CASE HISTORIES INVOLVING FATIGUE AND FRACTURE MECHANICS

A symposium
sponsored by ASTM
Committee E-24 on
Fracture Testing
Charleston, SC, 21–22 March 1985

ASTM SPECIAL TECHNICAL PUBLICATION 918
C. Michael Hudson, NASA-Langley Research
Center, and Thomas P. Rich, Bucknell
University, editors

ASTM Publication Code Number (PCN)
04-918000-30

 1916 Race Street, Philadelphia, PA 19103

Library of Congress Cataloging-in-Publication Data

Case histories involving fatigue and fracture
  mechanics.

  (ASTM special technical publication: 918)
  Includes bibliographies and indexes.
  "ASTM publication code number (PCN) 04-918000-30."
  1. Metals—Fatigue—Congresses.   2. Fracture
mechanics—Congresses.   I. Hudson, C. M.   II. Rich,
Thomas P.   III. ASTM Committee E-24 on Fracture
Testing.   IV. Series.
TA460.C334   1986   620.1'126   86-17208
ISBN 0-8031-0485-5

NOTE

The Society is not responsible, as a body,
for the statements and opinions
advanced in this publication.

Printed in Baltimore, MD
October 1986

Foreword

The symposium on Case Histories Involving Fatigue and Fracture Mechanics was held 21–22 March 1985 in Charleston, South Carolina. ASTM Committee E-24 on Fracture Testing sponsored the symposium. C. Michael Hudson, NASA-Langley Research Center, and Thomas P. Rich, Bucknell University, served as symposium cochairmen and coeditors of this publication.

Related
ASTM Publications

Automated Test Methods for Fracture and Fatigue Crack Growth, STP 877 (1985), 04-877000-30

Fracture Mechanics: Sixteenth Symposium, STP 868 (1985), 04-868000-30

Elastic-Plastic Fracture Test Methods: The User's Experience, STP 856 (1985), 04-856000-30

Fracture Mechanics: Fifteenth Symposium, STP 833 (1984), 04-833000-30

Design of Fatigue and Fracture Resistant Structures, STP 761 (1982), 04-761000-30

A Note of Appreciation
to Reviewers

The quality of the papers that appear in this publication reflects not only the obvious efforts of the authors but also the unheralded, though essential, work of the reviewers. On behalf of ASTM we acknowledge with appreciation their dedication to high professional standards and their sacrifice of time and effort.

ASTM Committee on Publications

ASTM Editorial Staff

Helen M. Hoersch
Janet R. Schroeder
Kathleen A. Greene
Bill Benzing

Contents

Introduction

The field of fracture mechanics has followed a distinct evolutionary pattern over the past 40 years. The period from the fifties through the sixties was a time of intense basic research from both the mechanics and materials points of view. During the seventies increasing transfer occurred from research into application with a corresponding increase in fracture mechanics based standards and specifications. Today while research into such areas as nonlinear fracture behavior and fracture/fatigue in new materials is still significant, the field of fracture mechanics has matured into a useful tool for engineering structural design, maintenance and inspection, and post failure analysis work.

Ever since the early days of research in fracture, numerous books and journals have been available to document the advances in the basic understanding of fracture and fatigue mechanisms and theory. However, the avenues for widely distributed information on experience gained in applying fracture mechanics to specific problems of hardware design and operation have been sorely lacking. With this in mind, the E-24 Committee on Fracture Testing of ASTM sponsored a Symposium on Case Histories Involving Fatigue and Fracture Mechanics in Charleston, South Carolina, on the 21st and 22nd of March, 1985. The response was extraordinary with over 20 case histories accepted for inclusion in the meeting. Many potentially good papers had to be excluded because of space and time limitations in the program.

The case history format was chosen for the meeting and resulting publication as the ideal medium for documenting experience. Authors were encourage to present clear and detailed examples of how fracture and fatigue concepts and data were applied to actual engineering components and systems. Sufficient information was sought to specify the geometrical, material, loading, environmental, and crack characterizations required to perform each case history investigation.

The results can be found in the case histories of this book. Major areas of application are pressure vessels, power generation equipment, structures, aircraft, manufacturing equipment, and bio-medical devices. In all cases, the authors are relating first hand experience on how they have applied fracture mechanics to engineering components in real life situations and time frames. Keep in mind that, as such, a case history is not put forward as "the best" way to tackle any given situation, but only as "a reasoned" way to accomplish stated objectives. They are to be learned from, built upon, and modified where necessary to enable the production and operation of safer, more reliable engineering products in the future.

C. M. Hudson

NASA-Langley Research Center, Hampton, VA 23665; cochairman and coeditor.

T. P. Rich

Bucknell University, Lewisburg, PA 17837; cochairman and coeditor.

Klaus Rahka[1]

Use of Material Characterization to Complement Fracture Mechanics in the Analysis of Two Pressure Vessels for Further Service in a Hydrogenating High-Temperature Process

REFERENCE: Rahka, K., "**Use of Material Characterization to Complement Fracture Mechanics in the Analysis of Two Pressure Vessels for Further Service in a Hydrogenating High-Temperature Process,**" *Case Histories Involving Fatigue and Fracture Mechanics, ASTM STP 918*, C. M. Hudson and T. P. Rich, Eds., American Society for Testing and Materials, Philadelphia, 1986, pp. 3–30.

ABSTRACT: Cracks of significant size were found in two chemical process pressure vessels after around 100 000-h service. The finding warranted fracture mechanics evaluation of further service fitness. This was done in five steps: fracture toughness measurements including determination of aging effects, loading limit identification including estimation of further aging and hydrogen effects, stress-intensity factor calculations, material and cracking mechanism investigation, and recommendations of precautions. The principal reason for crack formation was found to have been hydrogen attack due to local insufficient alloying in repair welds. Sufficient safety margins for further service were provided by a combination of precautions during startup and shutdown operations, low-working stresses, and sufficient fracture toughness as well as a detailed service record. Recommendations for further precautions included more frequent and intensified recurrent inspection, inert gas pressurization, and limitation of process pressures at ambient and lower temperatures during startups and shutdowns.

KEY WORDS: fracture mechanics, fitness for purpose, high temperature, hydrogen attack, pressure vessels.

Nomenclature

K_I Stress-intensity factor, MPa$\sqrt{m}$

σ_m Membrane stress

σ_b Bending stress

M_m Membrane stress correction factor

M_b Bending stress correction factor

[1] Senior research officer, Technical Research Centre of Findland (VTT), SF-02150 Espoo, Finland.

3

Q Flaw shape parameter
a Flaw size (edge crack size $= a$; internal crack size $= 2a$)
σ_a Axial stress
p Internal gas pressure
D Pressure vessel diameter
t Wall thickness

Fracture mechanics is a discipline within engineering, which has evolved through a need to rationalize the effect of crack-like discontinuities on the failure resistance of structures. Applications of fracture mechanics include both failure analysis and evaluation of cracked structures for continued use. Applications of fracture mechanics with the latter scope have economic as well as safety value, while the former serves the same purposes for a more distant future. Since the safety risks of flaw acceptance may be significant, a formal fracture mechanics calculation must be supported by detailed material characterization. In our case, two 2.9-m-diameter process vessels with 150 mm wall thickness were found to contain cracks up to 26 mm in depth and 430 mm in length. The vessels contain a process medium including gaseous hydrogen and hydrogen sulfide at an elevated temperature and high pressure.

Because a repair of the vessels would have included an unscheduled lengthy downtime and loss of production, a fracture mechanics evaluation was undertaken in order to find out whether the vessels would be fit for continued service. This continued service would allow smooth planning and implementation of repairs. In combination with this evaluation, an investigation was conducted to identify the mechanism of crack formation. This investigation used material that was locally removed from the vessel wall. Fracture toughness was determined using the same class of material from a "retired" end closure of a fellow vessel. The retirement of this end closure had been caused by brittle fracture due to an excessively low pressurizing temperature during a hydrotest.

Material Requirements for Elevated Temperature and Hydrogen Service

Service reliability of steel in elevated temperature hydrogenating environments is set by its resistance to hydrogen attack and creep resistance. Reliable process startup, shutdown, and hydrotesting also require brittle fracture precaution measures.

Limiting compositions of steels, which have shown satisfactory resistance to hydrogen attack, are given by Nelson diagrams [1]. In these diagrams, the hydrogen partial pressure and process temperature are used to characterize the severity of the environment. Based on the Nelson diagram in Fig. 1, it is concluded that in the process environment encountered in our case (H_2 partial pressure $\sim$ 10 MPa, temperature 420 $\pm$ 20°C) an iron car-

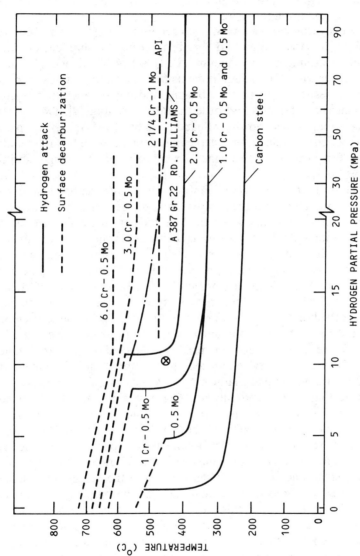

FIG. 1—*Nelson diagram for determining compositional levels for steels for potential hydrogen attack resistance* [1]. *Line for A 387 Grade 22 inserted according to data in Ref 2.*

bide is expected to undergo hydrogen attack. This attack involves the combining of hydrogen and carbon from the iron carbide to form methane gas. An alloy steel containing alloy rich iron, chromium, and molybdenum carbides will be chemically stable and therefore resistant to hydrogen attack in the conditions of interest here.

Based on hydrogen-attack-resistance requirements, a 2¼Cr-1Mo steel according to the ASTM Standard Specification for Pressure Vessel Plates, Alloy Steel, Chromium Molybdenum (A 387-82), A 387 Grade 22, is most suitable for our case. Other requirements for high temperature service are set by creep resistance to prevent bulging and high-temperature fracture [3] and aging characteristics for brittle fracture precaution [4]. Twenty years ago, creep strength was known and handled in practice. Aging, and particularly temper embrittlement or creep embrittlement, were less fully understood. For the purposes of this paper, I shall focus on the aging-embrittlement aspect only.

Aging embrittlement works in two ways: it will raise the brittle to ductile transition temperature and thus increase the danger for brittle fracture at low temperature, and it may lower the high-temperature resistance to fracture and thus increase the chance for cracking at elevated temperatures. In practice, the effect of aging is seen as a shift in transition temperature on a Charpy-V impact toughness curve. The Charpy-V impact toughness test is standardized: ASTM Standard for Notched Bar Impact Testing of Metallic Materials (E 23-82). Impact toughness is expressed either as the amount of energy absorbed or the percent of ductile fracture. The amount of embrittlement is expressed as the amount in shift of the toughness curve along the temperature scale. For aged material it is detected by regeneration heat treatment, Fig. 2 [4]. Service-induced aging embrittlement involving impurity effects is reversible to a large extent. While aging simulation is almost trivial for testing the susceptibility to aging of virgin materials, the reversibility of aging makes it possible to approximate—by testing—the actual aging that has occurred in material that has been in service. The "virgin" properties are regenerated by heat treatment of a specimen from the aged material, by, for example, a 1 h anneal at 650°C followed by water quench. In practice some uncertainties are anticipated in the aging embrittlement shift measured with regenerated material. For the purpose of estimating material fracture toughness, this uncertainty, together with postulated effects of future aging, are accounted for by shifting the toughness curves for the aged material further "upward" along the temperature scale by an amount equal to that obtained by regeneration in the "downward" direction.

Fracture Mechanics for Service Fitness Analysis

Apart from bulging by creep and embrittling by hydrogen attack, brittle fracture during hydrotesting or startup and shutdown operations is the most hazardous failure mode for pressure vessels of the kind in this paper. Brittle

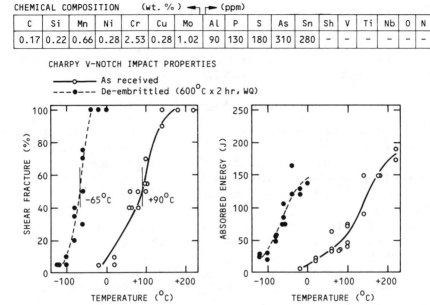

CHEMICAL COMPOSITION   (wt. %) ⟵ ⟶ (ppm)

C	Si	Mn	Ni	Cr	Cu	Mo	Al	P	S	As	Sn	Sh	V	Ti	Nb	O	N
0.17	0.22	0.66	0.28	2.53	0.28	1.02	90	130	180	310	280	–	–	–	–	–	–

CHARPY V-NOTCH IMPACT PROPERTIES

———o——— As received
----•---- De-embrittled (600°C x 2 hr, WQ)

FIG. 2—*Example of aging influence on the impact toughness of 2¼Cr-1Mo steel as detected by regeneration anneal (2 h at 690°C + water quench)* [4].

fracture at stresses well below the yield strength occurs at low or ambient temperatures under domination of elastic deformations. Consequently, it is often appropriate to apply linear elastic fracture mechanics to these cases. The principles listed in the ASME Boiler and Pressure Vessel Code, Section XI [5], therefore, were adopted. Accordingly, the stress-intensity factor, calculated using appropriate factors for geometrical, crack size, and load influence, should remain less than the structural fracture toughness. The material fracture toughness is the estimated value of a stress-intensity factor which produces failure. Usually, a safety margin is required to keep the applied stress-intensity factor at some level well below the material fracture toughness. This safety margin provides a buffer against uncertainties in the procedures or representativity of the data.

Fracture Toughness Measurement

Fracture toughness was determined using the unloading compliance method for early crack growth detection, and the ASTM Standard Test Method for J_{Ic}, a Measure of Fracture Toughness (E 813-81). Backup testing was done using the standard Charpy-V test. The material used for testing came from a heat-exchanger end closure from a fellow vessel. The manufacture of this end closure dates back to the same time as that of the vessels, and similitude in properties, therefore, was assumed. This end closure had failed by brittle fracture during a cold-hydro test a few years earlier. Investigation of this failure had been reinitiated through adoption of novel fracture-toughness measurement methods employing J_{Ic} methodology and considerations of aging. The principle for including the effect of aging into the integrity analysis is given in Fig. 3. Results including virgin properties obtained by regeneration treatment are shown in Fig. 4.

Determination of Actual Material Fracture Toughness

The material fracture toughness was estimated based on the results from testing specimens by allowing for extra aging effects and related uncertainties. This was done by fitting the ASME Section XI fracture toughness curves to the results from testing, and making an additional 60°C shift of the curves according to the regeneration result. The result is shown in Fig. 5.

The 60°C shift is somewhat arbitrary. The accuracy of the estimated material fracture toughness is improved, however, if one considers it to be probably higher than the highest stress-intensity factor ever sustained in Vessel A by the ligament in front of an even larger flaw (originally A4) than those now analyzed. This flaw had existed in the wall of Vessel A until the performance of the present analysis and was most probably first detected by ultrasonic inspection nine years earlier in 1973 (but not definitely identified as a crack until 1982). It then measured 80 mm in depth and 400 mm in width. It lay in the same horizontal girth weld as the rest of the cracks, which were now analyzed.

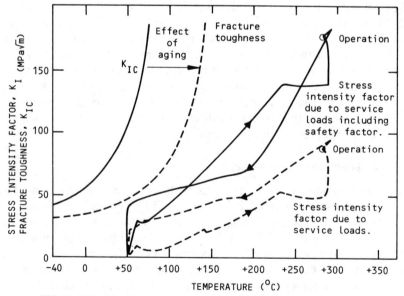

FIG. 3—*Principle for fracture prevention by fracture mechanics* $K_I < K_{Ic}$.

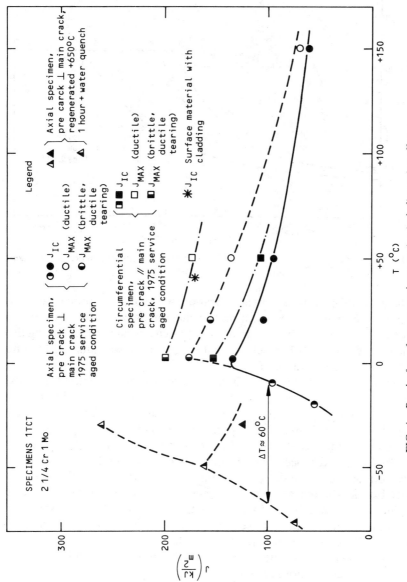

FIG. 4—*Results from fracture toughness testing including aging effect.*

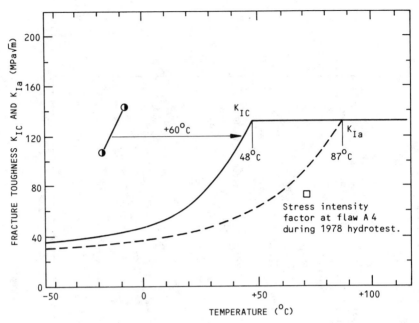

FIG. 5—*Choice of structural fracture toughness by using fracture toughness measurements, aging effects and ASME XI* [5] *fracture toughness curves.*

During a 1978 hydrotest crack indication A4 remained dormant as measured by pre- and post-test ultrasonic evaluation as well as *in situ* acoustic emission monitoring. The maximum stress-intensity value of A4 during this hydrotest was regarded to set a possible ultimate lower limit of fracture toughness for the material of the vessels.

A particular concern in the estimation of the material fracture toughness was the possible hydrogen entrapped within the vessel wall itself. This hydrogen takes some time to diffuse. This concern was relieved by recommending careful shut down operations to allow time for hydrogen to diffuse out of the vessel wall, and by avoiding operations causing thermal stresses. A related concern was the effect of hydrogen on high-temperature fracture toughness. The latter question was answered directly by a recent paper by Landes and McCabe [6]. Their paper showed that the fracture toughness of aged 2¼Cr-1Mo at high temperature is sufficient to prevent high-temperature fracture or even significant subcritical crack growth. Testing times up to six months [6] showed that this steel resists subcritical cracking sufficiently to allow at least six months of inspection interval. Pressure testing was not recommended.

Stress-Intensity Factor Calculation

Because precautions for reducing thermal stresses during startups, shutdowns, and operations were recommended, a simplistic stress and stress-intensity factor analysis was used. This analysis was based on the Inservice Inspection Section of the ASME Boiler and Pressure Vessel Code [5]. According to this code, the stress-intensity factor due to pressure, thermal, and shape-induced secondary stresses is calculated according to

$$K_I = M_m\sigma_m \sqrt{\frac{\Pi a}{Q}} + M_b \, \sigma_b \sqrt{\frac{\Pi a}{Q}} \tag{1}$$

where

σ_m and σ_b = membrane and bending stress components at the location of the crack, respectively;

M_m and M_b = membrane and bending correction factors, respectively, taking into consideration flaw eccentricity and shape;

Q = a flaw shape parameter, which takes into account flaw "topography" that is, ellipse ratio of the flaw and stress level relative to yield stress; and

a = actual flaw depth in the case of a surface flaw, for an embedded flaw a is half of the smaller axis of the ellipse that approximates the extremes of the actual flaw (the ligament, d, to the nearest free surface, must be $> 0.4 \, a$).

Location of Cracks and Significant Stress

From a stress calculation point of view, the cracks are all located in the simplest possible way. A graph of the pressure vessels in Fig. 6 includes the crack locations and orientation. All cracks were found in the center beltline cirumferential weld seam. They are, therefore, subject to crack opening due to axial membrane stresses only, that is

$$\sigma_m = \sigma_a = \frac{pD}{4t} \tag{2}$$

therefore

$$K_I = M_m\sigma_a \sqrt{\frac{\Pi a}{Q}} \tag{3}$$

Recurrent Inspections

The process vessels have been subjected to regular recurrent nondestructive examinations since 1973. In the 1973 examination, a large discontinuity was identified in one of the vessels (A) with the ultrasonic equipment. It extended to a depth of $a = 80$ mm and was 400 mm wide. Metallographic

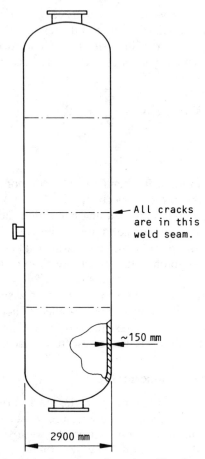

All cracks
are in this
weld seam.

~150 mm

2900 mm

FIG. 6—*Main dimensions of vessels and location of cracks.*

examination of the inside surface (cladded with stainless steel) of the vessel revealed no cracking in 1973, and the indication was then judged to be due to a material inhomogeneity. In subsequent regular yearly inspections, minor changes in the nature of this "inhomogeneity" were "observed." Motivated by suspicions as to the real nature of this discontinuity, a pressure test was performed in 1978. Acoustic emission techniques were employed to trace any cracking activity during this test. No indications of further cracking were detected during acoustic emission monitoring. The result of this pressure test was used in this evaluation to support the estimation of the material fracture toughness. By the time the initial fracture mechanics calculation was made in 1982, it was decided to investigate this discontinuity by removing material covering it. It became evident that the discontinuity was a crack and was numbered A4. In 1982, flaw A4 was removed from the vessel. The resulting cavity was ellipsoidal. During this shape modifi-

cation of the vessel, material adjacent to this flaw was removed and subjected to detailed chemical analysis and metallography. The ellipsoidal cavity was validated by stress analysis in a separate task, and it was left unrepaired.

Metallographic Observations of Defect Causes

The defect was identified as grain boundary cracking confined to weld
metal, Fig. 7. Metallography included scanning electron micrography, fractography (Fig. 7e) and transmission electron diffractometry to identify carbides in the steel (Fig. 8). The chemical composition was determined from
several locations in order to detect inhomogeneities, Table 1.

Results indicated that the local weld metal either contained no carbide
forming elements, or insufficient elements (see Cases a and b in Table 1).
The base metal was found to fulfill the specifications. The weld metal on
the outside surface of the vessel fulfilled the requirements as well. A micrograph of the sound weld metal is shown in Fig. 9.

Crack Inventory

Topographical maps of the center beltline weld seams are given in Fig.
10a and b, to show relative locations, approximate lengths, and identification
numbers of the cracks found.

A series of sections parallel to the axis of the Vessel B across the center
beltline weld at flaw (B4) is shown in Fig. 11. These sections show a typical
cross-section profile of a flaw. Intuitively one may conclude that the detected
flaws are most probably associated with repair of the weld root of the original
single sided V-groove weld. The quality control documentation from the
manufacturing also reports on weld repairs.

Stress-Intensity Factors

Stress-intensity factors were calculated using Eqs 2 and 3, assuming influence of membrane stress only. Factors M_m and Q were taken from the
appropriate graphs in Ref 5. The flaw sizes and aspect ratios were determined by enclosing the projected actual flaw within a rectangle according
to the ASME XI procedure. The dimensions of the rectangle are given by
the positions of the flaw extremities, Table 2a and b. Results of the stress-
intensity factor calculations are also reproduced in Table 2.

After the initial fracture mechanics evaluation in 1982, a shorter inspection interval of approximately 6 months was recommended. Also, inspection
was intensified and a lower reporting threshold was adopted. This resulted
in finding new flaws as well as growth of flaws that had already been found.
Strangely also, some flaws had "diminished" in size. Together with the
intensified inspections, a new coordinate and flaw numbering system was
adopted for Vessel A.

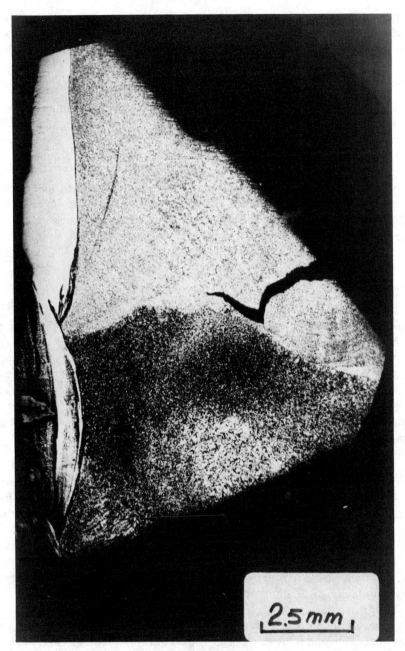

FIG. 7a—*Macrosection of specimen removed from the inside surface adjacent to indication A4. Cracking in confined to weld metal. Cladding is displayed.*

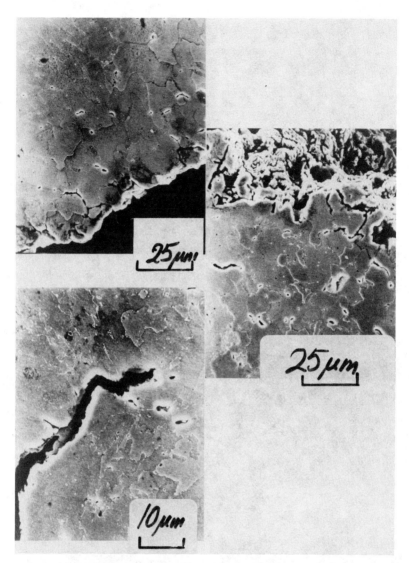

FIG. 7b—*Micrograph of material adjacent to main indication of flaw A4 grain boundary crack.*

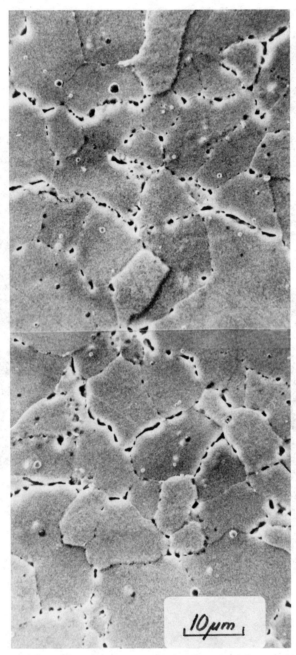

FIG. 7c—*Micrograph of material adjacent to side indication of flaw A4. Massive hydrogen attack.*

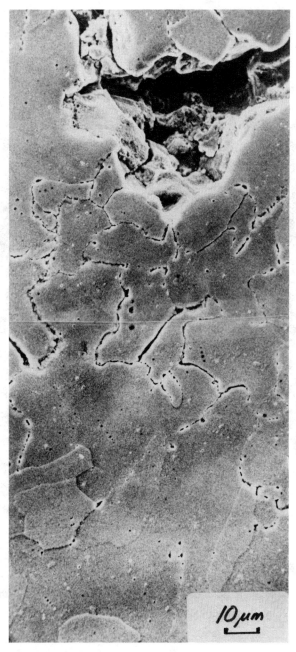

FIG. 7d—*Side indication of flaw A4.*

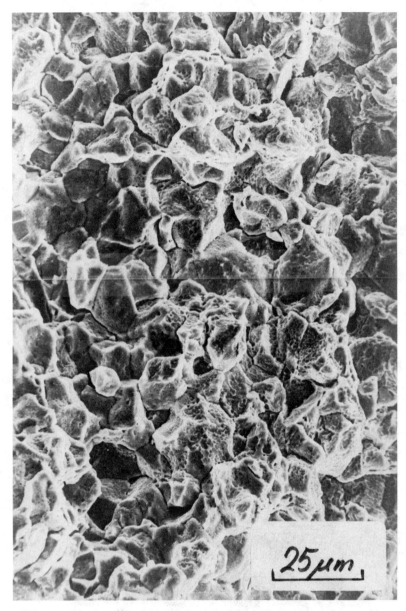

FIG. 7e—*Fracture surface of side indication of flaw A4. Grain boundary cracking.*

FIG. 8a—Bright field transmission electron micrograph of area containing chromium-carbide and electron diffraction pattern for determination of carbide type.

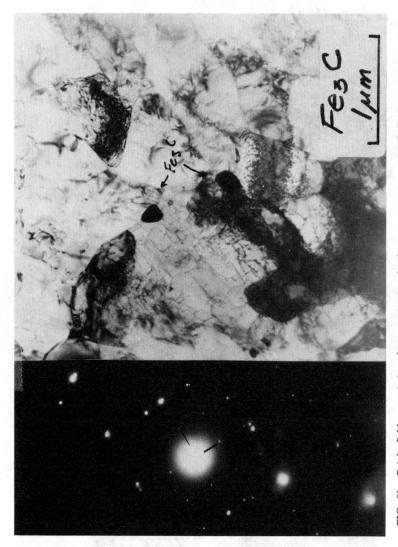

FIG. 8b—*Bright field transmission electron micrograph of area containing potentially hydrogen attack sensitive iron carbide and electron diffraction pattern to determine carbide type.*

TABLE 1—Chemical compositions determined for weld material at different locations.
(a) Material adjacent to main indication of Flaw A4.
(b) Material adjacent to side indication of Flaw A4.
(c) Material adjacent to Flaw A1.
(d) Weld material on the outside of the vessel.

Case	C	Si	Mn	S	P	Cr	Ni	Mo	Cu	Al	W	V	Ti	Co	Sn	As
a	0.04	0.30	0.45	0.027	0.014	1.6	0.10	0.78	0.15	0.01	0.01	0.02	0.00	0.02	0.019	0.033
b	0.02	0.30	0.67	0.022	0.019	0.05	0.06	0.02	0.04	0.005	0.00	0.02	0.01	0.02	0.005	0.04
c	0.12	0.48	0.66	0.040	0.022	2.25	0.21	1.15	0.39	0.01	0.01	0.02	0.00	0.02	0.02	0.04
d	0.07	0.39	0.84	0.029	0.028	2.5	0.12	1.15	0.28	0.02	0.01	0.02	0.00	0.02	0.023	0.028

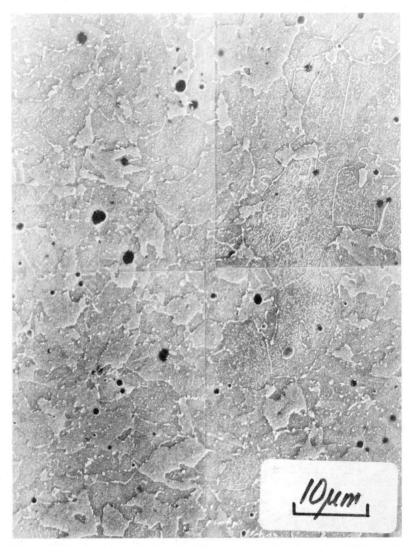

FIG. 9—*Weld material surrounding flaw A1 (numbering according to pre 1983 system).*

The final result of comparing fracture toughness and applied stress-intensity factors is given in Fig. 12. This figure shows a sufficient safety margin, provided no undue pressurization occur during shutdown and startup. In practice, pressurization with hydrogen gas was not done below ~150°C.

Discussion

Three points of the present case deserve some discussion. These are: stress level versus allowable stress; reasons for the repair action; and quality

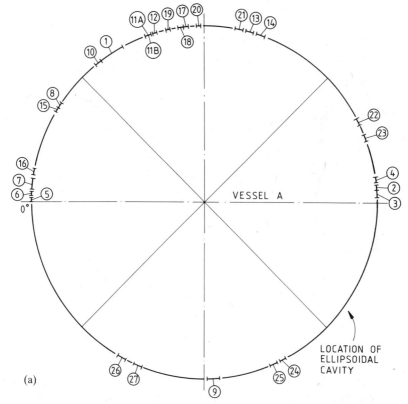

VESSEL A

LOCATION OF
ELLIPSOIDAL
CAVITY

(a)

0°

FIG. 10—*Flaw maps for center beltline weld for* (a) *Vessel A and* (b) *Vessel B. Approximate flaw length and flaw number are indicated.*

assurance deficiencies in the repair and manufacturing processes. The original choice of weld groove design is an item for consideration. The "lower than specified chromium and molybdenum" weld material is a central item for discussion. This deficiency is related to the original quality assurance procedures on the shop floor during manufacture. The adequacy of this analysis will be discussed bearing in mind the long period of safe service, the low service stresses, the shortened inservice inspection interval, and the intensified inspection procedures.

Stress Level

The allowable membrane stress at the peak operating temperature equals two thirds of the minimum guaranteed yield strength. Specifically this amounts to 180 MPa for the circumferential beltline membrane stress and 90 MPa for the axial membrane stress. The flaws are subject to opening mode by the axial stress.

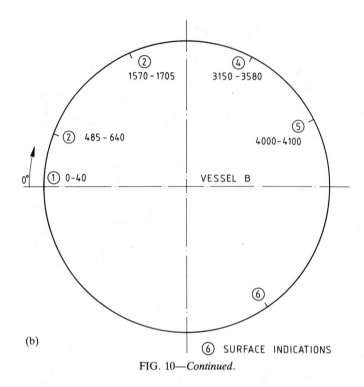

FIG. 10—*Continued*.

The wall thickness exceeds by 50% the value that would bring the wall stresses to their allowable values. Thus, the original design was quite conservative. With respect to possible creep of the vessel, the lower stress level is also conservative and contributes to an overall positive feeling of safety.

Reasons for the Repair

The reasons for the cracking raises other kinds of questions. A special concern is the local repair weld material, which did not meet specifications. On the other hand, the initial weld design was partly responsible for the original cracking. The latter is a welding design and technique related question, and is most probably connected to the choice of groove shape, filler material, preheat, and order of laying the beads. The former, or failure to use welding rods having the proper composition, is difficult to explain at this point. The problems that were met with at the analysis stage were connected to the unknown time dependence of hydrogen attack. Little information on this dependence was found during the limited time available to this analyst. Further, the amount and distribution of lower-than-specification weld material as well as alloy distribution was unknown. Some suggestive information on this distribution was found in the detailed flaw profile sketches that the nondestructive evaluation (NDE) personnel provided.

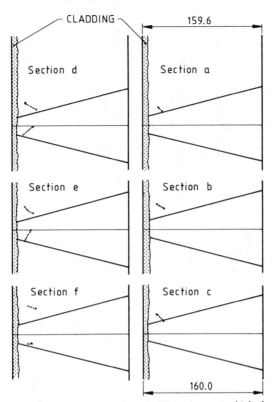

FIG. 11—*Six sections of flaw B4 approximately 50 mm/s apart which show probable asso-ciation of indications with weld root repair.*

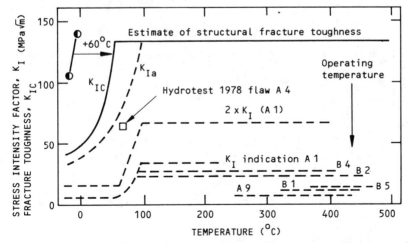

FIG. 12—*The estimated structural fracture toughness and stress-intensity factors caused by operating pressure and flaws found in Vessels A and B. Fracture toughness estimate is based on fitting the ASME XI [5] reference fracture toughness curves to the test data and allowing for aging. Upper shelf fracture toughness was estimated based on elevated temperature test data including effects of hydrogen [6].*

TABLE 2a—*Indications in Vessel A center beltline weld XX surface flaw assumption.*[a]

Indication Group	Ind	Record Date	Old	New	Flaw Location, mm Circumferential	Depth	K_I MPa$\sqrt{m}$	Note
I	A1	8/82	2		2445 to 2508[x]	30[px]	15.11[xx]	internal flaw length increased since 8/82
	A5	10/83		9	6715 to 6910	10 to 20	8.28 / 16.50[px]	
II	A1	8/82	3		2660 to 2710[x]	40[px]	15.11[xx]	small internal indications ($K_I \ll 10$ MPa$\sqrt{m}$)
	A5	10/83		25				
	A5			24				
III	A1	8/82	5		8 to 2_0^x	60[px]	18.51[xx]	small internal indications ($K_I \ll 10$ MPa$\sqrt{m}$)
	A5	10/83		5				
				6				
				7				
				16				
IV	A1	8/82	6		−1490 to 1515[x]	90[px]	23.79[xx]	additional point defect at 100 mm depth flaw 10 is small, extent in depth direction 97 to 104 mm
			7		−1340 to 1360[x]	90	23.79	
	A3	10/82		1	1360 to 1580	100	29.73	
	A4	4/83		1	1420 to 1690	85	37.99	
	A5	10/83		1	1420 to 1690	104		
				10	1305 to 1325			
V	A2	9/82	8	2	length 200 mm[xx]	50[px]	23.71[xx]	
	A3	10/82		3	length 165 mm[xx]	47[xx]	20.79[xx]	
	A4	4/83		2a	4345 to 4475			2b pointlike
				2b	4370			
				3	4366 to 4505			
				4	4340 to 4373			
	A5	10/83		2a	4345 to 4475	19 to 23	20.19	2b separate pointlike
				2b	4370	85		
				3	4350 to 4505	18 to 37		
				4	4300 to 4325	10 to 20		

VI	A5	10/83	22a	3785 to 3845	40 to 54	9.08	
			22b	3790 to 3835	39 to 48	7.83	
			23	4060 to 4095	20 to 30		
VII	A5	10/83	21	2710 to 2750	31 to 40	15.63	flaws combined 1 = 2750 to 2670 = 80
			13	2575 to 2605	76 to 80		
			14a	2670 to 2740	15 to 33		
			14b	2710 to 2743	29 to 41		
VIII	A5	10/83	11a	1780 to 1800	40 to 50		
			11b	1773 to 1800	42 to 48		
			12	1840 to 1860	35 to 43		
			19	2075 to 2095	82 to 92	8.95	17 and 19 combined
			18	2160 to 2178	76 to 84		
			17	2110 to 2120	76 to 84		$a = 8$, $b = 45$ internal flaw
			20	2145 to 2160	79 to 81		
					83 to 85		
					99 to 100		
IX	A5	10/83	8	700 to 750	89 to 105		small ($K_I \ll 10$ MPa$\sqrt{m}$)
			15	650 to 700	94 to 96		
X	A5	10/83	26	7620 to 7650	40 to 55		small separated flaws ($K_I \ll 10$ MPa$\sqrt{m}$)
			27a	7125	33 to 37		
			27b	7325	33 to 36		
			27c	7205	14 to 42		

[a] X = old coordinate system.

TABLE 2b—Indications in Vessel B center beltline weld.

NDE No.	Record Date	Indication	Location, mm		K_I MPa$\sqrt{m}$	Note
			Circumferential	Depth		
B1	10/82	1	0 to 40	17 to 30	13.09	
B2	4/83	1	0 to 40	notrepentations		
B3	10/83	1	0 to 40	12 to 35 / 54 / 70	14.84	pointlike / pointlike
B1	10/82	2	485 to 640	12 to 42	19.1	
B2	4/83	2	485 to 640	12 to 37		
B3	10/83	2	485 to 640	10 to 57	21.41	depth increased by ~20 mm
B1	10/82	3	1570 to 1680	14 to 26	17.45	
B2	4/83	3	1570 to 1680	8 to 32		
B3	10/83	3	1570 to 1705	13 to 37	18.96	length increased by ~25 mm
B1	10/82	4	3150 to 3580	15 to 37	22.13	
B2	4/83	4	3150 to 3580	13 to 35		
B3	10/83	4	3150 to 3580	12 to 39	25.45	
B1	10/82	5	4000 to 4100	22 to 47	16.74	
B2	4/83	5	4000 to 4100	16 to 47		
B3	10/83	5a	4000 to 4100	23 to 41	16.73	inter planar distance ~56 mm
		5b	4000 to 4100	45 to 65		flaw 5B is new
B1	10/82	6	5800 to 6180	surface		
B2	4/83	6	5800 to 6180	surface		
	10/83					ground

Extent of Lower than Specified Weld Filler and Adequacy of This Analysis

An indication of the extent of weld repair was obtained from the profiles of the ultrasonic indications, Fig. 11. These indications covered the major parts of the respective repairs. Accordingly, only limited further hydrogen attack was anticipated. However, since crack growth was the prime concern of this analysis, other modes of failure such as creep crack growth was covered as well. Some relevant literature was screened (for example, [*1,2,6*]). This screening indicated a limited possibility of further cracking. Such cracking would not happen in the fully alloyed material, however. It was concluded that the shorter inspection interval should bring this aspect into adequate control. Based on the stress-intensity factor calculations, several tens of millimetres of further crack growth could have been allowed before the safety margins were exceeded.

Conclusions

Detailed investigation of the vessels revealed significant deviations from material specifications. These deviations were the mechanistic reason for the crack formation mechanism. Under conditions of careful startup and shutdown and because operating stresses were low, the resulting stress-intensity factors were low as well. Thus sufficient safety margins were found for further service fitness. Uncertainties in the amount of underalloyed repair-weld filler material were compensated by a shortened inspection interval. Although further growth of some flaws were later detected, the margins of safety were not violated before the vessels were withdrawn from service.

Acknowledgments

Several colleagues of the present author contributed to the success of this investigation. They include Aki Valkonen, Pertti Nenonen, and Markku Kemppainen, who were involved in the work. Helpful suggestions were given by Dr. H. Hänninen.

References

[*1*] Nelson, G. A. and Effinger, R. T., *Welding Journal, Research Supplement,* Jan. 1955. pp. 12–21.
[*2*] Williams, R. O., "The Effect of Microstructure on the Susceptibility of Low Alloy Steels to Hydrogen Attack," ORNL 5781, Oak Ridge National Laboratory, Oak Ridge, Aug. 1981.
[*3*] Simmons, W. F. and Cross, H. C., *Report on Elevated Temperature Properties of Chromium-Molybdenum Steels, ASTM STP 151,* American Society for Testing and Materials, Philadelphia, 1955.

[4] Murakami, Y. and Ishikawa, N., "Properties of C-Mo and Cr-Mo Pressure Vessel Steels After Long Time Service at the Temper Embrittlement Temperature Range," Muroran Plant, Research Laboratory Paper PV-78-8-232, The Japan Steel Works, 21 Aug. 1978.
[5] ASME Boiler and Pressure Vessel Code; Part XI, Inservice Inspection Nonmandatory Appendix A, American Society of Mechanical Engineers, New York, 1980.
[6] Landes, J. D. and McCabe, D. E. in *Application of 2¼Cr-1Mo Steel for Thick-Wall Pressure Vessels. ASTM STP 755*, G. S. Sangdahl and M. Semchyshen, Eds., American Society for Testing and Materials, Philadelphia, 1982, pp. 68–92.

C. William Smith[1]

Cracking at Nozzle Corners in the Nuclear Pressure Vessel Industry

REFERENCE: Smith, C. W., **"Cracking at Nozzle Corners in the Nuclear Pressure Vessel Industry,"** *Case Histories Involving Fatigue and Fracture Mechanics, ASTM STP 918*, C. M. Hudson and T. P. Rich, Eds., American Society for Testing and Materials, Philadelphia, 1986, pp. 31–45.

ABSTRACT: Cracks in nozzle corners at the pressure boundary of nuclear reactors have been frequently observed in service. These cracks tend to form with radial orientations with respect to the nozzle central axis and are believed to be initiated by thermal shock. However, their growth is believed to be primarily due to a steady plus a fluctuating internal pressure. Due to the impracticality of fracture testing of full-scale models, the Oak Ridge National Laboratory instituted the use of an intermediate test vessel (ITV) for use in fracture testing which had the same wall thickness and nozzle size as the prototype but significantly reduced overall length and diameter. In order to determine whether or not these ITVs could provide realistic data for full-scale reactor vessels, laboratory models of full-scale boiling water reactors and ITVs were constructed and tested. After briefly reviewing the laboratory testing and correlating results with service experience, results obtained will be used to draw some general conclusions regarding the stable growth of nonplanar cracks with curved crack fronts which are the most common precursors to fracture of pressure vessel components near junctures. Use of linear elastic fracture mechanics is made in determining stress-intensity distributions along the crack fronts.

KEY WORDS: stress intensity factors, nuclear pressure vessels, nozzle corner cracks, intermediate test vessels, boiling water reactors, photoelasticity

The results described in this paper are drawn from studies motivated by the discovery of a large number of cracks at re-entrant nozzle corners in the coolant conduits of operating boiling water reactors (BWR) in the early and middle 1970s in the United States. The cracks were believed to be of thermal shock origin, but there was concern over the growth of the cracks due to fluctuations in the internal pressure in the vessels.

Due to the prohibitive costs involved in full-scale testing of reactor vessels, engineers at the Oak Ridge National Laboratory had devised a model known as the intermediate test vessel (ITV) by reducing the diameter and length of the reactor vessel by an order of magnitude but maintaining the same

[1] Alumni professor, Department of Engineering Science and Mechanics, Virginia Polytechnic Institute and State University, Blacksburg, VA 24061.

wall thickness and nozzle dimensions. A photo of an ITV is shown in Fig. 1. In designing the ITV, the Oak Ridge engineers included a conservative variation from the full-scale reactor geometry by making the nozzle corner root radius one half of that in a scale model of the reactor vessel. When, however, the ITV concept was introduced to the international technical community, the question was raised as to whether or not the thick-walled ITV could be used to predict crack growth in the thin-walled reactor vessel at nozzle corners, and whether or not the ITV was an adequately severe test.

Around 1975, the Delft Technical University sponsored a series of frozen stress photoelastic experiments at Virginia Tech on flat plates with nozzles which showed that cracks grown under monotonically increasing load in the photoelastic models were identical in shape to those grown under fatigue loading with stress ratios as low as 0.1 in reactor steel models [1]. However, the crack shapes, normally assumed by designers to be quarter circular at that time, were more complex.

Based upon these results, the Oak Ridge National Laboratory sponsored an extensive program at Virginia Tech to apply the frozen stress method to a study of nozzle corner cracks and accompanying stress-intensity factor (SIF) distributions in both ITV and BWR models and to compare results.

After briefly reviewing the frozen stress method and its use, this paper will focus upon the findings of this latter study and its relation to nozzle corner crack design rationale.

FIG. 1—*Photograph of an intermediate test vessel.*

Frozen Stress Method

There are two classes of basic considerations involved in the development of the "frozen stress" method for estimating SIFs. They are (a) experimental and (b) analytical. Experimental considerations include the "frozen stress" method and the deviations from two-dimensional behavior produced by through cracks in bodies due to local constraint variation. Analytical considerations include the utilization of near-tip field theory to compute quantities measured photoelastically.

Experimental Considerations

We shall first review concepts associated with the frozen stress photoelastic method.

This method involves a technique which dates back to the studies of Oppel [2]. The term "frozen stress" is a misnomer, in the sense that one does not obtain frozen in macrostresses. Instead the deformations and stress fringes are the frozen or fixed values. In order to accomplish stress freezing, one begins with a transparent material exhibiting certain diphase optical and mechanical properties. At room temperatures, these materials can be characterized as Kelvin-like in their response to load (Fig. 2). That is, mechanical loads tend to generate optical and mechanical creep or flow as well as elastic response. Above a certain temperature, however (called the critical temperature), the material's viscous component is suppressed, and its response

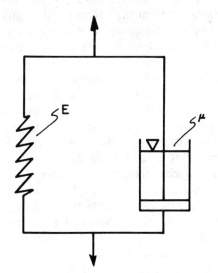

E = YOUNG'S MODULUS

μ = FIRST COEFF. OF VISCOSITY

FIG. 2—*Kelvin material.*

becomes linearly elastic and virtually time independent. Thus, one can adopt the following procedure for producing "frozen stress" models:

1. Heat model in unrestrained form to above critical temperature.
2. Apply live loads.
3. Cool to room temperature under load. [Upon removal of the loads, the stress fringes and deformations produced above critical temperature will be retained. The recovery effect due to unloading is negligible since the material fringe constant (or stress per unit fringe) above critical is an order of magnitude below the room temperature value and the value of Young's modulus (E) above critical is two orders of magnitude below that at room temperature. Since no macrostress results, the model may be sliced without altering the fringe or deformation patterns.

There are two basic limitations to the frozen stress method in general. The first is that it applies only to linear elastic behavior, and the second is that all stress freezing materials are essentially incompressible above critical temperature (that is, Poisson's ratio, $v \simeq 0.5$).

The practical significance of the first limitation is that when one introduces a sharp crack into the body (theoretically a singularity) one expects a local region of nonlinearity very near the crack tip. Such a region, in fact, is observed photoelastically. This region may be due to nonlinear material behavior, crack-tip blunting, nonlinear strain-optic relations or some combination of these effects [3]. The important point is that this region must be identified and excluded from the SIF determination. The second limitation leads to an important consideration especially when verifying the accuracy of the method by comparing with two-dimensional analytical solutions.

The effect of the elevated value of Poisson's ratio in cracked body problems may be described as follows: When a through the thickness crack occurs in a body of finite thickness, a region of high constraint develops around the crack tip characterized by the root radius of the crack tip and the plate thickness, the latter always being orders of magnitude larger than the former. This constraint dissipates rapidly with increasing distance from the crack tip and away from the crack tip generalized plane stress then prevails in a "two-dimensional" problem. The presence of the constraint variation, however, is a three dimensional effect which is present in all real bodies, and when measurements are made near the crack tip, they will be associated with a state of nearly plane strain, while measurements away from the near tip zone, such as compliance measurements, will be for a zone in which generalized plane stress exists. Irwin and others have suggested [4,5] that near tip measurements in such problems can be "converted" to two-dimensional results through the multiplying factor $(1 - v^2)^{1/2}$. This effect, while small for $v \simeq 0.3$, is substantial when $v \simeq 0.5$. The author and his associates have extended these ideas and suggested the factor $\{(1 - 0.3^2)/(1 - 0.5^2)\}^{1/2}$ for converting plate type frozen stress data for comparison with "two-dimensional" results where $v \simeq 0.3$. This method has been found to be remarkably accurate for plate shaped bodies. Details of this study are found in Ref. 6. When cracked bodies are not two-dimensional (as with surface flaws) this effect cannot be estimated in the same way, but estimates of the author and his associates indicate a maximum elevation in the SIF from near field measurements of the order of the experimental error (that is, about 5%).

Analytical Considerations

Mode I Analysis—It has been shown by Sih and Kassir [7] that the singular elastic field surrounding the tip of an elliptically shaped flaw border can be expressed in the same form as for the plane case if a local moving rectangular cartesian coordinate system is employed. For Mode I loading, these stresses can be written in the form

$$\sigma_{nn} = \frac{K_{\mathrm{I}}}{(2\pi r)^{1/2}} \cos\frac{\theta}{2}\left(1 - \sin\frac{\theta}{2}\sin\frac{3\theta}{2}\right) - \sigma_{nn}^{O}$$

$$\sigma_{zz} = \frac{K_{\mathrm{I}}}{(2\pi r)^{1/2}} \cos\frac{\theta}{2}\left(1 + \sin\frac{\theta}{2}\sin\frac{3\theta}{2}\right) - \sigma_{zz}^{O} \qquad (1)$$

$$\sigma_{nz} = \frac{K_{\mathrm{I}}}{(2\pi r)^{1/2}} \sin\frac{\theta}{2}\cos\frac{\theta}{2}\cos\frac{3\theta}{2} - \sigma_{nz}^{O}$$

where the notation, adapted to surface flaw problems is pictured in Fig. 3. The first terms represent the singular part of the stress field and the σ_{ij}^{O}, (following the Irwin approach for the plane problem), represent the contribution of the regular stress field in the form of the first terms of Taylor series expansions of the regular stress components near the crack tip. Assuming that any flaw border can be represented as locally elliptic in shape [8], Eqs 1 can be considered as applicable to such flaw border shapes.

Observation of Mode I fringe patterns (Fig. 4) reveals that the fringes tend to spread in a direction approximately normal to the flaw surface. Thus best fringe discrimination and accuracy is expected along $\theta = \pi/2$ in Eq 1. Evaluating Eq 1 along $\theta = \pi/2$, computing

$$\tau_{\max}^{nz} = \frac{1}{2}\left[(\sigma_{nn} - \sigma_{zz})^2 + 4\sigma_{nz}^2\right]^{1/2} \qquad (2)^2$$

---

[2] Also, from the Stress Optic Law $\tau_{\max}^{nz} = n'f/2t'$ where n' = stress fringe order, f = material fringe value, and t' = slice thickness.

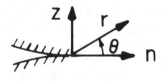

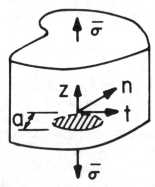

FIG. 3—*General problem geometry and notation.*

FIG. 4—*Near tip Mode I stress fringes.*

and truncating to the same order as Eq 1, one obtains

$$\tau_{max}^{nz} = \frac{A}{r^{1/2}} + B \tag{3}$$

where

$A = K_I/(8\pi r)^{1/2}$ and
$B = $ a constant containing σ_{ij}^O.

Rewriting Eq. 3 in a normalized form, we have

$$\frac{K_{Ap}}{q(\pi a)^{1/2}} = \frac{K_I}{q(\pi a)^{1/2}} + \frac{B(8)^{1/2}}{q} \left(\frac{r}{a}\right)^{1/2} \tag{4}$$

where $K_{Ap} = \tau_{max}^{nz}(8\pi r)^{1/2}$ is defined as an "apparent" stress-intensity factor. Equation 4, when plotted as $K_{Ap}/q(\pi a)^{1/2}$ versus $(r/a)^{1/2}$ yields a straight line which when extrapolated across the nonlinear zone to the origin will yield $K_I/q(\pi a)^{1/2}$, the normalized SIF. q is a nonlocal load parameter, such as internal pressure. A typical example of the extrapolation using data for the problem which follows is given in Fig. 5.

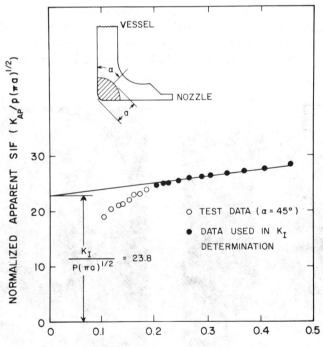

FIG. 5—*Estimating the SIF from stress frozen slice data.*

The foregoing method depends crucially upon the ability of the experimentalist to accurately locate the linear zone represented by the black data points in Fig. 5. Since Eqs 1 are the same form as the two-dimensional near tip equations for a crack in an infinite body, one expects that the linear zone may be constricted, sometimes severely from the outside due to complex boundary conditions and three-dimensional effects other than the crack surfaces themselves. A recent study [9] by the author and his colleagues suggests that the inner boundary of the linear zone along $\theta = \pi/2$ for Mode I is almost invariant with $(r/a)^{1/2} \approx 0.2$. However, the outer boundary of the linear zone may vary from $(r/a)^{1/2} \approx 0.4$ to 0.7 or more. In actual dimensions, this gives a linear zone from $r \approx 0.1$ to $r \approx 1.0$ mm. Thus, in practice, the investigator seeks linear zones to the right of $(r/a)^{1/2} \approx 0.2$ which are common to all slices along a given flaw border. In this way, the proper linear zone location is insured.

The Experiments

Experimental Procedure—Model sections for both ITV and BWR models were cast from suitable stress freezing materials. A starter crack was inserted into each nozzle in a plane normal to the vessel hoop stress direction by striking a sharp blade held normal to the reentrant corner with a hammer. Model parts were then glued together to form scale models of ITVs, each containing one nozzle and BWRs each containing two diametrically opposite nozzles. Photos of each are shown in Fig. 6 showing assembly glue lines.

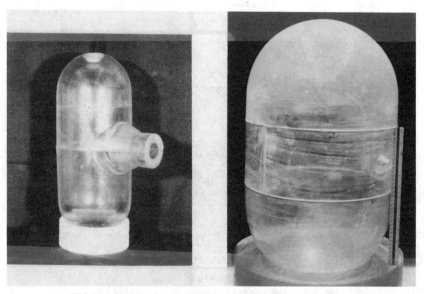

FIG. 6—*Photos of photoelastic models of an ITV and a BWR.*

These vessels were then placed in a stress freezing oven and heated to well above the critical temperature. After a thermal soak to achieve a uniform temperature distribution, the vessels were pressurized to a sufficient level to enlarge the flaws to the desired sizes. Upon reaching the desired sizes, the pressure was reduced to stop flaw growth, and the models were cooled under load. Pertinent data on dimensions and crack sizes are found in Table 1 using the notation of Fig. 7. Finally, slices were removed parallel to the *nz* plane at intervals along the flaw border and analyzed in a crossed circular polariscope at ×10 using white light and the Tardy method to read tint of passage.

Results—The flaw shapes obtained for the ITVs are shown in Fig. 8 and may be regarded as typical of both the ITVs and the BWRs. These shapes reveal two important features:

TABLE 1—*Local model and crack dimensions.*[a]

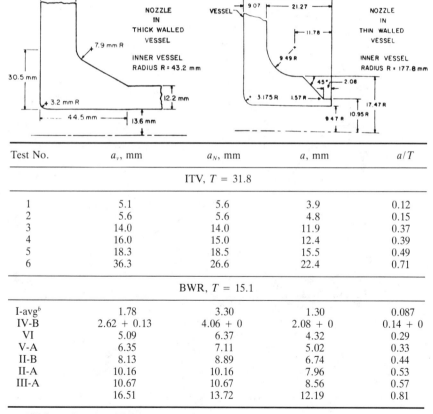

Test No.	a_v, mm	a_N, mm	a, mm	a/T
		ITV, $T = 31.8$		
1	5.1	5.6	3.9	0.12
2	5.6	5.6	4.8	0.15
3	14.0	14.0	11.9	0.37
4	16.0	15.0	12.4	0.39
5	18.3	18.5	15.5	0.49
6	36.3	26.6	22.4	0.71
		BWR, $T = 15.1$		
I-avg[b]	1.78	3.30	1.30	0.087
IV-B	2.62 + 0.13	4.06 + 0	2.08 + 0	0.14 + 0
VI	5.09	6.37	4.32	0.29
V-A	6.35	7.11	5.02	0.33
II-B	8.13	8.89	6.74	0.44
II-A	10.16	10.16	7.96	0.53
III-A	10.67	10.67	8.56	0.57
	16.51	13.72	12.19	0.81

[a] All dimensions in millimetres.
[b] Average of IA, IB, and VB.

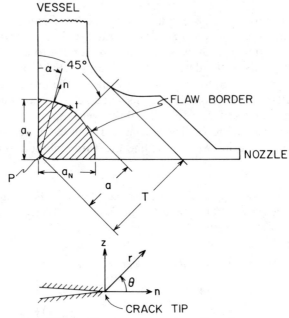

FIG. 7—*Nozzle corner cracks and notations.*

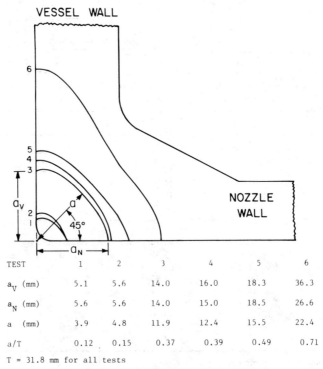

TEST	1	2	3	4	5	6
a_V (mm)	5.1	5.6	14.0	16.0	18.3	36.3
a_N (mm)	5.6	5.6	14.0	15.0	18.5	26.6
a (mm)	3.9	4.8	11.9	12.4	15.5	22.4
a/T	0.12	0.15	0.37	0.39	0.49	0.71

T = 31.8 mm for all tests

FIG. 8—*ITV Model flaw geometries.*

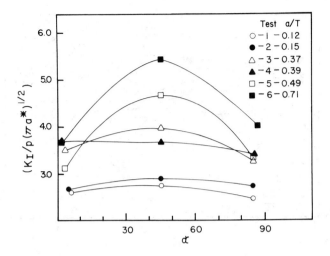

NOTE: $a^* = 11.9$mm, the value of a for Test 3

FIG. 9—*ITV SIF distributions.*

1. For the small flaws $a_N > a_V$ but for the deep flaws $a_V > a_N$.

2. As a/T increases, the flaw growth in the central part of the flaw is retarded relative to the outer portions producing a flattening of the flaw border in the center. It is thus clear that flaw growth was not self-similar.

Normalized SIF distributions for the ITVs are given in Fig. 9 and for the BWRs in Fig. 10. Ordinates were normalized with respect to a common crack length in each case to show the increase in SIF level with flaw depth. For the shallowest flaw in the BRW, only one slice (and hence one SIF value) was obtained. The concave shapes exhibited by the shallow BWR flaws did not occur with the ITV flaws because the smaller $r_f/R_O - R_i$ employed in the ITVs[3] caused an ITV crack border of the same a/T to be relatively further from the corner than a BWR flaw border of about the same a/T. This is illustrated in Fig. 11. It is suggested that, if the same value of $r_f/R_O - R_i$ were used in both geometries, similar SIF distributions would result.

Taken collectively, these results show that the ITVs produce flaw shapes and SIF distributions similar to those in BWRs. However, because of the smaller radius ITV nozzle fillet, the SIF values in the ITVs for about the same a/T values are significantly higher than those found in the BWRs. This result tends to support the use of the ITV as a conservative code verification test for a BWR.

[3] r_f = fillet radius.

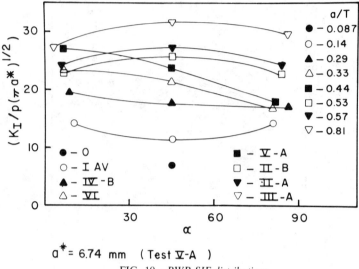

$a^* = 6.74$ mm   (Test V-A)

FIG. 10—*BWR SIF distributions.*

Assessment of Design Rationale

A number of numerical solutions have been proposed for estimating SIF values for nozzle corner cracks in thin-walled pressure vessels. These include numerical estimates [10] for very shallow flaws of prescribed simple flaw shape, as well as for quarter circular [11–13] straight front [14] and quarter elliptical [15] flaws over larger flaw depth ranges. Some analyses [11] predicted concave SIF distributions along flaw borders and others [14] predicted convex SIF distributions. One analysis predicted both types of distributions. As seen in Fig. 10, the distributions can be reversed as a result of changes in the flaw size and shape during flaw growth.

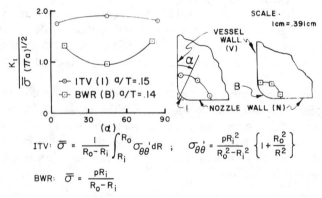

$$\text{ITV:} \quad \bar{\sigma} = \frac{1}{R_o - R_i} \int_{R_i}^{R_o} \sigma_{\theta\theta}{}' dR \quad ; \quad \sigma_{\theta\theta}{}' = \frac{pR_i^2}{R_o^2 - R_i^2} \left\{ 1 + \frac{R_o^2}{R^2} \right\}$$

$$\text{BWR:} \quad \bar{\sigma} = \frac{pR_i}{R_o - R_i}$$

FIG. 11—*Effect of reduced ITV fillet radius $r_f/R_o - R_i$ on SIF distribution.*

Two of the preceding analyses have received considerable attention from the reactor technology community in the United States. The first one [12], due to Gilman and Rashid, employs quarter circular flaw shapes, assumes self-similar flaw growth and uses a compliance like finite-element approach in order to compute an average SIF for a given flaw size. The second analysis [13], due to Besuner and his associates, utilizes an influence function-boundary integral approach in order to estimate average SIF values for the same flaw geometry studied by Gilman and Rashid.

In Figure 12, we compare the average experimental BWR results with those just cited. By plotting results in this manner, the crack size is normalized out and the usual increase in SIF with increasing crack size is not seen. It is clear from Fig. 12 that both analyses are conservative for shallow flaws when compared with the results of the present study.

Only midpoint or average values from the experimental results were used in the preceding comparisons since the experiments revealed complex crack shapes and nonself-similar flaw growth which were not included in the analyses cited. To the writers knowledge, the most comprehensive studies of flaw shape and flaw growth in this problem to date have been carried out by Broekhoven [15]. Other important studies on flaw shapes and non-self-similar flaw growth have been conducted by Sommer, Hodulak, Kordish, and Kunzelmann [16–18].

Conclusions

A method which combines the frozen stress photoelastic method with the local stress field equations of LEFM for use in estimating SIF distributions along the border of cracks in three-dimensional problems is used to augment a case study of cracking at nozzle corners in reactor vessels. The method produces flaw shapes believed to be very close to those generated by stable fatigue crack growth. In applying the method to study nozzle corner cracks in ITV and BWR models, the following results are noted:

1. Flaw shapes in ITV models are similar to those in BWR models but SIF distributions are significantly higher (that is $\approx 2x$).
2. Flaw shapes were not simple curves, and flaw growth was not self-similar in either model geometry.

The method of analysis used is not without its own limitations: the results are constrained to the elastic behavior of incompressible material. However, for problems where stable flaw growth, usually in fatigue, is controlled by geometric effects and small-scale yielding only occurs, the writer conjectures that engineering accuracies (say ±5%) can be expected in flaw shapes and SIF distributions for homogeneous, isotropic materials in problems such as presented here.

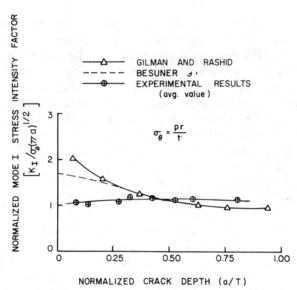

FIG. 12—*Comparison of experimental results for BWR geometry with analyses used in design rationale.*

Acknowledgments

The author wishes to acknowledge the contributions of his colleagues as noted in the references and the support of the Oak Ridge National Laboratory under Contract No. W-7405-Eng-26 and the National Science Foundation under Grant No. MEA 832-0252.

References

[1] Smith, C. W. and Peters, W. H., "Prediction of Flaw Shapes and Stress Intensity Distributions in 3D Problems by the Frozen Stress Method," *Preprints of the Sixth International Conference on Experimental Stress Analysis,* Munich, Sept. 1978, pp. 861–864.
[2] Oppel, G., "Photoelastic Investigation of Three Dimensional Stress and Strain Conditions," NACA TM 824 (Translation by J. Vanier), 1937.
[3] Smith, C. W., McGowan, J. J., and Peters, W. H., *Journal of Experimental Mechanics,* Vol. 18, No. 8, Aug. 1978, pp. 309–325.
[4] Irwin, G. R., "Measurement Challenges in Fracture Mechanics," William Murray Lecture, SESA Fall Meeting, Indianapolis, Oct. 1973.
[5] Srawley, J. R., Jones, M. H., and Gross, B., "Experimental Determination of the Dependence of Crack Extension Force on Crack Length for a Single Edge Notch Tensile Specimen," NASA-TN-D2396, National Aeronautics and Space Administration, May 1964.
[6] Smith, C. W., McGowan, J. J., and Jolles, M., *Journal of Experimental Mechanics,* Vol. 16, No. 5, May 1976, pp. 188–193.
[7] Sih, G. C. and Kassir, M., *Journal of Applied Mechanics,* Vol. 33, No. 3, Sept. 1966, pp. 601–611.
[8] Sih, G. C. and Liebowitz, H., "Mathematical Theories of Brittle Fracture," *Fracture VII Mathematical Fundamentals,* 1968, pp. 68–188.

[9] Smith, C. W., Olaosebikan, O., and Epstein, J. S., "An Analytical-Experimental Estimation of Elastic Linear Zone for Photoelastic Mode I SIF Extraction" (in press), *International Journal of Theoretical and Applied Fracture Mechanics*, 1985.

[10] Hellen, T. K. and Dowling, A. R., *International Journal of Pressure Vessels and Piping*, Vol. 3, 1975, pp. 57–74.

[11] Reynen, J., "On the Use of Finite Elements in the Fracture Analysis of Pressure Vessel Components," ASME Paper No. 75-PVP-20, American Society of Mechanical Engineers, June 1975.

[12] Rashid, Y. R. and Gilman, J. D. in *Proceedings*, First International Conference on Structural Mechanics in Reactor Technology, Vol. 4, Reactor Pressure Vessels, Part G, Steel Pressure Vessels, Sept. 1971, pp. 193–213.

[13] Besuner, P. M., Cohen L. M., and McLean, J. L., "The Effects of Location, Thermal Stress and Residual Stress on Corner Cracks in Nozzles with Cladding," *Transactions*, Fourth International Conference on Structural Mechanics in Reactor Technology, Vol. G, Structural Analysis of Steel Reactor Pressure Vessels, Paper No. G 4/5, Aug. 1977.

[14] Schmitt, W., Bartholome, G., Gröstad, A., and Miksch, M., *International Journal of Fracture*, Vol. 12, No. 3, June 1976, pp. 381–390.

[15] Broekhoven, M. J. G., *Proceedings*, Third International Conference on Pressure Vessel Technology, Part II, Materials and Fabrication, April 1977, pp. 839–852.

[16] Sommer, E., Hodulak, L., and Kordisch, H., *Journal of Pressure Technology, Transactions*, American Society of Mechanical Engineers, Vol. 99, Series J, No. 1, February 1977, pp. 106–111.

[17] Hodulak, L., "Development of Part Through Cracks and Implications for the Assessment of the Significance of Flaws," Paper No. C89/78, *Proceedings*, Institution of Mechanical Engineers, 1978, pp. 115–120.

[18] Hodulak, L., Kordisch, H., Kunzelmann, S., and Sommer, E., *International Journal of Fracture*, Vol. 14, 1978, pp. R-35-R-38.

Mitchell P. Kaplan,[1] Terrence Willis,[1] and Ralph L. Barnett[2]

A Pressure Vessel Hatch Cover Failure: A Design Analysis

REFERENCE: Kaplan, M. P., Willis, T., and Barnett, R. L., **"A Pressure Vessel Hatch Cover Failure: A Design Analysis,"** *Case Histories Involving Fatigue and Fracture Mechanics, ASTM STP 918,* C. M. Hudson and T. P. Rich, Eds., American Society for Testing and Materials, Philadelphia, 1986, pp. 46–64.

ABSTRACT: This case study involves a hatch cover located on a pressurized barge unloading concrete. The hatch cover exploded as a workman was kicking off one of the locking cams holding the cover to the deck. The cover, attached to the barge by a hinge, utilized six locking cams to hold the pressure seal to the compartment. The cover was 0.51 metres in diameter and fabricated from A356-T6 aluminum. The pressure to off-load the cement was 206.8 kPa. A finite element study was performed to determine the stresses in the cover as a function of locking geometry. Fracture toughness tests to determine K_{Ic}, a nondestructive inspection to determine crack size and crack population, and a design study to determine pressure vessel design characteristics were undertaken. A failure analysis was also conducted. Results of the studies indicate that the design choices (internal piping, hatch cover attachment, and material selection) allowed a potentially catastrophic situation to develop. Recommendations for design improvements are given.

KEY WORDS: pressure vessel, design, failure analysis, fracture mechanics, finite element analysis, mechanical properties, aluminum, fractography

In August 1975, a barge on the Mississippi River was off-loading cement into a receiving warehouse. The cement unloading process had begun approximately 2 h earlier. A hole developed in the main discharge line allowing cement to leak. The compressors driving the pneumatic off-loading system were switched off and gate valves were opened to release the pressure. As the depressurization was occurring rather slowly, the hatch covers to the compartments storing the cement were partially opened. This task was done by workers standing on the hatch and kicking the locking cams or dogs with their feet. After kicking off the second dog on a particular hatch cover, that hatch cover exploded throwing both it and the individual over the side of the barge into the water.

---

[1] Consulting engineers, Willis, Kaplan & Associates, Inc., Arlington Heights, IL 60004.
[2] Chairman, Triodyne, Inc., Niles, IL 60648.

The hatch cover was not recovered; therefore, no inspection, testing or analysis could be conducted on that component. However, four other hatch covers were obtained from the same vessel. As all of the hatch covers were with the vessel, it was assumed that tests conducted on the exemplar hatch covers would be indicative of the results that may have been obtained from the hatch cover that had failed. Two of the hatch covers supplied, Nos. 1 and 3 had failed hinges while the remaining two had the hinge intact.

The four hatch covers that were obtained were photographed in the "as-received" state, cleaned by sand blasting, and inspected to ascertain internal defects or cracks.

The hatch covers were machined to obtain tensile and fracture mechanics coupons. Tension and fracture mechanics tests were conducted, a metal-lurgical specimen was examined to determine the microstructural charac-teristics, and the fracture surfaces were examined using a scanning electron microscope.

Function/Design

Barge

The barge that was being utilized to transport the cement was manufac-tured in the late 1960s. The vessel was 57.3 m (188 ft) long, 15.24 m (50 ft) wide, and had a draft of 3.66 m (12 ft). It has a capacity of 2.545 10^6 kg (2800 tons) of cement and off-loads cement by pneumatic discharge. The barge has two parallel, cylindrically shaped tanks (hopper bottomed) each divided into three separate compartments as shown in Fig. 1.

As this barge has no onboard compressor, a compressor from another vessel was being utilized as the power source for unloading. All of the six compartments had been utilized to store cement. The off-loading operation was approximately 2 h old (cement was being unloaded from compartment 3), when a leak developed in the main airline just beyond the modulator valve. The pressure utilized to empty the cement from the barge was 207 kPa (30 psi). This pressure had to be reduced before repairs could be conducted on the air conveyor system. Consequently, it was necessary to wait for the pressure to abate before the leak could be fixed.

A section view of the barge, shown in Fig. 2, shows two of the six com-partments on the barge. Each of the individual compartments are loaded or unloaded through a common line. During the loading operation the top of the tanks are vented through a 0.305 m (12 in.) pipe, and during un-loading, the cement from each compartment is discharged through a 0.203 m (8 in.) line into a common 0.305 m (12 in.) discharge header.

An additional problem at this time was that the butterfly shut-off valves were not functioning properly. This was attributed to the caustic nature of cement and the fineness of the powder. As a result, there was a pressure

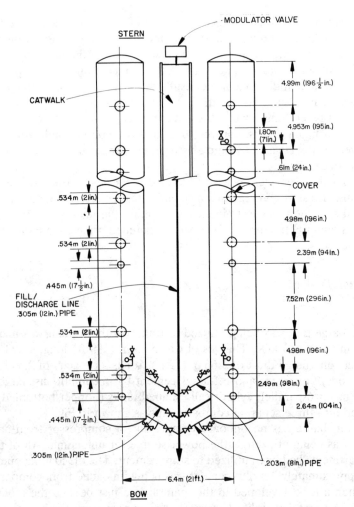

FIG. 1—*Plan view of barge.*

build up in the other compartments. To repair the leak, it was necessary to depressurize all compartments.

Gate valves located on compartments 5 and 6 were opened to relieve the pressure. This process was rather lengthy. To expedite matters, the operators began releasing the two dogs adjacent to the hinge on the hatch on each of the six compartments. The seals on these hatch covers being broken, the pressure in the compartments would dissipate more quickly. A number of compartments were vented in this manner. However, on one of the hatch covers on compartment 3, failure occurred immediately after an employee kicked off the second dog.

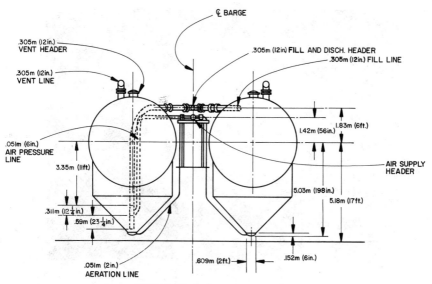

FIG. 2—*Typical transverse section.*

Hatch Cover

The hatch covers were manufactured from A356-T6 aluminum, and were 0.508 m (20 in.) in diameter and approximately 0.0064 m (0.25 in.) thick. Refer to Fig. 3 for the dimensions. Figure 4 shows a photograph of a representative hatch cover in the as-received condition. Figure 4*a* shows

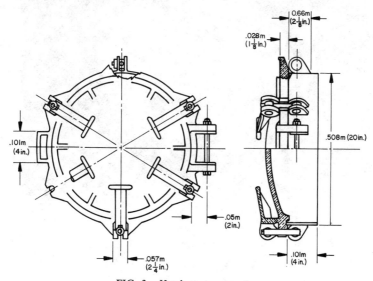

FIG. 3—*Hatch cover geometry.*

a)   HINGE INTACT

b)   HINGE   FAILED

FIG. 4—*Exemplar hatch covers.*

the hatch cover with the hinge intact, and Fig. 4*b* shows the hatch cover with the hinge broken.

There is a neoprene pressure seal on the hatch cover as shown in Fig. 4. As can be seen, the hatch cover is fastened to the hatch by six dogs, 60° apart. The attachment mechanism is also noted in Fig. 3.

Analytical Program

Analysis Procedure

The hatch cover was analyzed using a finite element analysis. The code utilized in this particular study was NASTRAN MSC-60. It was determined that one half the hatch cover need be modeled due to its geometric configuration and loading symmetry. This can be seen in Fig. 3. For the model, 111 elements were defined. These consisted of 14 bar elements, 7 rod elements, 60 quadrilateral elements, and 30 triangular elements. Figure 5 shows the model and the various elements.

To determine the stresses of the hatch cover as a function of pressure loading and partial engagement, five cases were analyzed. These were a baseline case, that is, all dogs attached. The internal pressure was defined as 276 kPa (40 psi). The additional four cases were at uniform pressures of 103.4 kPa (15 psi) and 275.8 kPa (40 psi). All four cases had one dog unattached in the model (two on the full hatch cover), and for each pressure condition the hinge was intact in one case and not intact in the other. Table 1 shows the conditions run.

Because of the symmetry of the hatch cover when the locking cams were engaged, it was necessary to determine the stresses of the various locations of interest only at one pressure due to stress linearity. However, when complex loading conditions were introduced, for example, broken hinges

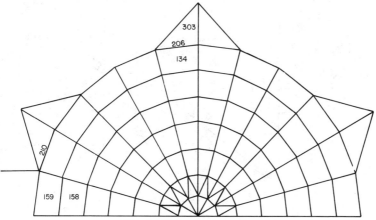

FIG. 5—*Finite element grid model.*

TABLE 1—*Test conditions run.*

Run No.	Pressure	Locking Cams	Hinge Condition
1	275.6 kPa (40 psi)	all attached	not intact
2	275.6 kPa (40 psi)	four attached	intact
3	103.4 kPa (15 psi)	four attached	intact
4	275.6 kPa (40 psi)	four attached	not intact
5	103.4 kPa (15 psi)	four attached	not intact

and open dogs or both, the stress response characteristics may deviate from linearity, and it would be necessary to bound the actual pressure. Interpolation could be then used to determine the characteristic stresses.

Figure 6 shows that the stress increase is noticeably faster as a function of pressure when the locking cams are opened. In this case, the attachment hinge has also failed. Figure 7 shows the maximum stress at a number of locations on the hatch cover when the attachment hinge is connected and two of the locking cams are opened. When the hatch is in this state, the stresses are higher in the zones adjacent to the hinge.

Analysis Correlation

A test program was conducted on a similar hatch cover in 1962 by Metal Control Laboratories. The test was accomplished in accordance with the

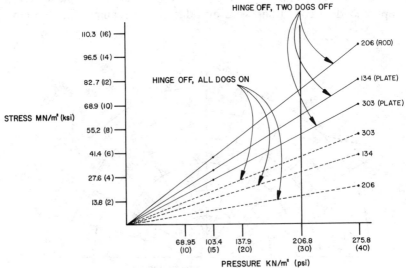

FIG. 6—*Maximum principal stress of selected elements.*

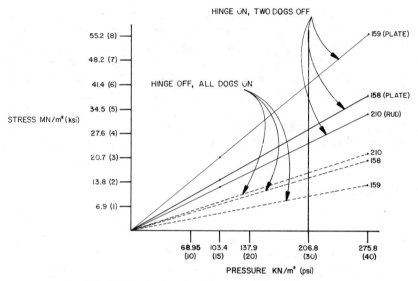

FIG. 7—*Maximum principal stress of selected elements.*

American Society of Mechanical Engineers (ASME) Unfired Pressure Vessel Code, Section VIII, paragraph UG101. In that test, Metal Control Laboratories placed strain gages on the hatch cover in the locations shown in Fig. 8. Prior to testing the cover was installed into a fixture simulating service conditions.

Pressure readings were taken on each of the five gages at 0 kPa (0 psi). The hatch cover was hydrostatically pressurized to 276 kPa (40 psi) and

NOTE: ALL DOGS ON, 275.8 KN/m²
(40psi) PRESSURE

GUAGE NO.	STRAIN MICRO METER/METER (MICRO INCH/INCH)	
	MEASURED	COMPUTED
1	1.40 (55)	1.17 (46.2)
2	1.45 (57)	1.14 (45.0)
3	1.45 (57)	
4	6.05 (238)	7.06 (278)
5	7.80 (307)	10.59 (417)

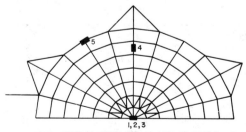

FIG. 8—*Finite element grid model with strain gage locations.*

readings were taken. Pressure was then reduced to 0 kPa (0 psi) and the strain gages were again read. The hatch cover was then repressurized to 413.7 kPa (60 psi), readings taken, and again the pressure was reduced to 0 kPa (0 psi) for determining drift or permanent deformation or both. This sequence was repeated in increments of 6.9 kPa (10 psi) until leakage of the gasket occurred at 1103 kPa (160 psi). Testing then ceased.

A comparison was made with the strain gage reading obtained from the test and the stresses obtained from test case 1 of the analysis. This comparison is shown in Fig. 8, and it shows reasonable correlation between test and analysis.

Experimental Procedure

Material

The material utilized for the hatch cover was A356-T6 aluminum. This was determined by analysis of the four additional hatch covers obtained from the barge. Table 2 lists the chemistry of the four hatch covers received, and in addition lists the requirements as described in ASTM Specification for Aluminum Alloy Sand Castings (B 26-82b). It should be noted that C-2 and C-3 met the requirements as described in ASTM B 26. C-1 was the cover that was most delinquent in meeting the chemistry.

Mechanical Tests

Both tension and fracture toughness tests were conducted on the hatch covers. Coupons were taken from two of the hatch covers. One tension test, in accordance with ASTM Specification for Tensile Testing of Metallic Material (E 8-82), and two fracture toughness tests, in accordance with ASTM Test Method for Plane Strain Fracture Toughness Testing of Metallic Materials (E 399-83), were then conducted. The test coupons for the tension tests were cylindrical test coupons with a 0.0508 m (2 in.) gage length and for the fracture toughness tests, they were 1/2 T compact specimens. Photographs of the specimens are shown in Fig. 9.

TABLE 2—Composition of exemplar hatch covers (weight percent).

Alloying Element	C-1	C-2	C-3	C-4	ASTM B 26
Cu	0.06	0.025	0.01	0.15	0.25 max
Zn	0.04	0.10	0.02	0.075	0.35 max
Fe	0.35	0.38	0.08	0.62	0.6 max
Si	3.95	6.80	7.40	7.00	6.5 to 7.5
Mn	0.12	0.15	0.002	0.24	0.35 max
Mg	0.93	0.30	0.34	0.50	0.2 to 0.4
Ti	0.01	0.06	0.15	0.025	0.25 max
Al	Re	Re	Re	Re	Re

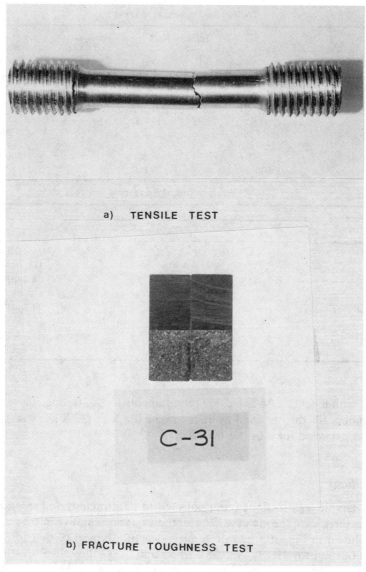

a)   TENSILE TEST

b) FRACTURE TOUGHNESS TEST

FIG. 9—Mechanical test coupons.

The load deflection curves from the tension tests were run in accordance with ASTM Specification E 8-82. The results of these tests are shown in Table 3. It should be noted that neither specimen met the requirements of ASTM Specification B 26-82b in either tensile strength or elongation. The fracture toughness tests were run in accordance with ASTM Method E 399-83. The K_{max} and K_Q values are also shown in Table 3. Specimens C-31 and

TABLE 3—*Mechanical test results.*

Tension Tests

Specimen Number	Tensile Strength	Yield Strength	Elongation 0.05 m gage length (2 in.)
C-3	137.9 MPa (20 ksi)	131 MPa (19 ksi)	<0.5%
C-4	161.7 MPa (23.5 ksi)	. . .	<0.2%
ASTM B 26	207 MPa (30 ksi)	138 MPa (20 ksi)	3%

Fracture Toughness Tests

Specimen Number	K_Q	K_{max}
C-31	16.2 MPa $\sqrt{m}$ (14.7 ksi $\sqrt{in.}$)	16.6 MPa $\sqrt{m}$ (15.1 ksi $\sqrt{in.}$)
C-32	11.9 MPa $\sqrt{m}$ (10.8 ksi $\sqrt{in.}$)	14.6 MPa $\sqrt{m}$ (13.3 ksi $\sqrt{in.}$)
C-41	12.5 MPa $\sqrt{m}$ (11.4 ksi $\sqrt{in.}$)	14.3 MPa $\sqrt{m}$ (13.0 ksi $\sqrt{in.}$)
C-42	14.2 MPa $\sqrt{m}$ (12.9 ksi $\sqrt{in.}$)	15.6 MPa $\sqrt{m}$ (14.2 ksi $\sqrt{in.}$)

C-42 both meet the ASTM E 399 requirements for K_{Ic} testing. The fracture toughness for this material averages 15.2 MPa $\sqrt{m}$ (13.8 ksi $\sqrt{in.}$) which is that expected for a brittle material.

Metallurgy

A sample was sectioned and polished to determine the microstructural characteristics of the material. These studies, representative microstructures are shown in Fig. 10, indicate a microstructure typical of that for ASTM Specification for Heavy-Walled Carbon and Low-Alloy Steel Castings for Steam Turbines (A-356-77) sand castings. There is evidence of shrinkage porosity.

Failure Analysis

The four hatch covers obtained from the barge were examined by radiographic (X-ray) and dye penetrant methods. Two of the four hatch covers, C-1 and C-2, showed cracking. C-1 had two edge cracks emanating from

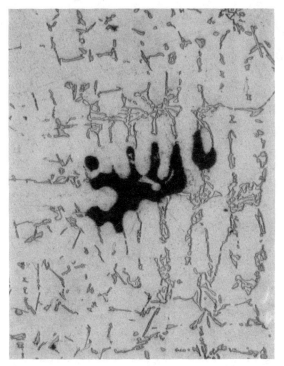

FIG. 10—*Typical microstructure showing shrinkage porosity (X100).*

the area of the broken hinge. the longer of these cracks, as measured on the radiograph, was 0.108 m (4.25 in.). This cracked area was removed from the hatch cover and is shown in Fig. 11. C-2 had an internal crack, probably due to shrinkage porosity, approximately 0.010 m (0.42 in.).

To determine the morphological characteristics of the edge crack, a study utilizing a JEOL Model JSM-U3 scanning electron microscope (SEM) utilizing an accelerating voltage of 25 kV was conducted. Prior to investigating that crack, a study to catalog the visual characteristics of overload and fatigue failure on the A356-T6 was performed. This study utilized the fracture toughness specimens discussed earlier.

The results of that study, shown in Figs. 12 and 13, indicate both fatigue and overload failure in this alloy. Figure 12a shows a low magnification of the entire area, whereas Fig. 12b shows the striations at a higher magnification.

Upon investigation of the fracture surface shown in Fig. 11, it was not possible to determine if fatigue was a contributing factor in the extension

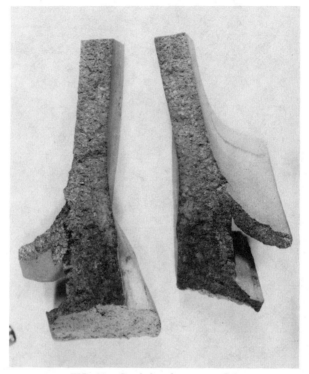

FIG. 11—*Crack face from cover C-1.*

of that crack. The photographs obtained, shown in Fig. 14, indicate corrosion product and areas of porosity (shrinkage cavity).

Discussion

Barge Design

As discussed earlier, the loading and unloading piping system of the barge was designed with butterfly valves separating the different compartments. As cement powder is extremely abrasive, degradation of the interior of the flow system occurred. This degradation, in time, would be significant enough to allow leakage, and this leakage would prevent the isolation of the various compartments.

In addition to the abrasive characteristics of cement, when it was in an aerated flow condition, it could easily pack itself into crevices and corners thus hampering valve movement. Again, in the case of a butterfly type shutoff valve, this is a serious problem as pressure leaks may become substantial.

Furthermore, the lack of sufficient gate valves on the vessel led to a

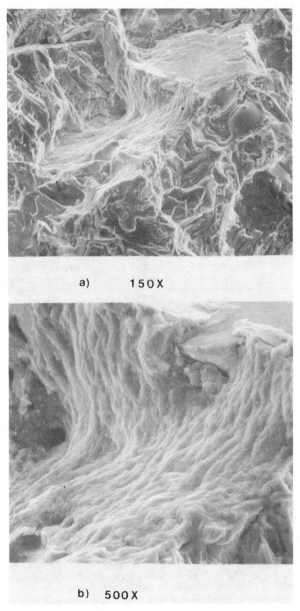

a)   150X

b)   500X

FIG. 12—*Fatigue region, compact specimen.*

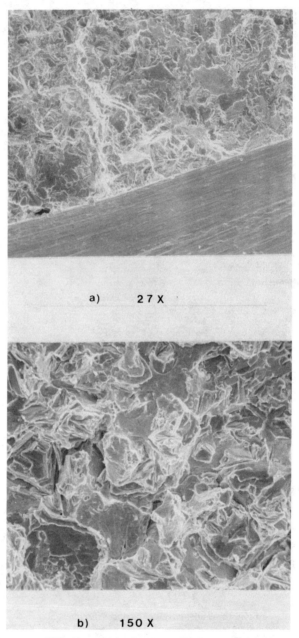

FIG. 13—*Overload region, compact specimen.*

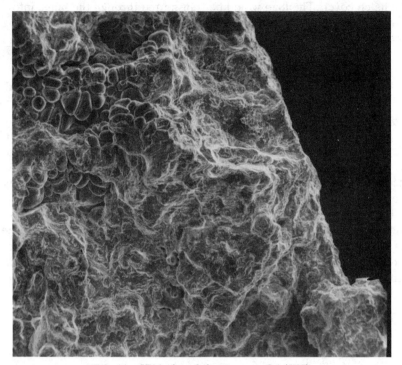

FIG. 14—*SEM of crack face in cover C-1 (X27).*

situation where depressurization was not occurring at a rate rapid enough for work to progress in a manner acceptable to the employees on the barge. If more gate valves had been present, they may have been opened and the depressurization rate enchanced.

Finally, there was a lack of sufficient pressure gages within the system. If the employees knew the pressure at various locations on the barge, preventative maintenance regarding the leakage may have taken place that would have hastened the depressurization process. If this had occurred, it would not have been necessary to partially open each hatch.

It should be noted that the particulars of the piping system and the flow system will not be discussed as they are beyond the scope of this paper.

Hatch Cover Design

The hatch covers were mounted in such a way on the barge that when they were in an open position they were resting neither on the deck surface nor in a protective cradle. Because of the geometry of the hatch when open, that is, cantilever, a load was imposed upon the hinge and its supporting structure. Furthermore, when traffic in the area of the hatch cover was heavy, equipment, cables, ropes, carts, personnel, etc., could easily contact the hatch cover. The load would be transmitted through the hinge into the cover. These impact type loads could place a significant stress in the area of the hinge. The hinge may crack, fail, and cause a crack to propagate into the hatch itself. It should be noted that at the time of this accident, 11 hinges on the 18 hatch covers on the barge were fractured.

The hatch cover opening mechanism was designed in a manner such that the locking cams could be open regardless of internal pressure. When the dogs were released and an internal pressure was present, the resulting stress on the hatch cover at the next attachment location increased as shown in Figs. 6 and 7. Designs exist that prevent opening of the hatch when there is a positive pressure on the hatch cover. If these types of designs had been incorporated into the hatch cover, it would not have been possible to release the pressure in the manner chosen.

The hatch cover, manufactured from A356-T6 aluminum, is a sand cast aluminum alloy. Although this alloy may be utilized where pressure tightness is required, a shortcoming is its low ductility and fracture toughness. Therefore, special care must be taken when using this alloy in an environment subjected to cyclical loads. Because the critical crack length is small, a program should be established that allows a periodic inspection of the critical areas.

Failure Mechanism

Figure 15 shows the difference in calculated stress-intensity for three loading configurations. These are: all dogs attached, four dogs attached with the hinge intact, and four dogs attached with the hinge failed. The stress intensity was determined by using the edge crack formula [1].

$$K = 1.12\sigma \sqrt{\pi a}$$

It can be seen in Fig. 15 that failure of the hatch cover would not occur if an edge crack were located adjacent to the failed hinge. Service experience indicates this is correct as failure did not occur on hatch cover 1 even though a crack of 0.108 m (4.25 in.) was present.

However, if a crack were located in the area of the locking cam 60° from the hinge, if the hinge had been broken, and the dogs opened as in this instance, the analysis indicates that a crack of approximately 0.009 to 0.083 m (0.35 to 3.25 in.) would cause failure.

Since the failed hatch cover was lost, it was not possible to categorically state where the crack was present. The analysis results indicate that the crack probably was adjacent to the next attached dog, and the hinge was not present. Thus, when the employee kicked open the dog, the increase in stress was sufficient to cause component failure.

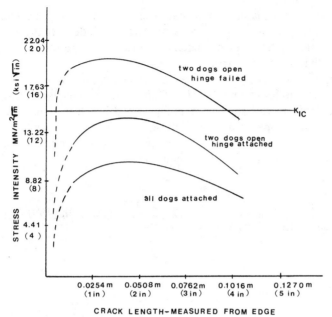

FIG. 15—*Failure diagram for hatch cover.*

Alternative Designs

There are a number of changes to the design of the barge, the hatch cover attachment, the hatch cover design, and the material that would have rendered this occurrence significantly less probable. These include:

1. Barge
   (*a*) Increased number of pressure gages.
   (*b*) Increased number of gate valves.
   (*c*) Utilization of more durable pressure isolation system.
2. Hatch design
   (*a*) Higher durability hatch attachment system.
   (*b*) Nonintegral hatch attachment system.
   (*c*) Increased protection of hatch cover when opened.
   (*d*) Nonopening locking cams when the compartment is pressurized.
   (*e*) Less fracture critical material.
      (i) Leak before failure criteria.
      (ii) Failure design to incorporate hatch cover deformation.

Conclusion

It can be seen from this paper that during the design process, the engineer must take into consideration a number of significantly different criteria if the product that is manufactured based upon his analyses and test results is useful, economically feasible and safe. These include both product design and material selection. In this instance there were a number of individual problems that led to catastrophic results.

A design based upon fracture mechanics techniques, if used, could have rendered this accident improbable. Thus, fracture mechanics can play a significant role in design and manufacturing. Fracture mechanics is not limited to highly sophisticated design such as aircraft, nuclear power plants and modern naval vehicles. It should be part of the design process wherever structural response plays an important role.

References

[1] Tada, H., Paris, P., and Irwin, G., *The Stress Analysis of Cracks Handbook,* Del Research Corporation, Hellertown, PA, 1973.

Hugh S. Pearson[1] and R. G. (Jesse) Dooman[2]

Fracture Analysis of Propane Tank Explosion

REFERENCE: Pearson, H. S. and Dooman, R. G., **"Fracture Analysis of Propane Tank Explosion,"** *Case Histories Involving Fatigue and Fracture Mechanics, ASTM STP 918,* C. M. Hudson and T. P. Rich, Eds., American Society for Testing and Materials, Philadelphia, 1986, pp. 65–77.

ABSTRACT: A truck-mounted propane tank exploded. The failure occurred through a girth weld, causing a section of the dome to fly from the tank. Initial evaluation of the failed tank showed a crack existed in the girth weld, for several years. The preliminary conclusion of this investigation was that the crack condition was the cause of failure; however, the authors suggested that a fracture mechanics analysis be conducted.

Four compact fracture toughness specimens were fabricated from weld metal in the tank and tested. It was shown that the failure could not have occurred from the existing crack and the tank internal pressure, unless the internal pressure far exceeded the maximum allowable 1.72 MPa (250 psi) pressure.

Further investigation led to the discovery that the relief valve of the tank was corroded so badly that it would not open, thus allowing excessive pressure in the tank. There was evidence that the driver had overfilled the tank. Heat from the sun and from the hot truck tailpipe caused the tank to become liquid full. Then, the pressure exceeded 6.9 MPa (1000 psi), causing the tank to explode. The fracture mechanics analysis thus showed that the explosion was not due only to the existing crack, but also to the stopped-up relief valve and excessive liquid propane in the tank.

KEY WORDS: fracture mechanics, failure analysis, tank explosion, pressure vessel, weldments, propane tank explosion

A truck-mounted propane tank (propane service truck) was filled, then driven to the driver's home, where he had lunch. Witnesses said the truck was left running. Shortly afterward, the tank exploded, causing considerable property damage and one death. The tank was constructed of one rolled steel plate (essentially 1030 steel), closed by a longitudinal weld and two domes joined to the rolled plate by girth welds. The failure occurred through a girth weld, causing a section of the dome to burst from the tank.

---

[1] President, Pearson Testing Laboratories, Inc., Marietta, GA 30067.
[2] Versitech, Inc., Norcross, GA 30071.

Figure 1 (*center*) shows the remains of the truck and the approximate location at the time of the failure. Figure 1 (*upper*) shows the separated section of the dome of the tank and its location after failure. Figure 1 (*lower*) shows the remaining portion of the tank.

Initial examination of the tank showed a crack existed in the subject girth weld, essentially all around it. The crack evidently existed for a long time (several years estimated), as indicated by slow, crack growth characteristics (crack branching). Based on this portion of the investigation it was preliminarily concluded that the crack was the cause of failure. However, a subsequent fracture mechanics analysis indicates that the failure was the result of excessive pressure in the tank. This paper presents the approach, procedure, analysis, and conclusions reached as a result of the investigation.

Visual Examination of Tank

The failed end of the tank was torch cut from the tank and brought into the laboratory for investigation as shown in Fig. 2. Also a small section of the circumferential weld opposite the failure was removed. This section is marked Exhibit 3 in Fig. 3. Also the section that separated from the tank was brought for examination and is shown in Fig. 3 as Exhibit 2.

The visual examination of the tank revealed:

1. The fracture had originated in the center portion of the circumferential weld, shown in Fig. 2, and propagated in both directions for a total of approximately 2000 mm (80 in.) before leaving the weld area and entering the parent material of the dome. The portion of the tank remaining is shown in Fig. 2. The size of the failure can be noted from the white numbers marked along the failure surface. The numbers are in inches from one side of the weld failure.

2. A visual examination of the welding revealed no reason to suspect defective welding.

3. The entire weld fracture was very flat indicating brittle fracture and the possibility of pre-existing cracks.

4. As a result of the visual examination specimens were cut off for closer microscopic, physical, and destructive analysis.

Metallurgical Evaluation

It was originally intended that impact tests be conducted of the circumferential weld that did not fail. Several blanks were laid out for specimens. When the specimens were being machined it was noted that cracks existed in each specimen in the center of the weld. At this point impact testing was abandoned in favor of a metallurgical examination.

All the sections examined contained cracks that originated at the "V" formed by the joint as shown in Fig. 4. During fabrication the longitudinal

FIG. 1—*Close view of the separated dome portion of tank* (upper); *View showing truck remains* (middle); *and view showing remaining tank portion* (lower).

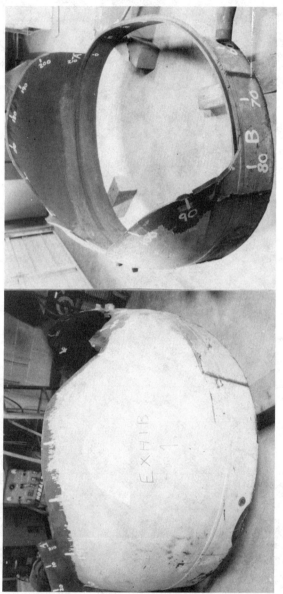

FIG. 2—End of failed tank.

FIG. 3—*Portion of tank that blew out (Exhibit 2). Also shown is circular area from other end of tank (Exhibit 3).*

section of the tank had been upset inward, the dome slipped over it and the circumferential weld made (this joint is an American Society of Mechanical Engineering approved joint type for pressure vessels). The cracks formed and propagated toward the outside of the tank through the weld area. Several features of the cracks led to suspicion of stress corrosion. Crack branching, as shown in Fig. 5, was found. In numerous instances, these crack branches propagated parallel to each other and subsequently rejoined, Fig. 6.

A section of the failed weldment was studied in the scanning electron microscope. Secondary cracking typical of stress corrosion was found.

Based on these findings it was concluded that numerous cracks had originated and propagated for several years in the welds of the tank. Furthermore, as a result of the above findings, it was concluded by some investigators involved that the vessel had failed during normal operation as a result of the existing cracked weldment.

Fracture Mechanics Evaluation

These investigators warned that the previously stated opinion could only be proven or disproven by a fracture mechanics evaluation which was approved by the tank manufacturer.

Plane-strain fracture toughness specimens (4 each) were made per ASTM Plane Strain Fracture Toughness of Metallic Materials (E 399-78) from the

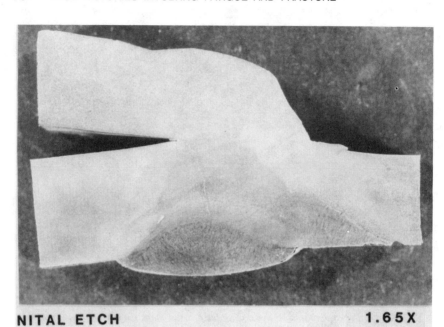

NITAL ETCH   1.65X

NITAL ETCH   6.4X

FIG. 4—*Typical cracks found in weld joint.*

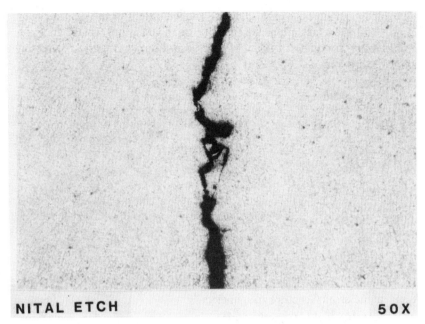

NITAL ETCH   50X

FIG. 5—*Crack branching found in cracks.*

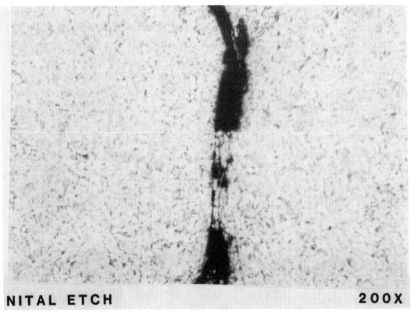

NITAL ETCH   200X

FIG. 6—*Parallel cracks.*

circular piece (Exhibit 3 of Fig. 3), with the weldment in the center of the specimen and the crack plane directed through the weld. Tests (and pre-cracking) were conducted on an MTS closed-loop servohydraulic testing machine with results as follows:

Specimen No.	Toughness K_Q,	
	MPa $\sqrt{m}$	(ksi $\sqrt{in.}$)
1	69.7	(63.4)
2	82.7	(75.3)
3	68.8	(62.6)
4	40.2	(36.6)

From the relationship in ASTM E 399-78

$$B \geqq 2.5 \ (K_{Ic}/\sigma_{ty})^2 \tag{1}$$

where

K_{Ic} = plane-strain fracture toughness,
σ_{ty} = material yield strength, and
B = material thickness.

It was determined that a valid K_{Ic} up to 38.5 MPa $\sqrt{m}$ (35 ksi $\sqrt{in.}$) could be determined from the available thickness (σ_{ty} = 550 MPa (80 ksi) deter-mined from hardness). While the value for Specimen 4 should have ap-proximated K_{Ic}, the failure did not appear different from the other tests; also, the test record was lost, and this number was considered questionable. From the tests (using the results of all tests) it was concluded that as a minimum a K_{Ic} of 38.5 MPa $\sqrt{m}$ (35 ksi $\sqrt{in.}$) or more likely a K_Q of approximately 73.6 MPa $\sqrt{m}$ (67 ksi $\sqrt{in.}$) (average of Specimens 1, 2, and 3) would best characterize the material in the weldment.

Next, a stress must be calculated for the tank. Using the data from Table 1 it was decided that we would calculate the working stress and the proof stress (1.5 times working stress) based on the working pressure. Then relate these numbers to a toughness level.

Using the standard equation [1]

$$\sigma = \frac{PR}{2t} = \frac{215 \times 33}{2 \times 0.475} \tag{2}$$

where

σ = tensile stress,
P = pressure in psi,
R = tank radius in inches, and
t = thickness in inches (thicker section).

TABLE 1—*Tank data.*

Design pressure: 1.72 MPa (250 psig)
Vapor pressure: not in excess of 1.48 MPa (215 psi) at 37.8°C (100°F)
Tank shell: 11.3 mm (0.443 in.)
Tank head: 12.1 mm (0.475 in.) (thicker at weld joint)
Length: 3710 mm (146 in.)
Outside diameter: 1680 mm (66 in.)
Water capacity: 7570 L (2000 gal)
Surface area: 22.3 m² (240 ft²)

We found the tensile stress equal to 51.7 MPa (7.5 ksi). Then, adding bending to account for the difference in wall thickness and dome thickness

$$\sigma_B = \frac{MC}{I} = \frac{6 \times 0.032 \times 215 \times 33/2/2}{(0.475)^2} \tag{3}$$

where 0.032 is the difference (in inches) in the centroids of the two materials.

We found the bending stress equal to 10.3 MPa (1.5 ksi). Adding, the working stress is 52 MPa (7.5 ksi) + 10 MPa (1.5 ksi) or 62 MPa (9 ksi) and the proof stress would be 93 MPa (13.5 ksi). Next, we selected the Irwin Part-Through Crack Equation [2] (also published in Naval Research Laboratory report in 1956).

$$K_I = 1.1 \, \sigma \sqrt{\frac{\pi a}{Q}} \tag{4}$$

where

K_I = stress intensity factor,
σ = applied stress,
a = crack depth, and
Q = flaw shape parameter.

We assume a crack depth of 12.7 mm (0.5 in.) (assuming a 19.1 mm (0.75 in.) thickness at the weld center), and a crack length ($2c$) of 127 mm (5 in.). (This resulted in an $a/2c$ ratio of 0.1.) The depth was assumed based on the fact that cracks approximately 10 mm (0.4 in.) deep were found. A little extra allows for conservatism. A length ($2c$) of 127 mm (5 in.) is larger than we believed possible, although some of the cracks found had joined along the surface.

Thus

$$K_I = 1.1 \, \sigma \sqrt{\frac{\pi \times 0.05}{1.09}} = 1.45 \text{ MPa } \sqrt{m} \text{ (1.32 ksi } \sqrt{\text{in.}})$$

or

$$\sigma = 0.758 \, K_I$$

Next, using the 62 MPa (9 ksi) working stress and 93.1 MPa (13.5 ksi) proof stress we find the K_I level at working stress to be 13.1 MPa $\sqrt{m}$ (11.88 ksi $\sqrt{in.}$) and at proof stress to be 19.6 MPa $\sqrt{m}$ (17.82 ksi $\sqrt{in.}$).

These values for stress-intensity factor are significantly less than the material toughness of between 38 and 73 MPa $\sqrt{m}$ (35 and 67 ksi $\sqrt{in.}$). Consequently, the tank could not have failed from even the proof pressure, much less the working pressure.

Using the equations to determine what the operating pressure must have been at failure, we find

$$\sigma_F = 0.758 \, K_I$$

$$\sigma_F = 183 \text{ MPa (26.5 ksi) (for } K_I = 38 \text{ MPa } \sqrt{m} \text{ (35 ksi } \sqrt{in.}\text{))}$$

or

$$= 350 \text{ MPa (50.8 ksi) (for } K_I = 73 \text{ MPa } \sqrt{m} \text{ (67 ksi } \sqrt{in.}\text{))}$$

Converting to pressure using the relation that 62 MPa (9 ksi) stress corresponds to a 1.48 MPa (215 psi) working pressure, we find the failure pressure to be between 4.36 MPa (633 psi) and 8.36 MPa (1213 psi). These values equate to approximately 3 to 6 times the working pressure.

Explanation for the Pressure in the Tank

At first it appears that the calculations cannot be correct, since the tank contains a relief valve set approximately at the working pressure. Further investigation found that the relief valve, shown in Fig. 7, was rendered inoperative by rust and corrosion. This explains how no gas could escape, but, if the tank had been filled properly, excessive pressure would simply liquify the gas. A properly filled tank does not contain more than 82.4% liquid by volume at 15.6°C (60°F) [3]. There was evidence, however, that the tank had been overfilled with liquid.

Further investigation leads to the conclusion that in addition to the stopped-up relief valve, the tank had been almost filled with liquid. Still, it was difficult to understand how simply heating by sunlight could cause enough heat to generate pressures like those calculated.

On inquiry, it was found that the driver had apparently left the truck running while he ate lunch. The heat from the muffler must have heated the tank. A test was conducted on a 1970 Chrysler where thermocouples were placed on the muffler, the frame, and a room temperature reference. (The room temperature reference remained at 20°C (68°F).) The engine

FIG. 7—*Corroded relief valve removed from tank.*

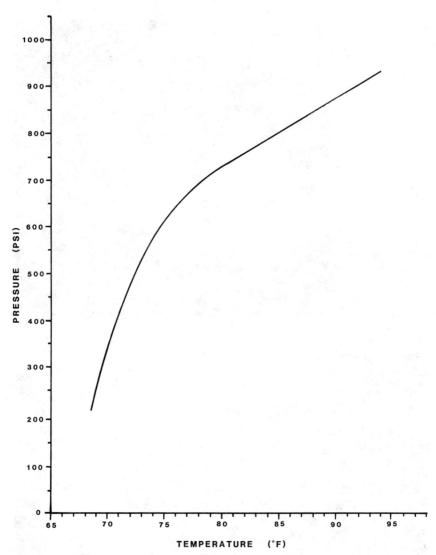

FIG. 8—*Pressure rise for temperature increase in liquid full tank.*

was started and allowed to idle for 6000 s (1 h and 40 min). After 1800 s (30 min) the muffler temperature stabilized at 102°C (215°F) and the frame temperature stabilized at 43°C (110°F). It was then concluded that a large amount of heat is available from an operating engine.

Next, a series of calculations including the swelling of the tank and accounting for the properties of propane (Ref *4*) resulted in the graph shown in Fig. 8. The significance of this figure is that pressures in excess of 4.1 MPa (600 psi) can result from a temperature increase of only 2.7°C (5°F) after the tank was liquid full. This temperature rise could be easily reached from muffler heat.

Conclusions

The failure was the result of multiple interacting factors.

1. Overfilling.
2. Heat generated by the exhaust system and solar radiation.
3. Inoperative safety relief valve.
4. Cracks in the weldment.

The existence of the cracks in weldments clouded the primary causes for this failure and would have resulted in merely dismissing the failure with insufficient investigation as an in-service failure.

Acknowledgments

We thank J. C. Tomlinson, welding consultant and J. Hubbard, electronmicroscopist, for their assistance in the preliminary evaluation.

References

[1] *Elements of Strength of Materials*, Timoshenko & MacCullough, VanNostrand, New York, 1940.
[2] Irwin, G. R., "Crack Extension Forces for a Part-Through Crack in a Plate," *Journal of Applied Mechanics*, Volume 84E, No. 4, Dec. 1962.
[3] *Hand Book of Butane-Propane Gases*, Chilton Publisher, Philadelphia, 1962.
[4] Katz, C. et al, *Handbook of Natural Gas Engineering*, McGraw-Hill, New York, 1959.

C. K. (Peter) Chow[1] and Leonard A. Simpson[1]

Analysis of the Unstable Fracture of a Reactor Pressure Tube Using Fracture Toughness Mapping

REFERENCE: Chow, C. K. and Simpson, L. A., **"Analysis of the Unstable Fracture of a Reactor Pressure Tube Using Fracture Toughness Mapping,"** *Case Histories Involving Fatigue and Fracture Mechanics, ASTM STP 918,* C. M. Hudson and T. P. Rich, Eds., American Society for Testing and Materials, Philadelphia, 1986, pp. 78–101.

ABSTRACT: On 1 August 1983, a Zircaloy-2 pressure tube in Ontario Hydro's Pickering-2 reactor failed by the unstable propagation of an axial crack. The reactor was quickly and safely shut down, without the need for any emergency cooling, and an extensive investigation into the event was initiated. One factor requiring explanation was the apparent failure of the leak-before-break principle, that is, why was the initiating flaw not detected by leakage prior to growing to an unstable size? This paper describes part of a program to address this question, which involved mapping the fracture toughness of the pressure tube as a function of position over the cracked region. The results of the fracture toughness mapping, plus selected fractography, are used to explain the sequence of events that occurred during the tube failure.

KEY WORDS: zirconium, pressure tube, zirconium hydride, critical crack length, fracture toughness, *J*-resistance curve

The CANDU reactor design is unique in that the heavy-water coolant passes through the core in a lattice of pressure tubes made of either Zircaloy-2, or Zr-2.5Nb (Fig. 1). A more detailed view of a single fuel channel assembly is shown in Fig. 2. The pressure tube has an inside diameter of 103 mm and a wall thickness of either 5 mm (Zircaloy-2), or 4.2 mm (Zr-2.5Nb). It contains the uranium dioxide fuel and operates at a temperature of 240°C at the coolant inlet end, rising to 290°C at the outlet end. The pressure-tube lattice is immersed in a pool of heavy water at 75°C, which serves as the moderator for the nuclear reaction. To insulate the pressure tube from this potential heat sink, it is surrounded concentrically by a second Zircaloy-2 tube, called the calandria tube. There is a 9-mm-wide annulus

[1] Research officer and manager, respectively, Materials and Mechanics Branch, Whiteshell Nuclear Research Establishment, Atomic Energy of Canada Limited, Pinawa Manitoba, Canada R0E 1L0.

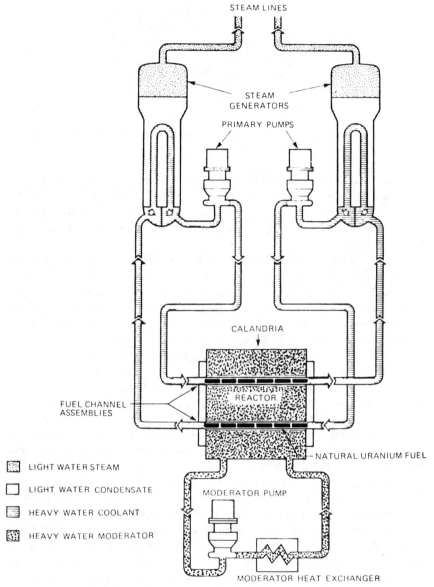

FIG. 1—*Cutaway view of a CANDU reactor core.*

between the two tubes, which is filled with a circulating gas, either carbon dioxide or nitrogen. The spacing between the two tubes is maintained by either two or four tightly wound zirconium alloy springs, called garter springs (Fig. 2).

If a crack should develop in a pressure tube and, prior to reaching an unstable size, penetrate the wall, it will leak moisture into the normally dry

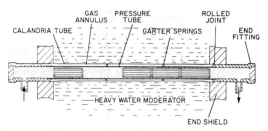

FIG. 2—*Schematic of a fuel channel for a CANDU reactor with pressurized heavy-water coolant.*

annulus region. This will be detected, enabling the reactor to be shut down prior to the crack reaching an unstable configuration. Thus, if it can be shown that any growing crack will penetrate the tube wall and leak prior to reaching an unstable size, a leak-before-break behavior has been established.

In fact, leak-before-break has been demonstrated in CANDU reactors during operation. In 1974 and 1975, several Zr-2.5Nb pressure tubes in Ontario Hydro's Pickering Generating Station Units 3 and 4 developed leaks at axial cracks near the end fittings where the pressure tube is joined to the external coolant circuitry. The reactors were shut down and inspected, and a number of cracked tubes were replaced. Examination of the tubes showed that semi-elliptical cracks had initiated on the inside of the tube wall and grown through the wall, the largest attaining a length of about 16 mm [1]. Since the critical length for a through-wall axial crack in a pressure tube at operating pressure ranges from 50 mm to over 100 mm for the range of anticipated material conditions during service, a large margin (in terms of crack size) apparently existed for leak-before-break. This was confirmed by an extensive research program that identified the crack growth mechanism as delayed hydride cracking and quantified the crack growth rates [2].

On 1 August 1983, a Zircaloy-2 pressure tube, identified as the G-16 tube from its core location in the Pickering Unit 2 reactor, failed by the unstable propagation of an axial crack. The reactor was quickly and safely shut down, without the need for emergency cooling, and an extensive investigation into the event was initiated. One factor requiring explanation was the apparent failure of the leak-before-break principle, that is, why was the initiating flaw not detected by leakage prior to instability? This paper describes part of a program to address this question, which involved mapping the fracture toughness of the G-16 pressure tube as a function of position over the cracked region.

Visual Examination

The cracked pressure tube was removed and examined at the Chalk River Nuclear Laboratories [3]. A photograph of the inboard (towards the interior of the reactor) end of the crack is shown in Fig. 3a. The crack was located at the bottom (the six-o'clock position) of the pressure tube. A number of white, blister-like, features were visible at points near the crack path and are indicated by letters in Fig. 3a. These were subsequently identified, using metallography, to be massive precipitates of zirconium hydride. Examination of the fracture surface near the center of the blister array revealed a flat brittle topography, intersecting several blisters (Fig. 3b). A change in coloration outlined a boundary that was suspected to be the stable size of the initiating flaw. It was approximately 100 mm long, but had not quite penetrated the wall, a thin shear lip being evident along the inner tube surface. From this point towards the outboard limit of the crack, the fracture surface acquired an increasing shear component, until it was 45° shear fully through the thickness. This end of the crack was arrested in the end fitting after travelling a distance of 1.7 m. In the inboard direction, the crack propagated approximately 0.3 m in a primarily brittle mode, and then bifurcated into two circumferential cracks (Fig. 3a), which were arrested after propagating about 50 mm in each direction.

The conclusions of the overall failure analysis were as follows [3]. The outlet garter spring moved about one metre inboard of its design location, probably during construction or commissioning. This eventually resulted in contact between the pressure tube and calandria tube, due to sag of the former by creep. The pressure tube was cooled at the contact points, causing hydrogen and deuterium to migrate by thermal diffusion and to precipitate there as zirconium hydride/deuteride.[2] This behavior was aggravated by a very high pickup of deuterium from the heavy-water coolant during operation. This resulted in a high concentration of hydrides near the blisters, which are known to lower the fracture toughness. A crack was initiated by an unspecified mechanism at one or more blisters and grew to an unstable size, possibly by delayed hydride cracking [2], without penetrating the inner surface of the tube.

A fracture mechanics analysis was clearly needed to quantify this behavior. Firstly, it was important to determine the fracture toughness of the tube near the suspected initiation site, to verify that it was consistent with the applied stress-intensity factor on the flaw boundary. Secondly, to understand why the unstable crack ran the course it did, it was necessary to determine the fracture toughness over the length of the crack and, particularly, at its extremeties where it bifurcated or was arrested. Before proceeding to the fracture mechanics analysis, it will be useful to review some of the important features of the metallurgy of the zirconium-hydrogen system that we will draw upon in our fracture mechanics assessment.

---

[2] Pressure tubes in service contain both hydrogen and deuterium. The former is present originally as a normal impurity, while deuterium is picked up from the heavy-water coolant during operation. As their respective effects on mechanical behavior are indistinguishable, when we speak of hydrides in this paper, we mean hydrides, deuterides, or both.

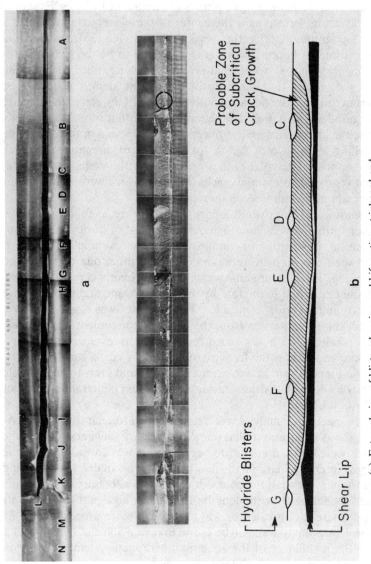

(a) External view of blistered region and bifurcation at inboard end.
(b) Photograph and drawing of fracture surface showing suspected initiating flaw.
(Circled region was examined in a scanning electron microscope.)

FIG. 3—*Crack in Pickering G-16 pressure tube.*

Zirconium-Hydrogen System

Hydrogen solubility in zirconium has been traditionally expressed in parts per million by weight or, more recently, in micrograms per gram. Because we are dealing with both hydrogen and deuterium in our system, the weight of deuterium is divided by two to give the "equivalent" hydrogen concentration. The solubility of hydrogen isotopes in zirconium ranges from less than 1 $\mu g/g$ at room temperature to approximately 60 $\mu g/g$ at 300°C [4]. When the solubility is exceeded, zirconium hydride precipitates in a platelet morphology. Under certain conditions [5], these platelets will fracture readily, and if they are aligned in the plane of an approaching crack, they can significantly lower the fracture toughness of the alloy.

A newly manufactured pressure tube contains approximately 10 $\mu g/g$ of hydrogen, which means that during reactor operation all hydrogen is in solution. However, during operation, deuterium is picked up, mainly via a corrosion reaction involving the heavy-water coolant [6]. In the Zircaloy-2 tubes in the Pickering Units 1 and 2 reactors, it was found [6] that the corrosion rate accelerated when the oxide film reached a critical thickness. This effect, which is enhanced by increasing temperature and irradiation, resulted in an unexpectedly high rate of ingress of deuterium into the outlet end of the pressure tube. Hydrogen concentrations (equivalent) were as high as 150 $\mu g/g$ near some parts of the crack in the G-16 tube. This finding, plus the discovery that other tubes in Pickering Units 1 and 2 were in the same condition as G-16, led to the decision to retube both reactors with Zr-2.5Nb, the alloy used in all other large CANDU power reactors. The Zr-2.5Nb alloy has a much lower pickup rate for deuterium than Zircaloy-2; a recent examination of a tube from Pickering Unit 3 showed less than 3 $\mu g/g$ hydrogen equivalent in the main body of the tube.

In both pressure-tube alloys, the normal orientation for the hydride platelets is in the circumferential plane of the tube, that is, the platelet normals are in the radial or through-wall direction. In this orientation, they have a negligible effect on the propagation of a through-wall, or part through-wall, crack propagating in the radial or axial direction. However, if the circumferential stress (hoop stress) in the tube exceeds a threshold value during hydride precipitation [7], the platelets will precipitate with their normals parallel to that stress, such that the platelets will lie in the radial-axial plane. A crack can propagate through the plane of the platelet with little resistance; hence, platelets in this orientation have a strong, deleterious effect on fracture toughness. For Zr-2.5Nb, this threshold stress is about twice the operating hoop stress. However, recent unpublished results from Atomic Energy of Canada Limited and Ontario Hydro suggest that, for Zircaloy-2, the threshold is very close to the operating hoop stress (90 MPa). Thus, in the G-16 tube, two factors were at work to lower the fracture toughness: the high ingress rate of deuterium and the precipitation of the resulting hydrides in the radial-axial plane.

The third factor that contributed to the failure was the contact between the pressure tube and the calandria tube. Hydrogen diffuses down a temperature gradient in zirconium by a process known as thermal diffusion [8]. If the point of contact is below the solvus temperature for the local hydrogen concentration, hydrides will precipitate, forming the nucleus of a blister, and serve as a sink for a steady-state diffusion of hydrogen. The result is a blister of zirconium hydrides surrounded by a two-phase region of hydride platelets and α zirconium (Fig. 4). A line of such blisters was found on the G-16 tube (see Fig. 3a), indicating that pressure-tube/calandria-tube contact had occurred at several points.

Experimental

To map the fracture toughness of the G-16 pressure tube, it was first necessary to develop a test method. Previous programs had been carried out to measure the effect of irradiation on pressure-tube material [9,10]. However, these had used compact specimens machined from flattened, unirradiated tubing that was subsequently irradiated in test reactors. Here we were faced with the problem of both producing and testing a specimen from a piece of highly radioactive tubing.

Because the fracture behavior was expected to range from brittle near the initiation region to very ductile at the crack extremities, we required a test method that would characterize the full range of crack propagation behavior. A method was chosen to take full advantage of the test procedures developed earlier for flat specimens [9,10]. This included determining the J-resistance curve for each specimen when stable crack growth occurred.

Flattening the material was not possible because of the irradiation embrittlement, and the concern that any dislocation movement would destroy the unique defect structure produced during irradiation. A procedure was, therefore, developed for testing curved specimens, spark machined from a piece of tubing in a single operation, using a "cookie cutter" electrode (Fig. 5). To ensure electrical conductivity for spark machining, the oxide film from the inside of the tube was first removed by sanding. The specimens were then formed by initiating sparking from the inside of the tube. Finally, the specimens were pickled in 45% nitric acid (HNO_3) 45% water (H_2O) and 10% hydrofloric acid (HF) for about 30 s to clean out the pinholes.

Figure 5 shows the dimensions of a typical specimen. Except for the thickness and the curvature of the tube, the in-plane dimensions of this specimen are in the proportions described for compact specimens in ASTM Standard Test Method for Plane-Strain Fracture Toughness of Metallic Materials (E 399-83). The two studs on the front edge of the specimen were used to attach the specimen to the excitation d-c current, to permit monitoring of the crack length by potential drop at the crack mouth. The potential drop (PD) pickup wires, made of Zr-2.5Nb, were spot-welded within 1 mm of either side of the specimen crack mouth.

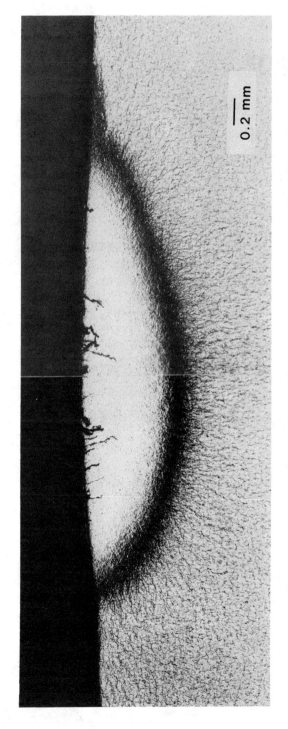

0.2 mm

FIG. 4—*Typical hydride blister in a Zircaloy-2 pressure tube arising from contact with calandria-tube.*

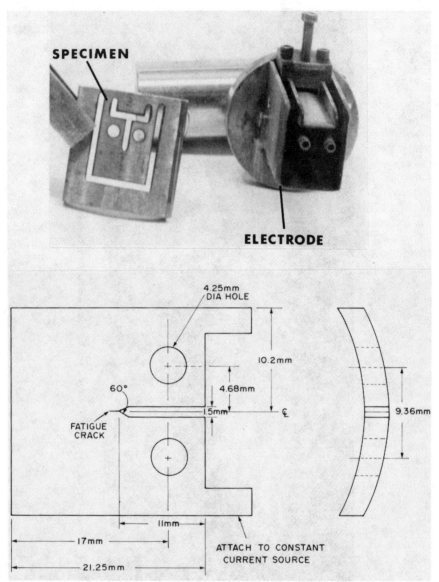

FIG. 5—*Curved compact specimen geometry and "cookie cutter" electrode for spark machining.*

Before proceeding with the testing of the G-16 material, a specimen was subjected to an extensive analytical and experimental program, to determine the errors involved in applying the flat-plate analysis to a curved specimen. This included comparative testing of sets of flat and curved specimens taken from identical material, and having a wide range of fracture toughness values. It was shown that the error involved in applying the flat-plate equations to these curved specimens was less than 10% [11].

Specimens were cut from the regions of the G-16 pressure tube shown in Fig. 6 and tested at the temperatures recorded in Table 1. The normal operating temperature of the tube is 280°C, while 230°C is the estimated, through-wall mean temperature in the contact region. The specimen, with the current and PD leads, was mounted in an Instron servo-hydraulic test machine in the hot cells at the Whiteshell Nuclear Research Establishment. Fatigue cracks were induced at room temperature by cycling between loads of 0.2 and 1.6 kN. The maximum load was reduced as the crack propagated. The maximum stress intensity factor was about 15 MPa · m$^{1/2}$ when the fatigue process stopped, with a crack length to specimen width ratio of 0.5. After fatiguing, the specimen was enclosed by a furnace and heated to the desired test temperature for fracture toughness testing. Care was taken that the temperature did not overshoot, as this might have changed the properties of the specimen, including the hydride morphology. When the temperature had been stabilized, the specimen was pulled at an extension rate of 0.25 mm/min.

During the test, time, load, potential drop, and load-point displacement were recorded by a data logger and tape recorder. The load-point displacement was measured using two linear variable differential transducers (LVDT) attached to opposite sides of the grips. The average of the LVDT readings minus the pin deflections gave the load-point displacement. The latter were determined by a preliminary calibration. If the specimen showed stable crack growth, the test was stopped when the potential drop indicated that

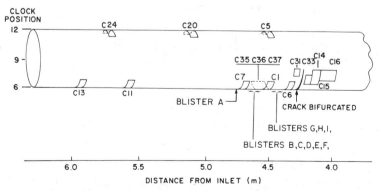

FIG. 6—*Locations of fracture toughness specimens.*

the crack had advanced about 4 mm. After unloading, the specimen was heat-tinted at the test temperature for 30 min. It was then broken open at room temperature. The fracture surface was photographed, and the crack lengths were measured by the nine-point average method.

Following testing, metallographic sections were prepared to examine the hydride morphology in selected specimens, and fracture surfaces were also examined by scanning electron microscopy. A segment of the suspected initiation region, outlined by the circle in Fig. 3b, was also examined to compare the features on either side of the alleged boundary between slow and unstable crack growth.

Data Analysis

For brittle specimens, the following equation was used to calculate the stress-intensity factor, as per ASTM Method E 399-83

$$K = \left(\frac{P}{B\sqrt{W}}\right)(2 + c)/(1 - c)^{3/2}$$
$$\times [0.886 + 4.64\,c - 13.32\,c^2 + 14.72\,c^3 - 5.6\,c^4] \quad (1)$$

where

P = applied load,
B = specimen thickness,
W = width of specimen,
a = crack length measured from the load line, and
c = a/W

The fracture toughness was taken as the critical value of Eq 1, at unstable propagation.

For ductile specimens, a crack-growth resistance curve based on the J-integral was calculated for each specimen. This J-resistance curve was calculated using equations derived by Ernst et al [12]

$$J_{i+1} = \left[J_i + \left(\frac{f(a_i/W)}{b_i}\right) \cdot \frac{A_{i,i+1}}{B}\right]\left[1 - \left(\frac{\gamma_i}{b_i}\right)(a_{i+1} - a_i)\right] \quad (2)$$

where

$\gamma_i = 1 + 0.76\,[(W - a_i)/W]$,
$f(a_i/W) = 2\,[(1 + \alpha)/(1 + \alpha^2)]$,
$\alpha = [4\,(a_i/b_i)^2 + 4\,a_i/b_i + 2]^{1/2} - [(2a_i/b_i) + 1]$,
$b_i = W - a_i$, and
$A_{i,i+1}$ = area under the load/load-point displacement curve between step i and $i + 1$.

Results

All of the specimens tested at 230°C showed brittle behavior, with a linear load deflection curve rising to a maximum coincident with unstable crack extension (see Table 1). Three specimens tested at 280°C are also classified as brittle, although they each showed limited (1 mm) stable crack growth prior to unstable failure. The remainder of the specimens listed in Table 1 were classified as tough because they showed several millimetres of stable crack growth during the test, and never became unstable. Typical load deflection curves for the two conditions are shown in Fig. 7. In general, the tough specimens did show some short-lived instabilities prior to maximum load (Fig. 7a). The potential-drop trace showed that these instabilities corresponded to sudden, short, crack-growth steps (pop-ins).

Table 1 also includes average tension test data for the two test temperatures, and these were used to calculate the minimum thickness, B, and crack length, a, for a valid, plane-strain, fracture toughness test according to ASTM Method E 399-83, using

$$a, B \geq 2.5 \ K^2/\sigma_y^2 \qquad (3)$$

where σ_y = yield strength.

The fracture toughness values for the brittle specimens were used to compute the critical crack length (CCL) in a pressure tube at operating pressure, which corresponds to a hoop stress of 90 MPa. The fracture tough-

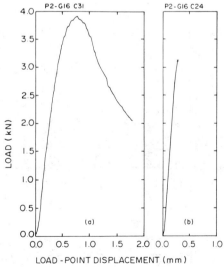

FIG. 7—*Typical load deflection curves for* (a) *ductile, and* (b) *brittle fracture behavior.*

TABLE 1—Summary of fracture toughness results and specimen properties.

Specimen No.	Test Temperature, °C	Tough or Brittle	Fracture Toughness, K_Q, MPa · m$^{1/2}$	Yield Stress, σ_y, MPa	[H] Equivalent, µg/g	Critical Crack Length, mm	2.5 K_Q^2/σ_y^2, mm
C5	280	T	54.0[a]	662	61	71	...
C7	280	T	63.6[a]	662	59	102	...
C13	280	T	69.7[a]	662	95	108(LB)	...
C14	280	T	56.9[a]	662	42	88	...
C16	280	T	63.6[a]	662	45	95	...
C31	280	T	67.3[a]	662	58	103(LB)	...
C33	280	T	66.0[a]	662	42	99(LB)	...
C35	280	T	55.4[a]	662	...	75	...
C1	230	B	48.7	719	72	49	11
C6	230	B	53.6	719	63	53	14
C11	280	B	75.4	662	110	70	14
C15	230	B	50.3	719	...	50	7.7
C20	280	B	46.0	662	109	47	13
C24	280	B	62.0	662	146	60	12
C36	230	B	40.0	719	81	41	19
C37	230	B	48.5	719	66	49	27

[a] Calculated from $K_Q = \sqrt{EJ_i}$, where J_i is read from the first pop-in on the J-resistance curve.

ness for a through-wall axial crack is given by [13]

$$K_Q = \left[\frac{8a_c\bar{\sigma}^2}{\pi} \cdot \ln \left(\sec \frac{\pi M\sigma_H}{2\bar{\sigma}} \right) \right]^{1/2}$$ (4)

where

$2a_c$ = critical crack length,
$\bar{\sigma}$ = 1/2 (yield strength + ultimate strength),
σ_H = applied hoop stress, and
M = stress magnification factor due to tube curvature [14].

Since M is a complicated function of a_c [14], this equation was solved graphically from plots of K versus crack length.

For the ductile specimens, typical J-resistance curves are shown in Fig. 8. All specimens showed some short instabilities or pop-ins in the early stages of the test, and the J-value corresponding to the first of these was defined as an initiation J-value, or J_i. (The application of ASTM Standard Test Method for J_{Ic}, a Measure of Fracture Toughness (E 813-81) to determine crack initiation does not yield reproducible results in irradiated zirconium alloys [9,10]). An equivalent initiation fracture toughness was calculated using

$$K_Q = \sqrt{EJ_i}$$ (5)

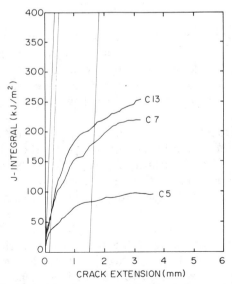

FIG. 8—*Typical J-resistance curves for ductile fracture behavior.*

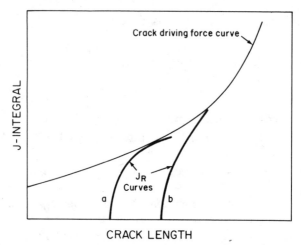

(a) Tangency method for complete J_R-curve.
(b) Lower-bound method for truncated J_R-curve.

FIG. 9—*Method for estimating through-wall, axial critical crack length using a crack-driving-force curve.*

The axial CCL for the ductile conditions was estimated by matching the J_R-curves to a crack-driving-force (CDF) curve. The latter was computed from Eq 4 but converted to units of J by Eq 5. An example of this method is shown in Fig. 9. Three of the J_R-curves did not reach a saturation level before the test was terminated, and the true tangency point may have been beyond the end of the J_R-curves. In this case, the CCL was read off when the top of the J_R-curve just touched the CDF curve and was, thus, a lower-bound estimate.

All but two of the fracture toughness specimens were analyzed for hydrogen and deuterium using a hot, vacuum extraction method. The results are given in Table 1 in terms of hydrogen equivalent. A significant observation is that the three specimens that showed brittle behavior at 280°C all had much higher hydrogen concentrations than the other specimens tested at that temperature.

The hydride morphology was examined in some selected specimens (Fig. 10). A large fraction of the hydrides were radial. The concentration was fairly uniform through the tube thickness, except for the blistered area where the hydrides were concentrated towards the outside of the tube. In this region the predominance of radial hydrides was also greatest at the outside of the tube.

Figure 11a shows the fracture surface of a brittle specimen and the flat cleavage regions corresponding to fractured hydrides. These regions are separated by ligaments of ductile void coalescence, which are created when the zirconium metal between the hydride platelets is separated. Figure 11b

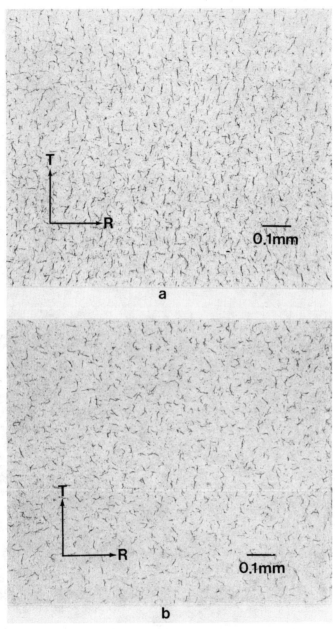

(a) Specimen C20: [H] = 109 μg/g.
(b) Specimen C7: [H] = 59 μg/g.

FIG. 10—*Typical hydride morphology in fracture toughness specimens showing large fraction of radial hydrides.*

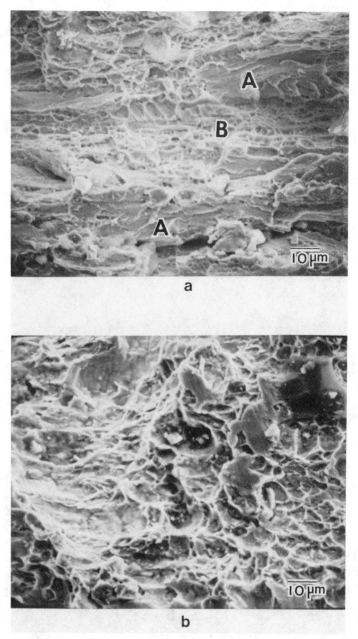

(a) Brittle fracture behavior showing cleavage of hydrides (A) and ridges of ductile tearing between hydrides (B).

(b) Ductile fracture behavior showing almost 100% void growth and coalescence.

FIG. 11—Typical fracture surfaces on fracture toughness specimens as seen in a scanning electron microscope.

shows the fracture surface of ductile specimens. The fracture surface consists almost entirely of ductile voids. Second-phase particles (intermetallic inclusions) can be seen inside some of the voids.

Discussion

Validity Criteria

Reference to Eq 3 and Table 1 reveals that none of the brittle fracture toughness tests satisfy the validity criterion for thickness and crack length required for a valid, plane-strain, fracture toughness measurement. Given the nonplanar geometry of the specimen, this observation is somewhat academic, but, in order to assess the failure behavior of the pressure tube, some feeling for the physical significance of the results is needed. The failure to meet the thickness criterion is irrelevant since the small specimen is the same thickness as the structure (pressure tube) being assessed. The requirement on crack length is close (our crack lengths were typically 8 to 9 mm) in some cases, but the lengths are small by up to a factor of two, or more. This means that the stress-intensity factor may not uniquely characterize the crack-tip, elastic stress field; therefore, the CCL prediction may be in error. However, we intend to use the CCL values only in an approximate sense and the estimates should be adequate for that purpose.

Similar concerns arise when we address the geometry independence of the J-resistance curve. The conditions for J-controlled crack growth (that is, when the J-integral uniquely describes the crack-tip, elastic-plastic deformation field) are based on the demonstration that proportional loading dominates the crack-tip field near an advancing crack tip [15]. Numerically, this means

$$\omega = \frac{W - a}{J} \frac{dJ}{da} >> 1 \tag{6}$$

Although there is no detailed justification, $\omega = 10$ is often taken as the critical value for validity in Eq 6. Application of this criterion to our data indicated that J-control was violated after 0.2 to 0.5 mm of crack growth. Thus, we cannot expect the J-resistance curves to be the same for the small specimen and the pressure tube, and the CCL predictions would be in error. Simpson [16] has shown that the use of J_R-curves from compact specimens underestimates the CCL in a tube, and hence gives a conservative value.

While there is some doubt about the absolute significance of the toughness values generated here, it is completely valid to use them in a comparative sense, to assess the variation of fracture toughness over the length of tube and the temperatures experienced. We shall show that this can reveal a great deal about the condition of the tube and the behavior of the crack.

Crack Initiation

The blistered region of the pressure tube was approximately 500 mm long, indicating that contact between the pressure tube and calandria tube existed over this region. Most fracture toughness tests on material from this region were performed at 230°C, which was the estimated mean through-wall temperature in the contact region. Specimens C1, C36, and C37 (Table 1) were taken very close to the suspected initiation site, near blisters C, D, E, and F (see Fig. 3). However since the specimen edge was just above the fracture surface, the crack plane was approximately 11 to 12 mm around the circumference of the tube. As this was away from the center of contact, which was a line running axially at the six o'clock position, the embrittlement could have been less severe in the fracture plane of the small specimens.

The fracture toughness in the region near the initiation site is in the range 40 to 50 MPa $\cdot$ m$^{1/2}$. This leads to an estimated CCL of 63 to 72 mm for a through-wall axial crack in a pressure tube with a 90 MPa hoop stress.

In some unpublished work, P. H. Davies of the Ontario Hydro Research Division applied the finite-element results of Raju and Newman [17] to the suspected, initiating crack boundary shown in Fig. 3b. They calculated a stress-intensity factor of 40 to 50 MPa $\cdot$ m$^{1/2}$ around the boundary of the crack for the reactor operating pressure. They also calculated that the thin uncracked ligament would have been in a state of plastic collapse, causing some load redistribution around the crack boundary. This would have caused the point of maximum stress intensity to shift from the point of deepest penetration through the wall towards the ends of the crack. Given these facts and the possible errors in the fracture toughness determinations, the results obtained here are wholly consistent with the suspected crack boundary being the limit of stable crack growth.

Further support for this part-through region being the initiating point is provided by the scanning electron micrographs taken on either side of this boundary. Figure 12a shows an enlargement of the region circled in Fig. 3b, which encompasses the boundary between the part-through crack and the alleged unstable crack surface. This boundary is drawn in Fig. 12a because the color contrast does not show up in the electron image. Photographs from region A (stable growth) and region B (unstable growth) are shown in Figs. 12b and 12c, respectively. Region A shows predominantly cleavage features with very few ductile tearing ridges. This would be characteristic of a heavily hydrided region, which might have grown by delayed hydride cracking [2] at subcritical K-levels. Region B also contains cleavage features, but these are connected by large regions of microvoid coalescence (ductile features), indicating that a significant transition in resistance to crack propagation occurred at the boundary in Fig. 3b. The features in region B are also quite similar to those seen in the unstable crack propagation region of the fracture toughness specimens (see Fig. 11a).

This change in fractographic appearance is consistent with a crack that initiated at one or more blisters, and grew by delayed hydride cracking at a relatively low K-value up to the boundary. It would not grow completely through the wall thickness because the equivalent hydrogen concentration at the inner surface was close to, or below, the limit of solubility for hydrogen, a consequence of the higher temperature at the coolant/pressure-tube interface. (Hydrogen content in excess of the solubility, that is, the presence of hydrides, is a prerequisite for delayed hydride cracking [2].) The crack continued to grow axially, in the normal stepwise manner characteristic of delayed hydride cracking. The stress intensity factor at the axial extremities of the crack was low, and a high density of hydride platelets was required to precipitate near the crack tip before it could advance, hence the absence of any ductile features in region A of Fig. 12a. As plastic collapse of the ligament spread outward from the point of deepest penetration, pop-through occurred, creating, suddenly, a through-wall crack approximately 100 mm in length, well in excess of the CCL for an axial crack (Table 1). The sudden transition in fracture morphology across that boundary is consistent with the sudden rise in stress intensity that would result from the conversion to an axial crack.

Ductile-Brittle Transition

Where the pressure tube was not in contact, its temperature during operation was 280°C. Thus, most of the fracture toughness tests away from the initiation region were carried out at this temperature. These specimens generally showed ductile fracture behavior, with little evidence on the fracture surfaces of fractured hydride platelets (Fig. 11b). Specimens C11, C20, and C24, however, were exceptions. They failed in a brittle manner much like the specimens tested at 230°C. The significant difference with these specimens was the much higher equivalent hydrogen concentrations. They were cut from the region of the tube where the combined effects of neutron flux and temperature yielded a maximum in the hydriding rate during the operation of this reactor [6].

Zirconium/zirconium hydride alloys undergo a remarkable ductile-brittle transition when the hydrides are radial. This is shown by some results on flat specimens of Zircaloy-2 containing 90 μg/g hydrogen as radial hydrides [18] (Fig. 13). Below the transition, at about 260°C in this case, the cracks propagate as much as possible by cleavage of hydride platelets, while above it the hydrides suddenly become innocuous with respect to the fracture process (even though they are still present). Davies and Stearns [19] have shown that this transition temperature increases with hydrogen content, rising from 90°C for [H] = 30 μg/g to 260°C for [H] = 90 μg/g. The hydrogen contents for the specimens showing brittle behavior at 280°C ranged from 109 to 140 μg/g. This is consistent with movement of the

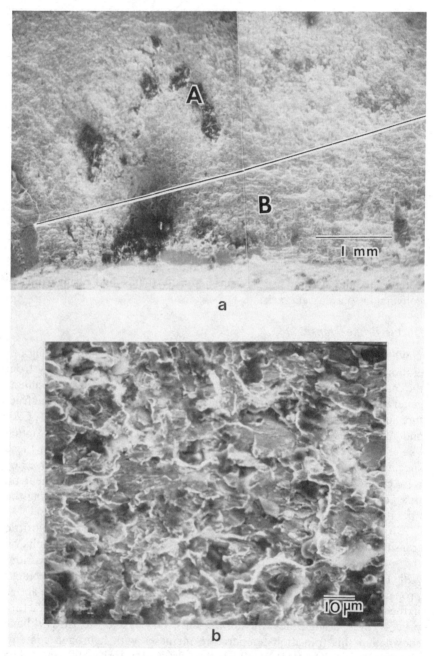

(a) Location of boundary and regions examined.
(b) Suspected region of stable crack growth (A).
(c) Suspected region of unstable crack growth (B).

FIG. 12—*Scanning electron microscopy of the region near the boundary of the suspected initiating flaw (see circle in Fig. 3b).*

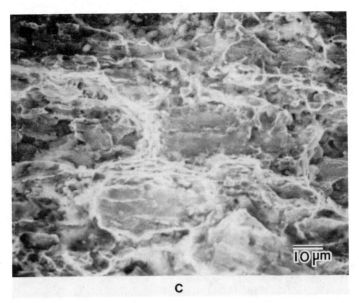

c

FIG. 12—*Continued*.

transition temperature to more than 280°C, as would be expected from the established data.

Two specimens from the initiation zone (C5 and C7) were tested at 280°C and showed ductile behavior, demonstrating that the ductile-brittle transition in this region lay between 230 and 280°C, consistent with the hydrogen concentration there (60 to 80 µg/g).

Specimen C13, with a hydrogen content of 95 µg/g, was taken only 23 cm towards the end fitting from the peak hydrogen region (specimen C24), where the hydrogen content was 146 µg/g. However, this drop in hydrogen concentration was sufficient to lower the transition temperature to below 280°C, and the fracture behavior was highly ductile, showing one of the steepest J_R-curves. It is clear, therefore, that the ductile-brittle transition played a significant role in the mode of crack propagation.

Bifurcation

There are several possible explanations for the bifurcation of the crack at the inboard end. As the crack propagated, the annulus region became pressurized due to the escaping coolant. This reduced the pressure drop across the crack and, therefore, the hoop stress. The axial stress on the tube arises from the end loads, which were not reduced. Eventually, the axial stress on the tube would have become the maximum principal stress and the crack could have tended to turn in response. There were blisters near the bifurcation point, and it is possible that one of them contained a

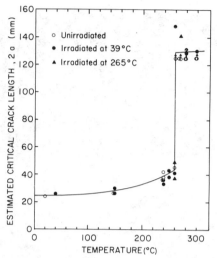

FIG. 13—*Results from fracture toughness tests on compact specimens containing 90 µg/g radial hydrides, showing brittle-ductile transition.*

circumferential crack that was rendered unstable under the dynamic stress field ahead of the propagating crack. A sudden change in axial fracture toughness distribution would have also encouraged bifurcation.

Specimens C31 and C33 (Table 1), tested at 280°C, were cut from opposite sides of the bifurcated crack. Their K_Q-values and CCL estimates were almost identical, indicating that no abrupt change in fracture toughness occurred in that region. Similar conclusions can be drawn from the results of specimens C6 and C15, which were tested at 230°C. Both of these specimens showed brittle behavior and nearly identical fracture toughness values. Thus, it would appear that the bifurcation was a consequence of either the blisters in the path of the advancing crack, or the change in the principal stress direction or both.

Summary and Conclusions

1. The curved compact specimen geometry has been used successfully to map the fracture toughness distribution along the path of a crack in a failed pressure tube.

2. The test specimens were ductile or brittle, depending on temperature and hydrogen content, consistent with current data on the ductile-brittle transition behavior of zirconium/zirconium hydride alloys.

3. Brittle behavior was related to cleavage of hydride platelets while ductile fracture was dominated by ductile tearing of the zirconium matrix.

4. Fracture toughness results and fractography support the identification of the critical initiating flaw.

5. The bifurcation of the crack was not induced by a sudden change in fracture toughness.

Acknowledgments

This paper describes only a small part of a many-faceted program to study the Pickering G-16 pressure-tube failure. We have drawn extensively on the results of that program in this paper. While it is not possible to mention all the individuals who have contributed, we must first acknowledge Ontario Hydro, who provided the funding for this study, via George Field and his failure analysis group. We are also grateful to Brian Cheadle and his colleagues at our Chalk River Nuclear Laboratories for providing their results prior to publication. Finally, we are grateful to our own technical staff, including Greg Brady, Dennis Stark, and Wayne Stanley, for their ingenuity and dedication in carrying out this work.

References

[1] Perryman, E. C. W., *Nuclear Energy,* Vol. 177, 1978, pp. 95–105.

[2] Simpson, L. A. and Puls, M. P., *Metallurgical Transaction,* Vol. 10A, 1979, pp. 1093–1105.

[3] Field, G. J., Dunn, J. T., and Cheadle B. A., "An Analysis of the Pressure Tube Rupture at Pickering NGS "A" Unit 2," *Canadian Metallurgical Quarterly,* Vol. 24, 1985, pp. 181–188.

[4] Kearns, J. J., *Journal of Nuclear Materials,* Vol. 22, 1967, pp. 292–303.

[5] Simpson, L. A., *Metallurgical Transactions,* Vol. 12A, 1981, pp. 2113–2123.

[6] Urbanic, V. F. and Cox, B., "Long-Term Corrosion and Deuteriding Behaviour of Zircaloy-2 Under Irradiation," *Canadian Metallurgical Quarterly,* Vol. 24, 1985, pp. 189–196.

[7] Hardie, D. and Shanahan, M. W., *Journal of Nuclear Materials,* Vol. 55, 1975, pp. 1–13.

[8] Sawatzky, A., *Journal of Nuclear Materials,* Vol. 2, 1960, pp. 321–328.

[9] Simpson, L. A., "The Effect of Irradiation and Irradiation Temperature on the Fracture Toughness of Cold-Worked Zr-2.5 Wt% Nb," Technical Report AECL 8368, Atomic Energy of Canada Ltd., Pinawa, Manitoba, Sept. 1984.

[10] Simpson, L. A. in *Mechanical Behaviour of Materials-IV,* J. Carlsson and N. G. Ohlson, Eds., Pergamon Press, Oxford 1984, Vol. 2, pp. 739–746.

[11] Chow, C. K. and Simpson, L. A., "Determination of the Fracture Toughness of Irradiated Reactor Pressure Tubes Using Curved Compact Specimens," 18th National Symposium on Fracture Mechanics, American Society for Testing and Materials, 1985, Boulder, Colorado (to be published).

[12] Ernst, H. A., Paris, P. C., and Landes, J. D. in *Fracture Mechanics: Thirteenth Conference, ASTM STP 743,* American Society for Testing and Materials, Philadelphia, 1981, pp. 476–502.

[13] Keifner, J. F., Maxey, W. A., Eiber, R. J., and Duffy, A. R. in *Progress in Flaw Growth and Fracture Toughness Testing, ASTM STP 536,* American Society for Testing and Materials, Philadelphia, 1973, pp. 461–481.

[14] Folias, E. S., *Engineering Fracture Mechanics,* Vol. 2, 1970, pp. 151–164.

[15] Hutchinson, J. W. and Paris, P. C. in *Elastic-Plastic Fracture, ASTM STP 668,* American Society for Testing and Materials, Philadelphia, 1979, pp. 37–64.

[16] Simpson, L. A. in *Advances in Fracture Research,* D. Francois, Ed., Pergamon Press, Oxford, 1982, Vol. 2, pp. 833–841.

[17] Raju, I. S. and Newman, J. C., Jr., *Journal of Pressure Vessel Technology,* Vol. 104, 1982, pp. 293–298.

[18] Simpson, L. A. and Chow, C. K., "The Effect of Metallurgical Variables and Temperatures on the Fracture Toughness of Zirconium Alloy Pressure Tubes," Seventh International Conference on Zirconium in the Nuclear Industry, Strasbourg, 1985, American Society for Testing and Materials (to be published).

[19] Davies, P. H. and Stearns, C. P. in *Fracture Mechanics Seventeenth Volume, ASTM STP 905,* American Society for Testing and Materials, Philadelphia, 1986, pp. 379–400.

Cedric N. Reid¹ and Brian L. Baikie²

Choosing a Steel for Hydroelectric Penstocks

REFERENCE: Reid, C. N. and Baikie, B. L., **"Choosing a Steel for Hydroelectric Penstocks,"** *Case Histories Involving Fatigue and Fracture Mechanics, ASTM STP 918,* C. M. Hudson and T. P. Rich, Eds., American Society for Testing and Materials, Philadelphia, 1986, pp. 102–121.

ABSTRACT: This case history concerns the design of the high pressure water pipes (penstocks) of the Dinorwig pumped storage power station. The first part describes an Open University teaching exercise which is based on the original assessment performed by Milne of the Central Electricity Generating Board. It involves the consideration of two contending steels, "A" and "B," and the selection and justification of a preferred choice. Since the hydraulic loading contains cycles of changing pressure, the fatigue lives of the penstock designs were compared. The possible presence of longitudinal surface cracks was considered, with a depth equal to the limit of ultrasonic detection (taken to be 6 mm). The Paris-Erdogan equation was used to calculate the length of time required to grow a fatigue crack from this depth up to the critical size. Comparing designs of equal cost, that in Steel B achieved a much longer "crack growth lifetime" and was the preferred choice. Finally, defect acceptance standards were drawn up for this design.

KEY WORDS: Dinorwig, pressure pipe, fatigue, crack growth lifetime, choice of material, defect acceptance standards

Nomenclature

σ_Y Yield strength
a Depth of semi-elliptical crack
$2b$ Length of semi-elliptical crack
K_I Plane-strain stress intensity
h "Head" of water
ρ Density of water
P Hydraulic pressure
R Mean radius of pipe
D Internal diameter of pipe
t Wall thickness of pipe
σ_R Residual stress

¹ Professor, Faculty of Technology, The Open University, Milton Keynes, U.K.
² Research officer, Central Electricity Generating Board, Manchester, U.K.

C Material cost of penstock
£   Cost per tonne of steel
N Number of loading cycles
G Crack growth parameter

The Central Electricity Generating Board (CEGB) operates the world's largest interconnected electricity supply system; it transmits electricity in bulk through the national grid to twelve Area Boards who in turn supply the consumers in England and Wales. The CEGB must have a large, quickly available reserve of power to meet sudden increases in demand, to safeguard the nation's electricity supplies in the event of a major failure of generating or transmission plant, and to maintain the frequency of the grid supply between prescribed limits. Hydroelectric and pumped-storage stations can be brought into operation much more quickly and reliably than other types of station. Hydroelectric power stations use water falling from a high-level source to produce electricity. The water drives turbines which, in turn, drive the generators that produce the electricity. Pumped-storage schemes, in effect, use water to store electrical energy. A pumped-storage station differs from a normal hydroelectric station because it has two reservoirs and uses the same water again and again. After driving the turbines the water is pumped from the lower reservoir back to the upper reservoir ready to be used again when needed. This is done mainly at night when demand for electricity is low and electricity for pumping is, therefore, cheaper, being provided by the most efficient, base-load power stations.

The pumped storage station at Dinorwig in North Wales is the subject of this case history. With a capacity of 1800 MW it is the largest such station in Europe, and it came into full operation in 1983. Water from the upper reservoir flows through hydraulic tunnels at a maximum rate of 3290 m^3/s. Inlet tunnels 3 km long, including a vertical shaft 440 m deep and 10 m in diameter, carry the water to the turbines (Fig. 1). In some pumped-storage schemes, the turbines and pumps are separate machines, but, at Dinorwig, the six turbine generators can also work in reverse, as motor pumps, to return the water to the top reservoir, powered by electricity from the national grid. The flow of water from the upper reservoir to the turbines is controlled by the main inlet valve. When the valve is closed the turbine is normally on "standby," spinning in air.

The Design Task

The task concerns the plain cylindrical sections of the upper and intermediate penstocks—the pressure pipes of internal diameter 2.5 m on either side of the main inlet valve (Fig. 2). Over part of their lengths, these pipes are not supported by surrounding rock and concrete, and consequently they have to be able to withstand the full loading of the water pressure throughout

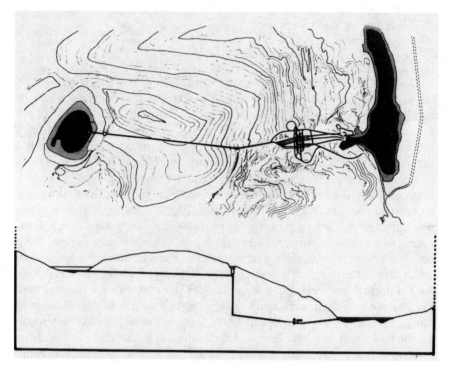

FIG. 1—*A map of the Dinorwig site and a vertical section through the tunnels.*

the design life of the system (55 years). The upper penstock is critical; since it lies above the main valve, it cannot be unloaded for internal inspection or repair (without draining the entire upper lake, high pressure shaft and tunnels).

The task is to select, from two candidate materials, the material which will provide safe penstocks with the required lifetime and the minimum cost.

Loading

The penstocks are loaded hydraulically. On the upper penstock with the main valve closed, the static internal pressure is that due to a "head" of water of 645 m, but the largest possible transient pressure is estimated to be 816 m of water, due to a "water hammer" effect which could occur under an unlikely coincidence of certain machine trip conditions (the "fault condition").

Changes in water pressure occur during operation of the system. For example (Table 1), when the station changes from pumping to generating

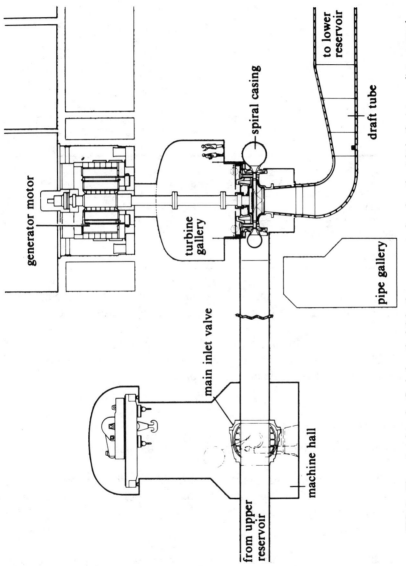

FIG. 2.—*The main inlet valve flanked on the left by the upper penstock and on the right by the intermediate penstock.*

TABLE 1a—*Upper penstock.*

Operation	Cycles per Annum, n_i	Pressure Change, h_i/m	
		First Cycle	Successive Cycles
Turbine load on	6500	115	95 <11
Turbine load off	6500	117	43, 36, 30, 26, 21, 18, 15, 13 and 11
Pump start	1000	27	<11
Pump stop	1000	238	199, 167, 140, 118, 99, 83, 70, 59, 49, 41, 35, 29, 24, 20, 16, 14, and 12
Turbine trip	4	323	262, 220, 185, 155, 130, 109, 92, 77, 65, 54, 45, 38, 32, 27, 23, 19, 16, 13, and 11
Pump trip	4	390	327, 275, 230, 194, 163, 137, 115, 96, 81, 68, 57, 48, 40, 34, 28, 24, 20, 17, 14, and 12

TABLE 1b—*Intermediate penstock.*

Operation	Cycles per Annum	Pressure Change/m
All mode changes starting and finishing with turbine spinning in air	7500	575

(or vice versa, a "mode change"), the main inlet valve is closed and opened, and the intermediate penstock is subjected to a pressure change equal to 575 m.

Pressure changes in the upper penstock are smaller but more complicated (see Table 1). Notice that the pressure changes following each operation are cyclic, with a steadily decreasing amplitude.

Candidate Materials

Two alternative materials were proposed for the penstocks:

(*a*) Steel A, a medium strength, low-alloy, quenched and tempered steel;
(*b*) Steel B, a carbon-manganese pressure vessel steel conforming to British Standards Specification BS1501:1964 (steels for fired and unfired pressure vessels).

The chemical compositions, mechanical properties, and prices of these steels are given in Table 2. It is to be assumed that the costs of fabricating the penstocks by welding are the same for the two steels. Notice that the yield strength of Steel A is more than twice that of Steel B, while its price is less than twice that of Steel B.

TABLE 2—*Material specifications.*

	Steel A	Steel B
C	0.15 to 0.21	<0.22
Si	<0.90	0.10 to 0.55
S	<0.04	<0.05
P	<0.04	<0.05
Mn	0.80 to 1.10	0.90 to 1.60
Cr	0.50 to 0.80	<0.25
Mo	0.25 to 0.60	<0.10
Ni		<0.30
Zr	0.05 to 0.15	
B	0.0005 to 0.0025	
σ_Y/MN m^{-2}	695	324 to 347
$\sigma_{tensile}$/MN m^{-2}	803	495 to 594
elongation	18%	26.5
K_{Ic}/MN m$^{-3/2}$	100	120[a]
Price of rolled plate/£ tonne^{-1}	542	295

[a] The mean value at 273K for various welded joints.

Design for Life

The objective is to choose one of these steels for the construction of each of the upper and intermediate penstocks and to specify the wall thickness for each penstock. Obviously, the wall must withstand the largest load applied (whether static or transient); in order to provide a "factor of safety," it is proposed that the largest principal stress difference in the material should not exceed two thirds of the yield strength σ_Y. The penstocks are fabricated like pressure vessels—by welding (longitudinally and circumferentially) shaped steel plates. Although these welds are to be stress-relieved, experience within CEGB suggests that a figure of 70 MN m^{-2} should be taken for the maximum residual hoop stress σ_R in the vicinity of the welds in both steels.

It is essential to consider the effects of cracks (both critical and subcritical) on the strength and lifetime of the structure. In so doing linear elastic fracture mechanics (LEFM) will be used. It is important to keep in mind the limits of applicability of LEFM, and the nature of any error (over- or underestimates) when LEFM is not strictly applicable.

As critical load-bearing components, the penstocks will be subjected to nondestructive inspection (NDI) after fabrication and during service. For the purposes of design it is proposed to assume that ultrasonic inspection is incapable of detecting cracks with a dimension in the through-thickness direction (*a* for a surface crack, 2*a* for an internal crack) of 6 mm or less (assuming that the other crack dimension, 2*b*, is longer).

In comparing the steels, it is further proposed that the presence of a hypothetical radial/axial surface crack should be considered on either inside or outside surfaces of the pipe. For the sake of simplicity and pessimism, the crack will be considered to have infinite length and a depth of a. The stress intensity factor K_I for this crack is given by Eq 1, where t is the thickness of the pipe wall and bending of the pipe wall (bulging) is not constrained.

$$K_I = Y\sigma\sqrt{(\pi a)} \tag{1}$$

where

$$Y = 1.12 - 0.231(a/t) + 10.55(a/t)^2 - 21.71(a/t)^3 + 30.38(a/t)^4$$

A crack on the inside surface of the pipe is also loaded by the hydraulic pressure P acting on its faces. The stress intensity due to this loading is [1]

$$K_I = P\sqrt{(\pi a)} \tag{2}$$

Since we are dealing with thin-walled tubes, $P \ll \sigma$, so this effect has been omitted.

Static Design of Penstock

Both the upper and intermediate penstocks are cylindrical thin-walled tubes with an internal diameter D of 2.5 m. Using the Tresca yield criterion and the prescribed "factor of safety"

$$\sigma_1 - \sigma_3 = \tfrac{2}{3}\sigma_Y \tag{3}$$

In this case, the radial stress (σ_3) is zero and the hoop stress (σ_1) is the sum of the applied and residual stresses

$$\frac{PR}{t} + \sigma_R = \frac{2}{3}\sigma_Y \tag{4}$$

Substituting for P and R

$$\frac{(\rho gh)[(D/2) + (t/2)]}{t} + \sigma_R = \frac{2}{3}\sigma_Y \tag{5}$$

Rearranging gives

$$t = \frac{3\rho ghD}{4\sigma_Y - 6\sigma_R - 3\rho gh} \tag{6}$$

Substituting the appropriate values

$$\rho = 1000 \text{ kg m}^{-3}$$
$$g = 9.81 \text{ m s}^{-2}$$
$$h = 816 \text{ m (the "fault condition")}$$
$$\sigma_R = 70 \text{ MN m}^{-2}$$

$$t = \frac{1}{6.68(\sigma_Y - 105 \times 10^6)10^{-8} - 1} \qquad (7)$$

Taking the yield stress of Steel A from Table 2, the required wall thickness is

$$t_A = 0.026 \text{ m} \qquad (8)$$

Using the minimum value of σ_Y for Steel B, the wall thickness is

$$t_B = 0.073 \text{ m} \qquad (9)$$

Which Static Design Has The Lower Cost?

The cost depends on the weight of steel used (Table 2). Since the diameter and length are fixed, the weight depends on the wall thickness, and the relative cost is

$$\frac{C_B}{C_A} = \frac{t_B}{t_A} \frac{\text{£}_B}{\text{£}_A} = \frac{0.073}{0.026} \times \frac{295}{542} = 1.53 \qquad (10)$$

On the basis of a static design, Steel B has costs 50% higher than Steel A. However, it is clear that the loading is not static, and the effect of dynamic loading must be considered.

Effect of Cracks on Design

Critical Cracks

Under changing loads, subcritical cracks may grow by fatigue up to the critical crack length. It is proposed to calculate the critical crack lengths, $a_c(A)$ and $a_c(B)$, of long surface cracks perpendicular to the hoop direction.

Recalling the expression for stress intensity (Eq 1) and putting in the critical values

$$K_{Ic} = Y\sigma\sqrt{(\pi a_c)} \qquad (11)$$

where σ is the design stress ($\frac{2}{3} \sigma_Y$).

Rearranging gives

$$a_c Y^2 = \frac{9K_{Ic}^2}{4\pi\sigma_Y^2} \qquad (12)$$

where Y is a function of a.
Recalling from Table 2 that

$$K_{Ic}(A) = 100 \text{ MN m}^{-3/2}, \qquad \sigma_Y(A) = 695 \text{ MN m}^{-2}$$

$$K_{Ic}(B) = 120 \text{ MN m}^{-3/2}, \qquad \sigma_Y(B) = 324 \text{ MN m}^{-2}$$

it follows that

$$a_c(A)Y^2 = 0.015 \text{ m} \qquad (13)$$

$$a_c(B)Y^2 = 0.098 \text{ m} \qquad (14)$$

Using Eq 1, these equations can be solved for a_c by trial and error, with the aid of Table 3.
The critical crack lengths in both penstocks under the maximum loading ($h = 816$ m) are

TABLE 3—Values of $[(a/t)Y^2]$ as a function of $[a/t]$.

a/t	Y (from Eq 1)	$(a/t)Y^2$
0.10	1.25	0.16
0.20	1.50	0.45
0.22	1.55	0.53
0.23	1.57	0.57
0.29	1.80	0.94
0.30	1.83	1.00
0.31	1.87	1.08
0.35	2.05	1.47
0.36	2.10	1.59
0.40	2.30	2.12
0.41	2.36	2.28
0.42	2.42	2.46
0.50	3.00	4.50
0.55	3.50	6.74
0.60	4.15	10.33

for Steel A,

$$a_c(A) = 6 \text{ mm} \tag{15}$$

for Steel B,

$$a_c(B) = 25 \text{ mm} \tag{16}$$

These are *upper* estimates; LEFM is not strictly applicable to these section thicknesses at values of K approaching K_{Ic} because the plastic zone at the crack tip is significant. If we had used a plastic zone correction factor we would have calculated somewhat *smaller* values of critical crack size.

Note that the critical crack length in Steel A is undetectable by NDI. This is unacceptable. To increase $a_c(A)$ the stress level must be reduced by increasing the wall thickness, thereby losing some of the cost advantage of Steel A. It is proposed to consider the limiting case where all the cost advantage of A is lost. The wall thickness t'_A at which this happens is given by

$$\frac{C_A}{C_B} = \frac{t'_A}{t_B} \frac{\pounds_A}{\pounds_B} = 1 \tag{17}$$

or

$$t'_A = t_B \frac{\pounds_B}{\pounds_A} = \frac{0.073 \times 295}{542} = 0.039 \text{ m} \tag{18}$$

The maximum hoop stress σ_{max} is equal to the sum of the applied hoop stress under the maximum head of water ($h = 816$ m) and the residual stress σ_R (70 MN m^{-2})

$$\sigma_{max} = \frac{PR}{t'_A} + \sigma_R \tag{19}$$

$$= \frac{(\rho gh)[(D/2) + (t'_A/2)]}{t'_A} + \sigma_R = 330.6 \text{ MN m}^{-2} \tag{20}$$

When the crack attains a critical length $a'_c(A)$, the stress intensity is K_{Ic} where

$$K_{Ic} = Y\sigma_{max}\sqrt{(\pi a'_c(A))} \tag{21}$$

or

$$a'_c(A)Y^2 = \frac{1}{\pi}\left[\frac{K_{\mathrm{Ic}}}{\sigma_{\mathrm{max}}}\right]^2 = \frac{1}{\pi}\left[\frac{100}{330.6}\right]^2 = 0.029 \text{ m} \qquad (22)$$

$$\frac{a'_c(A)Y^2}{t'_A} = 0.75 \qquad (23)$$

From Table 3, this implies that

$$\frac{a'_c(A)}{t'_A} = 0.26 \text{ or } a'_c(A) = 10.1 \text{ mm} \qquad (24)$$

(a crack detectable by NDI).

Fatigue Crack Growth

In this section is it proposed to calculate the time required to grow a crack by fatigue from the largest undetected length ($a = 6$ mm) up to the critical size. This will be done for each of the two penstocks, constructed from each of the two steels. The instantaneous rate of fatigue crack growth is given by the Paris-Erdogan [2] equation

$$\frac{da}{dN} = C(\Delta K)^m \qquad (25)$$

This case deals with cracks in ferritic steels growing in contact with water, for which the appropriate values are [3]

$$\frac{da}{dN} = 2 \times 10^{-11}(\Delta K)^3 \qquad (26)$$

(a in metres, N cycles, K in MN m$^{-3/2}$)

This represents an increase by a factor of two of the upper bound growth rate for low-alloy steels in air (Lindley and Richards [4]).

$$\frac{da}{dN} = 2 \times 10^{-11}(Y\Delta\sigma\sqrt{(\pi a)})^3 \qquad (27)$$

Separating the variables

$$dN = \frac{da}{2 \times 10^{-11}Y^3(\Delta\sigma)^3\pi^{3/2}a^{3/2}} \qquad (28)$$

Integrating for growth from crack length a_1 and a_2 gives

$$N = \frac{1}{2 \times 10^{-11}(\Delta\sigma)^3\pi^{3/2}} \int_{a_1}^{a_2} \frac{da}{Y^3 a^{3/2}} \tag{29}$$

Intermediate Penstock

Recalling Table 1, the cyclic loading of the intermediate penstock involves a change in the head of water Δh of 575 m. This causes a change in hoop stress of

$$\Delta\sigma = \frac{\Delta PR}{t} = \frac{(\rho g \Delta h)[(D/2) + (t/2)]}{t} \tag{30}$$

$$= 2.82 \left[\frac{2.5}{t} + 1\right] \text{ MN m}^{-2} \tag{31}$$

For Steel A, $t'_A = 0.039$ m, and

$$\Delta\sigma(A) = 183.6 \text{ MN m}^{-2} \tag{32}$$

For Steel B, $t_B = 0.073$ m, and

$$\Delta\sigma(B) = 99.4 \text{ MN m}^{-2} \tag{33}$$

Evaluating Eq 29 for Steel A, using the values $a_1 = 0.006$ m, $a_2 = 0.0101$ m and $\Delta\sigma(A) = 183.6$ MN m^{-2}, gives the life (expressed in loading cycles). Carrying out this integration numerically on a computer gives the result

$$N_A = 3344 \text{ cycles (about 5 months of operation)} \tag{34}$$

For Steel B, using the values $a_1 = 0.006$ m, $a_2 = 0.025$ m and $\Delta\sigma(B) = 99.4$ MN m^{-2}, the life is

$$N_B = 55836 \text{ cycles (about 7½ years of operation)} \tag{35}$$

These results clearly favor the use of Steel B over that of Steel A. Comparing Steels A and B on the basis of equal cost, it takes a period over 16 times longer for a crack in Steel B to grow from the maximum undetectable size up to the critical length. The shortness of this period (5 months) rules out Steel A for the intermediate penstock. The use of Steel B of thickness 73 mm is recommended for the walls of the intermediate penstock.

Upper Penstock

Recalling Table 1, the cyclic loading of the upper penstock is more complicated than that of the intermediate penstock. It consists of annual "blocks" of pressure cycles, each of which is described by the data in Table 1. The hoop stress range $\Delta\sigma_i$ due to a pressure change of Δh_i is given by Eq 30

$$\Delta\sigma_i = 4905\Delta h_i \left[\frac{2.5}{t} + 1\right]10^{-6} \text{ MN m}^{-2} \tag{36}$$

The change in hoop stress is proportional to Δh_i

$$\Delta\sigma_i = k\Delta h_i \tag{37}$$

where for Steel A

$$(t'_A = 0.039 \text{ m})k_A = 0.319 \text{ MN m}^{-3} \tag{38}$$

and for Steel B

$$(t_B = 0.073 \text{ m})k_B = 0.173 \text{ MN m}^{-3} \tag{39}$$

Combining Eqs 29 and 37 we get the number of loading cycles of range Δh_i which are accompanied by crack growth from length a'_1 to a'_2

$$n_i = \frac{1}{2 \times 10^{-11}(k\Delta h_i)^3\pi^{3/2}} \int_{a'_1}^{a'_2} \frac{da}{Y^3 a^{3/2}} \tag{40}$$

Rearranging, we can call the integral a *crack growth parameter*

$$G_i = \int_{a'_1}^{a'_2} \frac{da}{Y^3 a^{3/2}} = (2 \times 10^{-11})k^3(n_i\Delta h_i^3)\pi^{3/2} \tag{41}$$

To get the total crack growth per annum (from length a_1 to a_2), we sum the crack growth parameters for all the load cycles listed in Table 1

$$G = \sum_i G_i = \int_{a_1}^{a_2} \frac{da}{Y^3 a^{3/2}} = (2 \times 10^{-11})k^3\pi^{3/2}\sum_i(n_i\Delta h_i^3) \tag{42}$$

It is found that the summation gives

$$\sum_i(n_i\Delta h_i^3) = 6.08 \times 10^{10} \text{ m}^3 \tag{43}$$

which, for crack growth purposes, is equivalent to 7500 cycles per annum (the number of operation cycles of the main valve), each of pressure change $h = 200$ m (that is, $7500(200)^3 = 6 \times 10^{10}$ m^3). It follows that the period (in years) required to grow a crack from length a_1 and a_2 is

$$
t = \frac{1}{(2 \times 10^{-11})\pi^{3/2}(6.08 \times 10^{10})k^3} \int_{a_1}^{a_2} \frac{da}{Y^3 a^{3/2}}
$$

$$
= \frac{0.148}{k^3} \int_{a_1}^{a_2} \frac{da}{Y^3 a^{3/2}} \tag{44}
$$

For Steel A, putting $k_A = 0.319$ MN m^{-3}, $a_1 = 0.006$ m (the largest undetectable size) and $a_2 = 0.0101$ m (the critical crack length) this gives a "growth lifetime"

$$
t_A = 10.6 \text{ years} \tag{45}
$$

For Steel B, putting $k_B = 0.173$ MN m^{-3}, $a_1 = 0.006$ m and $a_2 = 0.025$ m, the growth life is

$$
t_B = 176 \text{ years} \tag{46}
$$

Actually this figure is an underestimation because at the initial crack length most of the loading cycles cause a change in stress intensity below the threshold value for fatigue, ΔK_{TH} (even taking a generous value of $\Delta K_{TH} = 10$ MN m$^{-3/2}$).

The growth lifetime using Steel A (10.6 years) is well below the design life of the station (55 years). This is unacceptable, because the upper penstock cannot be readily isolated for repair. On the other hand the predicted life using Steel B exceeds the design life, so this steel in a thickness of 0.073 is recommended for construction of the upper penstock.

Recommended Design

The task has been to choose between two steels: A, a high strength, high cost steel and B, in effect a mild steel. A design is sought which, after an initial inspection for defects, will carry the service loads safely throughout the design life of 55 years. Such a design can be made in either A or B; obviously, the one with the lower cost is to be preferred.

In designing to meet the largest conceivable load (the static design), the wall thickness of Steel B must be about three times greater than that of Steel A; the cost of such a design in Steel B would be about 50% higher than that in Steel A.

However the penstocks must also resist repeated loads. This carries the risk of causing the growth of fatigue cracks. The successful design of upper penstock must ensure that the largest undetected crack at installation cannot grow to the critical size during the service life. For the intermediate penstock, the requirement is less severe; this penstock *could* be repaired in service.

In terms of the rates of fatigue crack growth, da/dN, under a given load cycle (ΔK), Steels A and B are identical. To offset its higher cost, Steel A must be used in smaller thicknesses and at higher stresses than Steel B, and the rates of crack growth will be correspondingly higher. Furthermore, the critical crack size will be much smaller in Steel A than Steel B (because K_{Ic} is smaller and σ_{max} is larger). Accordingly, the crack growth lifetime of Steel A is much lower than that of Steel B. The possibility exists of fatigue crack growth leading to failure within the design life in a lower cost penstock made of Steel A. This risk is unacceptable, and therefore Steel B 73 mm thick is preferred for both penstocks. In the upper penstock, the largest undetected crack (length 6 mm) cannot grow to failure in Steel B, 73 mm thick, within 55 years; in the intermediate penstock, such a crack would take some 7½ years to grow to failure in Steel B, 73 mm thick, giving adequate time to detect the subcritical crack.

It was noted earlier that the use of LEFM at values of K approaching K_{Ic} is not strictly valid in either steel. As a consequence, the calculated values of critical crack length ($a_c(A)$ and $a_c(B)$) are overestimates. However, this error does not seriously affect the calculated crack growth lifetimes; such lifetimes are very sensitive to the *initial* crack length (when growth is slow) and relatively insensitive to the final crack length (when growth is fast).

The Adopted Design

The penstocks at Dinorwig use Steel B, 65 mm thick for the walls of the plain cylindrical section of the upper penstock and 80 mm thick for the cylindrical section of the intermediate penstock. Subsequent to this final design being adopted, fracture mechanics has been used to draw up defect acceptance standards for the penstocks.

Defect Acceptance Standards

The CEGB proposes that in-service inspection should be carried out on the upper and intermediate penstocks at frequencies which guarantee a residual life of up to five times the inspection interval, the precise figure being dependent upon the accuracy with which the operational stresses and defect sizes are known. The current view on inspection interval is that it should be linked to the maintenance interval of 6 years for the intermediate penstock and 20 years for the upper penstock. With this in mind and with a knowledge of the most recent fatigue crack growth rates and loading regimes, limits of acceptability have been determined for the complete range of possible defect shapes.

Crack growth is assumed to occur in accordance with the wet fatigue law stated earlier until the defect reaches the critical defect size, at which point failure of the component could occur if it was subjected to the fault condition. The failure size for all but through-thickness defects and infinitely long surface defects has been determined with the aid of the computer programme FRACPAC (Chell [5]) which is based upon the post yield fracture mechanics methodology outlined by Milne [6]. For this purpose the lower bound flow stress is taken to be

$$\left[\frac{\sigma_{tensile} + \sigma_{yield}}{2} \right] = 409 \text{ MN m}^{-2} \tag{47}$$

Within the FRACPAC program it is relatively simple procedure to obtain the failure defect size (at the fault condition) for a range of defect shapes from semi-circular to that approaching an infinitely long surface defect. This approach has been adopted, supplemented by analytical calculations for the through-thickness defect and infinitely long surface defect. Acceptable initial defect sizes were then obtained by considering crack shrinkage from the failure sizes determined previously, by employing analytical techniques and through the use of the computer programme FATPAC (Chell [7]). By specifying a range of initial defect sizes, aspect ratios, stresses, and cracked body geometry, FATPAC automatically accounts for changes in aspect ratio as the crack shrinks. In both approaches, reverse fatigue proceeds until a predetermined number of stress cycles has been used up. For example, where a residual life of 5 times the inspection interval (say 6 years) is required, the appropriate input is $5 \times 6 \times 7\,500 = 225\,000$ cycles.

Results are presented in this paper in graphical form. A common feature of all the diagrams is the linear ordinate (defect depth) versus logarithmic abscissa (defect length) format. In this form, the defect acceptance line can be regarded as a locus of defect sizes with a common residual life. The residual life is expressed in multiples of the inspection interval and each defect acceptance line is labelled with this factor. A factor of zero denotes the locus of failure defect sizes.

The acceptance curves for longitudinal/radial cracks (that is, cracks normal to the hoop stress) are shown in Figs. 3 and 4. Figure 3 refers to the intermediate penstock while Fig. 4 refers to the upper penstock.

In the upper penstock the axial component of stress undergoes large cyclic changes which arise from operations of the main inlet valve. Each operation changes the end-load on the upper penstock. Changes in end-load arise primarily as a result of the change in the pressure differential at the main inlet valve service seal and from the presence of the sliding joint between

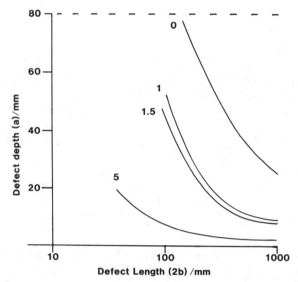

FIG. 3—*Acceptance lines for surface defects in the intermediate penstock welds, orientated normal to the hoop stress. The residual life represented by each line is expressed as a multiple of the inspection interval (6 years).*

the valve and the intermediate penstock. These changes in axial stress may cause growth of hoop/radial cracks in the upper penstock, and defect acceptance curves have been obtained for this case (Fig. 5). The change in axial stress is

$$\Delta\sigma_L = \frac{\Delta P^1 r_2^2}{r_1^2 - r_2^2} \tag{48}$$

where ΔP^1 is the change in pressure differential at the valve seal (56.5 bar or 580 m of water) and r_1, r_2 are the outer and inner pipe radii, respectively. 7500 such changes occur in a given penstock per annum. Figure 5 shows that the defect acceptance lines for hoop/radial cracks are in general shifted to smaller crack sizes than those in Fig. 4, reflecting the relative severity of the axial cyclic loading.

The prime requirement of the defect acceptance standards is that they provide accurate predictions of both defect stability and residual life. These predictions clearly depend upon the accuracy of the source data. For the most part, pressure-induced fatigue stresses are known with a high degree of certainty, and so too is the flow stress. In contrast, less is known about the level of the residual stress and fracture toughness. It may be argued, however, that precision is not required here; the residual stress affects the mean stress level, and this is known to have little effect on the rate of fatigue crack growth in the range described by the Paris-Erdogan equation [4]. The

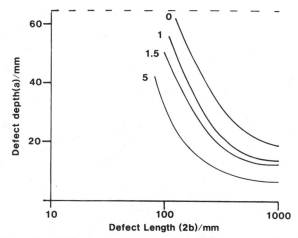

FIG. 4—*Acceptance lines for surface defects in the upper penstock welds, orientated normal to the hoop stress. The residual life corresponding to each line is expressed as a multiple of the inspection period (20 years).*

value of fracture toughness affects the critical crack length, but this has little effect on the fatigue life because crack growth is rapid at crack lengths approaching the critical value.

Results of the fracture mechanics analysis are presented in the standards in the form of "iso life" lines in the range 0 to 5I where I is the proposed nondestructive inspection interval. It is envisaged that all detected defects will be plotted as a point on the relevant standard. Where stresses are known accurately, a safety factor of 1.5 on life is considered to be a tolerable minimum. The 1.5I line, therefore, sets an upper limit to the allowable defect sizes within the welds for the proposed inspection frequency. Defects whose sizes are within these limits are assured of a residual life in excess of the inspection interval, in which case the position of the defect relative to the additional assessment lines enables the criticality of the defect to be determined. Defects which are found to be beyond the limits set should be referred for individual fracture mechanics assessment, where account can be taken of its precise location in the weld and a revised in-service inspection program developed.

Conclusions

1. This case history formed a successful teaching exercise based on the CEGB's original assessment which showed that Steel B is preferred over Steel A for the construction of both upper and intermediate penstocks. In a thickness of 73 mm, Steel B has a predicted lifetime for the growth of detectable fatigue cracks that exceeds the design life of the upper penstock.

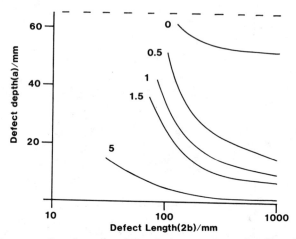

FIG. 5—*Acceptance lines for surface defects in the upper penstock welds, orientated normal to the axial stress. The residual life corresponding to each line is expressed as a multiple of the inspection period (20 years).*

A penstock made of Steel A with the same cost as that of Steel B has a predicted growth life of 10.6 years (only 19% of the design life).

2. Defect acceptance standards have been determined for the upper and intermediate penstock welds, based upon a 20 and 6 year inspection interval, respectively. In all cases, residual life factors in the range 1 to 5 were considered.

3. The data are presented in a form which enables the engineer responsible to derive alternative inspection intervals depending upon the results of the preservice inspection and the accuracy of the stress determination and nondestructive test results.

Acknowledgments

This paper is published by kind permission of the Director General of the CEGB, North Western Region. The materials selection teaching exercise is based on an original assessment by Milne [8] and other work by the Technology, Planning, and Research Division and the Generation, Development, and Construction Division of the CEGB. We also acknowledge the work of several of our colleagues, notably Dr. W. Laidler.

References

[1] Broek, D., *Elementary Engineering Fracture Mechanics*, Martinus Nijhoff, 1982, p. 76.
[2] Paris, P. C., and Erdogan, F., *Journal Basic Engineering*, Vol. 85, 1963, p. 528.
[3] Atkinson, J. D. and Lindley, T. C., Report RD/L/N117/79, Central Electricity Research Laboratories, Leatherhead, U.K., 1979.
[4] Lindley, T. C. and Richards, C. E., *Engineering Fracture Mechanics*, Vol. 4, 1972, p. 971.

[5] Chell, G. G., Report RD/L/N170/77, Central Electricity Research Laboratories, Leatherhead, U.K., 1978.
[6] Milne, I., *Fracture 1977*, Vol. 3, p. 419, University of Waterloo Press, Waterloo, Ontario, Canada.
[7] Chell, G. G., Report RD/L/N205/78, Central Electricity Research Laboratories, Leatherhead, U.K., 1979.
[8] Milne, I., Report RD/L/N132/75, Central Electricity Research Laboratories, Leatherhead, U.K., 1975.

Curt Christensen[1] and Richard T. Hill[2]

Applied Fracture Mechanics for Assessing Defect Significance in a Crude Oil Pipeline

REFERENCE: Christensen, C. and Hill, R. T., **"Applied Fracture Mechanics for Assessing Defect Significance in a Crude Oil Pipeline,"** *Case Histories Involving Fatigue and Fracture Mechanics, ASTM STP 918,* C. M. Hudson and T. P. Rich, Eds., American Society for Testing and Materials, Philadelphia, 1986, pp. 122–135.

ABSTRACT: The integrity and service life of a pipeline will depend, to a large extent, on the fracture toughness properties and subcritical crack growth resistance of pipes with defects either inherently present or introduced during construction or service. Unstable fracture from crack like defects can occur if the fracture toughness of the line pipe material is inferior, whereas subcritical crack growth can occur under sustained load and cyclically varying loads (fatigue). Both of these subcritical growth mechanisms can be strongly influenced by the internal pipe environment (hydrogen sulfide, carbon dioxide) and external environment (corrosion, cathodic protection).

A case history of the application of fracture mechanics for evaluating the static fracture toughness and subcritical crack growth resistance of an external longitudinal weld toe crack is discussed.

KEY WORDS: crack tip opening displacement, pipeline failure, weld toe crack, critical engineering assessment, unstable cracking fatigue, cathodic protection

Weld toe cracks in submerged arc welded pipes have been reported in several investigations [1–3]. In a recent study of a 508 mm outside diameter by 12.7 mm wall API 5LX 60 pipeline the toe cracks occurred exclusively at the external weld seam due to a process defect. If less than 1 mm deep such indications are not easily detected by either mill X-ray or ultrasonic inspection but can be disclosed by subsequent field magnetic particle inspection. As the weld toe crack would be exposed to varying hoop stresses from the service loads and a possible hydrogen charging effect from external cathodic protection coming into action at coating damages, a fitness-for-service analysis was performed.

An intensive nondestructive field inspection program supplemented by metallographic examination showed a maximum toe crack depth of 0.9 mm

[1] Metallurgist, Korrosionscentralen, Copenhagen, Denmark.
[2] Consultant, J. P. Kenny & Partners Ltd., London, U. K.

and a length of 5 to 20 mm in disconnected rows of up to 200 mm long. Based on this examination the worst case depth was assumed to be 1.5 mm.

To determine whether such defects were tolerable the following fracture mechanics program was implemented.

1. *Unstable fracture*—Establishment of the static fracture toughness properties to assess the risk of fast fracture during hydrostatic testing and define the maximum allowable flaw size.

2. *Full scale high stress testing*—Confirmation of the predicted failure stress level in a pipe section containing a weld toe crack.

3. *Fatigue testing*—Establishment of the subcritical crack growth behavior in air and in simulated service environment.

Unstable Fracture

The fracture toughness value can be obtained from fracture mechanics tests that measure the load required to cause a well characterized crack or flaw to propagate under a specific stress state. From previous reported failure analysis of weld toe cracks it appeared that the crack would propagate in truly radial direction, perpendicular to the principal hoop stress. Hence the area of interest in terms of fracture toughness was the cross-hatched zone shown in Fig. 1.

With very ductile line pipe steels having surface breaking defects, the plane strain fracture toughness, K_{Ic}, has no real meaning at the crack tip. Instead, analysis based upon measurement of the critical crack-tip opening displacement (CTOD), J-integral and J-resistance, (J_{Ic} and J–R curves) gives a more precise description of the elastic-plastic fracture toughness behavior. CTOD and J_{Ic} measure the resistance to crack initiation, that is, actual ductile crack extension, whereas J–R curves measure the materials resistance to crack growth. The physical meaning of CTOD, J_I and J–R curves are shown in Fig. 2.

The British Standards Institution PD-6493-1980 outlines a well developed and conservative method for defect assessment based on the principles of

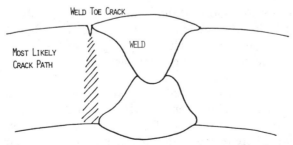

FIG. 1—*Position of weld toe crack and path of expected crack growth.*

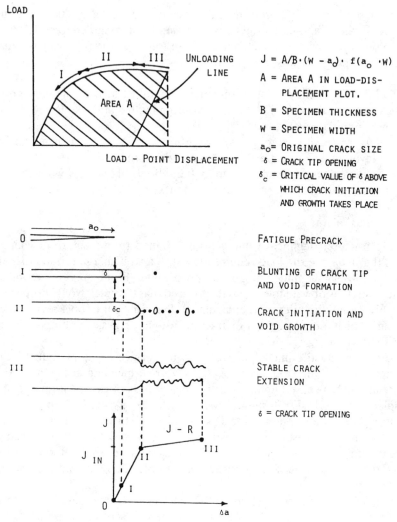

FIG. 2—*Schematic model of crack propagation in fracture toughness specimens* [9].

fracture mechanics. Specifically it allows the use of CTOD-values in defining the size of flaws that can safely be left in a structure without the risk of unstable fracture. The CTOD-values for the pipe material were obtained using single edge notched and fatigue precracked specimens machined from the pipe material as shown in Fig. 3.

When the test specimen is loaded in 3-point bending, crack extension, or fracture is assumed to occur when the CTOD (which relates to the crack-tip strain) exceeds a critical value. This criterion is equivalent to the K_{Ic} criterion in cases where plane-strain conditions prevails at the crack tip [4].

The dimensions of the specimen, position of notch, and fatigue-precrack length were in accordance with BSI 5762 as shown in Fig. 3. A total of 10 specimen were used to allow for a statistical representation of the values. As the material was very tough and ductile, slow stable crack extension was expected to occur prior to fracture. Indeed, the obtained load displacement records indicated slow stable crack extension as shown in Fig. 4. The critical CTOD-values evaluated at maximum load were: average value: 0.545 mm, minimum value: 0.500 mm, and maximum value: 0.600 mm.

Utilizing the minimum value of CTOD obtained the maximum allowable flaw size was then determined using the guidelines of BSI-PD-6493, 1980

$$a_m = C \frac{\delta_{crit}}{\epsilon_y}$$

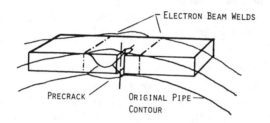

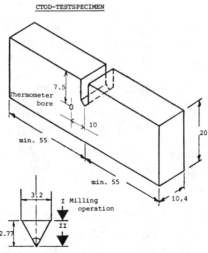

FIG. 3—*Single edge notched test specimen for CTOD measurement.*

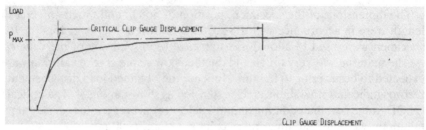

FIG. 4—*Load displacement plot of one actual CTOD test.*

where

a_m = tolerance defect parameter equivalent to the half crack length of a through wall defect,

δ_{crit} = critical CTOD value,

C = constant derived from $C = 1/[2(\epsilon/\epsilon_y - 0.25)]$,

ϵ = total strain (applied and residual), and

ϵ_y = yield strain.

The total elastic strain was derived from the total stress consisting of the maximum hoop stress during hydrostatic testing, that is, 72% of yield, and a weld residual stress equivalent to the yield strength of the pipe. This was done to further enhance the conservatism of the analysis, although in practice the weld residual stresses has most likely been partly relieved during pipe expansion and hydrostatic testing in the pipe mill.

Conversion of the through-wall defect parameter a to a surface breaking defect of depth t and length l is obtained through the relationship outlined in BSI-PD-6493.

Assuming a safety factor of 2 on strain and considering general yield or plastic collapse of the ligament below the defect the allowable flaw size was determined to be

	Depth, t	Length, l
or	5.2 mm	45 mm
	4.5 mm	infinity

This is approximately 5 times the deepest defect found in the toe region of the external longitudinal weld, and 3 times the anticipated worst case condition.

Full Scale High Stress Testing

To add confidence to the CTOD analysis a full-scale high-stress test was conducted on a pipe section containing a defect representative of the deepest indications found.

The pipe section was provided with end caps and pressurized stepwise to pressure levels equivalent to 35, 45, 75, 100, 130, and 139% of specified minimum yield stress (SMYS) with intermediate depressurization to 0 bar gram. Generation of acoustic emission (AE) from the defect was monitored by general AE activity transducers and an AE location system with particular reference to the area of the defect. During the testing only very low general activity was recorded. The AE location system did indicate single AE events related to the end cap welds, but none within the area of the weld toe indication.

The final pressurization reached 139% of SMYS before it was stopped because of significant general yielding. Had the test been continued the pipe may have burst at the defect thereby destroying the information concerning possible post yield stress stable crack extension.

The test pipe instrumented with AE equipment is shown in Fig. 5. Although the permanent deformation after testing was measured to be 1.5%, metallographic examination of the weld toe crack showed that there had been no measureable crack extension as shown in Fig. 6 and 7. This test confirmed the very high flaw tolerance of the pipe and showed that even when loaded to general yielding it was not in a state of imminent failure.

Fatigue Analysis

Crack growth behavior in air is characterized by 3 regions, as shown in Fig. 8. The initial and terminal region is the stress ratio or dependent on

FIG. 5—*Test pipe instrumented with acoustic emission transducers.*

FIG. 6—*Microsection of weld toe crack.*

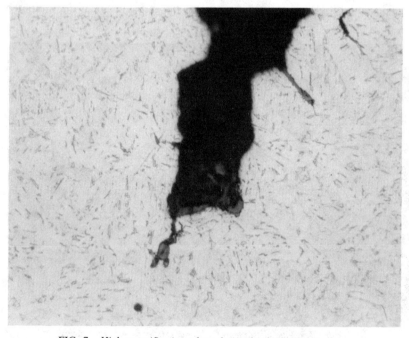

FIG. 7—*High magnification of crack tip after high stress testing.*

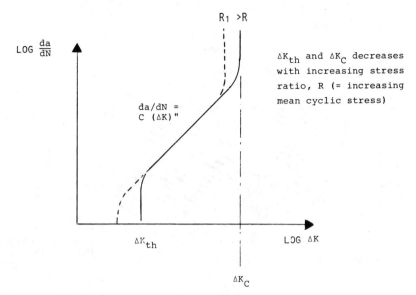

FIG. 8—*Schematic representation of fatigue crack growth behavior.*

cyclic ratio R = min/max = K_{min}/K_{max}. Increasing this ratio (that is, higher mean stress for unchanged maximum stress) decreases both the threshold ΔK for crack growth and the critical K for unstable fracture.

Within a limited range of ΔK-values the plot of da/dN versus ΔK falls on a straight line giving the relation known as the Paris law

$$da/dN = C(\Delta K)^n$$

but obviously the exponential relationship depend upon the position of the ΔK range considered (that is, high, intermediate, or low ΔK-values). In practice the relationship is a function of both ΔK and K_{max} and hence of the cyclic ratio R but for most technical problems the simpler Paris law will often be sufficient as the variations due to R-dependency lies within the normal experimental scatterband [4].

To gain a general view of the defect behavior the following equations was used to calculate stress intensity ranges for various defect depths

$$K_{min}/K_{max} = \sigma_{min}/\sigma_{max} = R$$

$$\Delta K = K_{max} - K_{min} = (1 - R)K_{max}$$

$$K_{max} = 1.1\sigma_{max}\sqrt{\pi a}$$

$$\Delta K = (1 - R)1.1\sigma_{max}\sqrt{\pi a}$$

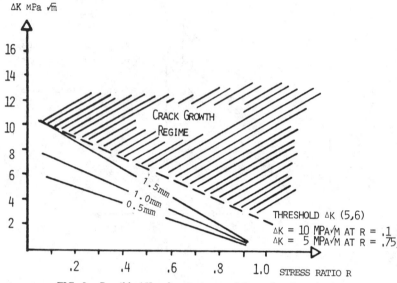

FIG. 9—*Possible ΔK-values acting on defects of various depths.*

Although the stress-intensity equation is valid for shallow cracks in a plate of infinite width it is also believed to be applicable, with reasonable accuracy, for shallow longitudinal weld toe cracks in thin wall large diameter pipes at membrane stresses well below the yield stress.

Figure 9 shows the possible ΔK-values acting on the defects during normal operation (maximum stress approximately 35% of SMYS). Also on the same figure is the threshold ΔK-value [5,6] which is valid for a carbon-manganese steel of similar strength. If testing was carried out at $\Delta K - R$ combinations below this line, crack growth may never be encountered. The program, therefore, made use of fracture mechanics test specimens tested at high ΔK-values for comparison with referenced values of other carbon-manganese steels. Additionally lower ΔK-values were used to establish a possible threshold ΔK-value and finally actual weld toe cracks were tested at comparable ΔK-value to compare the crack growth behavior of the actual defect with crack growth in the through thickness direction.

SEN specimens similar to those used for CTOD-testing were tested at a high ΔK using a triangular load and a frequency of 0.1 Hz, and at a low ΔK using a sinusoidal load at a frequency approximately 60 Hz. During the precracking of the specimens using a high frequency pulsator, Amsler HFP 10, the maximum K-value was reduced by 10% for every 36 000 cycles to approximately 80% of the value used in the beginning of the fatigue measurement. For the very low ΔK specimens though, the maximum K-value was about the same at termination of precracking and start of fatigue testing. The results, summarized in Fig. 10, showed that the crack growth behavior

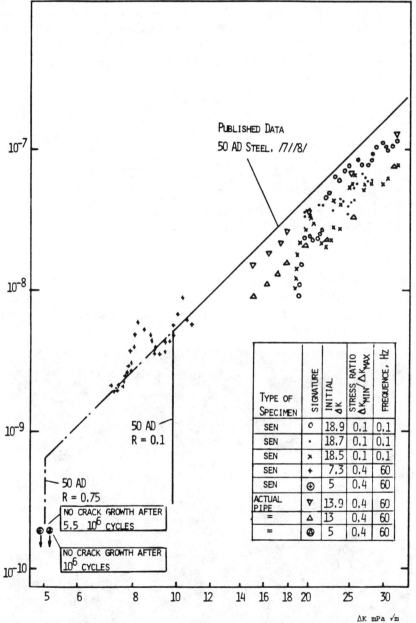

FIG. 10—*Fatigue test results. Crack growth rate versus stress intensity range.*

was similar to that from published data of similar type of steels for cracks growing in the same direction.

Crack growth in the through thickness direction was measured by also testing actual defects in a 3-point bend test as shown in Fig. 11. Although the test specimens could not be considered standard fracture mechanics specimens the linear elastic fracture mechanics stress intensity equation

$$K = \frac{6PL}{4 \cdot B \cdot w^{1/2}} \, a \left(1.99 - 2.47 \frac{a}{w} + 12.97 \left(\frac{a}{w}\right)2 \right.$$
$$\left. - 23.17 \left(\frac{a}{w}\right)3 + 24.8 \left(\frac{a}{w}\right)4 \right)$$

where

    P = load,
    L = distance between supports,
    B = specimen width,
    w = specimen height, and
    a = crack depth.

was used to allow comparison with the fracture mechanics tests. The crack growth was monitored by microscope examination of acetal replica's of the specimen sides taken at intervals during the testing. The actual crack growth

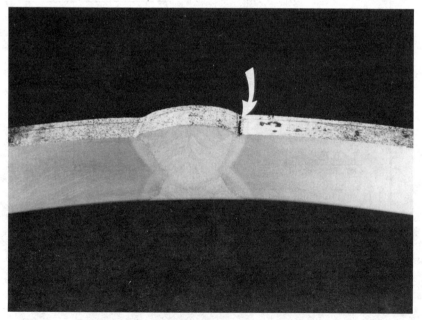

FIG. 11—*Test specimen containing actual defect.*

behavior is also shown in Fig. 10 and as can be seen the results are in fair agreement.

Using the Paris law constants established from the testing it was shown that it would take some 20 pressure fluctuations per day from full operating pressure to zero pressure to extend an initial flaw of 1.5 mm to a depth of 3 mm within the expected 30 years lifetime of the pipeline. Any expected pressure fluctuations having a frequency above 20 a day would be of a size that would result in stress intensity ranges below the fracture mechanics threshold value.

Although the above tests were conducted in air this represents a very likely environmental condition as the pipes were coated with a heavy polyethylene coating. However it could be argued that during the construction phase, water penetrates from the pipe end along the external weld reinforcement, and, as the coating, despite being of superior quality, does not yield complete diffusion tightness, cathodic protection might become effective. Also it would be inferred that coating damage may occur at a weld toe crack thus exposing the cracks to full cathodic protection. Thus in an anaerobic soil, hydrogen evolution might then occur.

From past experience [7], it could be surmised that cathodic protection would have a beneficial effect at low loads. As indicated in Fig. 12, reproduced from [8], the threshold stress-intensity range for possible crack growth

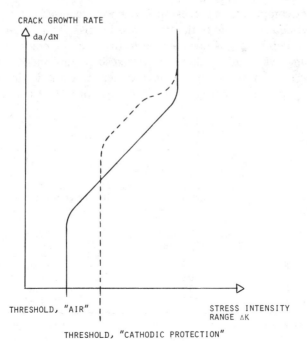

FIG. 12—*Schematic illustration of influence of cathodic protection on fatigue crack rate* [8].

is shifted towards higher (more beneficial) values. However, once exceeded the crack growth rate is likely to be accellerated by the influence of the hydrogen charging.

To clarify the possible influence of cathodic protection on crack growth behavior, a fracture mechanics specimen was fatigue tested in the load range corresponding to the established air fatigue threshold value using an impressed cathodic protection of −1100 mV saturated calomel electrode (SCE) in a simulated worst case water environment.

As the possible crack growth rate near the threshold value was around 10^{-9} m/cycle and the test frequency was approximately 0.1 Hz to allow for the influence of the environment, time did not allow for the specimen to crack through. Therefore the program was designed so that the load range was increased at 50 000 cycle intervals if the compliance monitoring of the specimen did not reveal any crack growth. It was believed this interval would suffice to detect the onset of crack growth, at least in the sense that if it had commenced just prior to termination of the 50 000 cycles, then the next higher load would certainly disclose it. Table 1 details the results of the actual test program. After a total of 504 000 cycles the crack had not yet started to grow, even at a relatively high stress intensity range. The number of cycles at the final stress range corresponds to more than 30 full depressurization from maximum operating pressure to zero bar gauge per day during 30 years of operation.

Following completion of the test specimen was fractured in liquid nitrogen to allow for determination of the true stress intensity value from the actual initial crack depth. It was not possible to disclose any crack extension by fractographic examination; thus, the test results confirmed that cathodic charging in the simulated pipeline environment was not detrimental in the expected operating load range of the pipeline. In fact, it may even result

TABLE 1—*Test conditions and results.*

Environment	Deionised water with 200 ppm NaCI; initial bulk pH 5 by addition of NaHSO₄
Test specimen	three-point bending single edge notched specimen
Cathodic protection	impressed current at −1100 mV SCE potentiostatically controlled

Consecutive Cycles	Stress Intensity Range, MPa $\sqrt{m}$, ΔK^a	R	Crack Growth	Bulk pH During Testing
0 to 50 000	6.7	0.4	none	3
50 000 to 100 000	10.0	0.2	none	3
100 000 to 150 000	13.4	0.1	none	3
150 000 to 504 000	16.7	0.1	none[b]	3

[a] The stress intensity range is calculated from the true depth of the initial crack, determined after final fracturing.

[b] No crack growth visible in fractographic examination of the freelaid initial crack.

in a beneficial effect with respect to an increase of the threshold stress intensity.

Concluding Remarks

Fracture mechanics is not just a sophisticated design tool but is also applicable to practical problems encountered within the pipeline industry. The BSI-PD-6493: 1980 outlines a straightforward procedure to follow when performing an engineering critical assessment of a defect based upon fracture mechanics concept. Using this procedure the weld toe crack of the above cited case history was determined to be insignificant to the continued safe operation of the pipeline system.

Acknowledgment

The CTOD testing and the acoustic emission analysis was carried out in the laboratories of the Danish Welding Institute by F. Jacobsen and Arved Nielsen, respectively.

References

[1] Smith, R. B. and Eiber, R. E., "Field Failure Survey and Investigations," AGA 4th Symposium on Line Pipe Research, Dallas, Texas, Nov. 1969.
[2] Vosikovsky, O. et al, "Allowable Defect Sizes in Sour Crude Oil Pipeline for Corrosion Fatigue Conditions," *International Journal of Pressure Vessels and Piping,* Vol. 13, No. 193, pp. 197–226.
[3] Krass et al, "Pipeline Technic," Mineral ol fernleitungen, *Verlag TUV,* Rheinland GmbH, Køln, 1979.
[4] Brock, David, *Elementary Engineering Fracture Mechanics,* Noordhoff International Publishing, Leyden, The Netherlands, 1974, Chapter 1 and 10.
[5] Morgan, H. G. et al, "An Investigation of The Corrosion Fatigue Crack Growth Behaviour of Structural Steels in Seawater," Conference on Steel in Marine Structures, Paris, Oct. 1981.
[6] Austen, J. M. et al, "Factors Affecting Corrosion Fatigue and Stress Corrosion Crack Growth in Offshore Steels," Conference on Steel in Marine Structures, Paris, Oct. 1981.
[7] Noppenau, H., "Korrosionsudmattelse af stål i havvand," *Thesis,* Instituttet for Metallaere, Danmarks Tekniske Højskole, Lyngby, Oct. 1980.
[8] Wei, R. P., "Fracture Prevention and Control," (D. W. Hoepner, Ed.), ASM No. 3, *Materials/Metal Working Technology Science,* American Society for Metals, 1974.
[9] Provenzano, V. et al, "A Comparison of J_{Ic}, Stretched Zone Size and Crack Opening Displacement in a Ferritic Steel," *Proceedings,* Fourth International Conference on Mechanical Behaviour of Materials, ICM 4, Stockholm, Sweden, August 1983.

Harold S. Reemsnyder[1]

Observations, Predictions, and Prevention of Fatigue Cracking in Offshore Structures

REFERENCE: Reemsnyder, H. S. **"Observations, Predictions, and Prevention of Fatigue Cracking in Offshore Structures,"** *Case Histories Involving Fatigue and Fracture Mechanics, ASTM STP 918,* C. M. Hudson and T. P. Rich, Eds., American Society for Testing and Materials, Philadelphia, 1986, pp. 136–152.

ABSTRACT: In the past, during transit of jack-up offshore drill rigs, vortex shedding caused the initiation of fatigue cracks at certain details. The problem was defined as an interaction of cyclic stresses introduced by column vortex shedding and tensile residual stresses in the details. The paper outlines two simultaneous approaches to solution of the problem: (1) elimination of vortex shedding through structural modifications, that is, addition of helical strakes or spoilers, and (2) determination, by experiment, whether thermal stress-relief was beneficial. Also reviewed are the development and experimental corroboration of a local-strain life prediction method applicable to this and other structural fatigue problems.

KEY WORDS: crack initiation, design, fatigue life prediction, fatigue of metals, fatigue tests, local strain, offshore platforms, residual stress, stress concentration, stress relief, vortex shedding

Nomenclature

a Material constant
b Fatigue strength exponent
c Fatigue ductility exponent
e Nominal strain
Δe Nominal strain range
E Young's Modulus
K' Cyclic strength coefficient
K_f Fatigue notch factor
K_σ Local stress concentration factor
K_ϵ Local strain concentration factor
K_t Theoretical stress concentration factor

[1] Senior research fellow, Research Department, Bethlehem Steel Corporation, Bethlehem, PA 18016.

n' Cyclic strain-hardening exponent
N_i Cycles to crack initiation
r Notch root radius
S Nominal stress
S_{max} Maximum nominal stress per cycle
S_m Nominal mean stress per cycle
ΔS Nominal stress range per cycle
S_y Monotonic yield strength
S_u Monotonic tensile strength
σ_f' Fatigue strength coefficient
σ_m Local mean stress per cycle
σ_{max} Local maximum stress per cycle
$\Delta\sigma$ Local stress range per cycle
σ_{res} Local residual stress before cyclic loading
ϵ_f' Fatigue ductility coefficient
$\Delta\epsilon$ Local total strain range per cycle
$\Delta\epsilon_e$ Local elastic strain range per cycle
$\Delta\epsilon_p$ Local plastic strain range per cycle

During the past 27 years, Bethlehem has designed and built over 80 mobile, mat-supported, jack-up drilling platforms for the offshore oil industry. These rigs, capable of operating in water depths up to 114 m (375 ft) consist of a buoyant platform, a cellular mat, three cylindrical legs, and a hydraulic jacking system, Fig. 1. The platform contains the derrick, drilling machinery, and living quarters. The mat rests on the ocean floor and serves as a foundation for the rig. The three cylindrical tubular legs or columns— approximately 3.6 m (12 ft) in diameter, 91 m (300 ft) tall, and fabricated from steel plate 38.1 to 63.5 mm (1.5 to 2.5 in.) thick—are integral to the mat and pass through hydraulic jacking systems located in jack-houses on the upper deck of the platform. Rows of rectangular holes, called pinholes, Fig. 1, along and around the column receive large structural pins to raise, lower, and support the platform in conjunction with the hydraulic jacking system.

The pinholes in the legs, approximately 254 × 406 mm (10 × 16 in.), are formed by drilling four 38.1 mm (1.5 in.) diameter holes at the four corners and joining them by gas cutting, Fig. 1. Gas cutting introduces tensile residual stresses on the order of one half the yield strength of the leg material in the drilled holes. (Experimental measurement of the residual stresses is described later.) In the past, some legs were thermally stress relieved while others are not. The stress concentration factors K_t (determined by both photoelastic and finite element analysis) on the gross and net longitudinal stresses (axial plus bending stress) at the pinhole corner were, respectively, 5.7 and 4.6.

FIG 1—*Mat-supported jack-up drill rig and pinhole.*

Operation

During a "wet" tow or transit to a drilling location, the platform serves as a buoyant hull from which the mat is suspended a short distance below the platform by the legs. The legs, in turn, are exposed to a height of approximately 79 m (260 ft) above the platform. The speed of a wet tow is on the order of 4 knots while typical periods of pitch and roll are, respectively, 17 and 23 s.

An alternate method of extended ocean tow is the "dry" tow where the rig is carried on a semi-submersible barge (the initial concept) or a semi-submersible ship. The lower towing resistance of the barge versus the floating rig results in a higher towing speed—from 6 to 8 knots. However, rolling resistance is reduced, and the period of roll is closer to the wave period, 6 to 9 s, with a concomitant increase in angle of roll. For a dry tow on a barge, it is common to remove the upper portion of the legs and to carry these leg sections on the deck of the rig platform.

More recently, special purpose heavy-lift semi-submersible self-propelled ships have become available. For example, two mat-type jack-up rigs were transported simultaneously 13 000 sea miles on a self-propelled semi-submersible ship at an average speed of 15 knots [1]. These special purpose ships experience a longer roll period and a smaller roll angle than a barge carrying a rig. Because of the improved roll response, rigs are transported with their legs intact aboard such ships.

Upon arrival at the work site, the mat is lowered to the sea floor, and the platform is then jacked up the legs to sufficient height for wave clearance and is locked to the legs.

The Problem

A few years ago, cracks 12 to 200 mm (½ to 8 in.) long were detected at pinholes in several rigs after transit by wet tow and in one rig after a dry tow. The cracks were concentrated in the lower one third of the legs just above the jack-house. The lengths of the cracks decreased as the distance of the pinhole above the jack-house increased. In the case of wet tows, the transit crews aboard the rigs observed significant cyclic deflections (transverse to the relative wind direction) of the legs even in the absence of rolling.

For the wet tows, the problem was defined as an interaction of cyclic stresses introduced by vortex shedding and tensile residual stresses in the pinholes. Vortices develop alternately on opposite sides of a cylindrical body immersed in a flowing fluid. These vortices are carried downwind, forming a double row of evenly spaced, staggered vortices (a vortex street). The staggered formation of the vortices results in an alternating pressure and concomitant cyclic displacement on the body and transverse to the direction of fluid flow. The period of the alternating force is approximately five times the ratio of the diameter (of the cylinder) to the relative velocity of the fluid. The dynamic, structural response of the legs to vortex shedding is that of vertical, elastic, cantilever beams with the lower ends fixed at the jack-houses.

Analysis showed that the period of vortex shedding equalled the leg natural period (approximately 1 to 2 s depending on the height of exposed leg) at a relative wind velocity of 38 kmph (24 mph). Towing logs showed that such wind velocities were experienced during the majority of the days at sea. The computed nominal bending stress range induced by the vortex shedding was about 106 MPa (15.4 ksi) near the fixed end of a leg. Subsequently, a local strain analysis (described later) estimated a life to crack initiation of 5×10^5 cycles for this stress range, that is, a 6- to 12-day duration of vortex shedding at the leg natural period (sea-times for wet tows range from one to three months.)

During a wet tow, the periods of pitch and roll are very large relative to the periods of waves likely to be encountered. Therefore, the resulting angles of pitch and roll and concomitant inertia bending stresses in the legs are small. The insignificance of these inertia stresses has been corroborated analytically by the rig designers. For the dry tow, however, the problem was identified as bending stresses induced by the inertia response of the legs to rolling and not to vortex shedding. This rig experienced very severe weather rounding the Cape of Good Hope. Also, the length of upper legs removed and carried on the deck of the rig platform was only two thirds of the length removed in dry tows that did not experience pinhole cracking.

The obvious solution to the fatigue problem was to lower the stresses in the critical areas by the elimination of vortex shedding through structural modifications and, for dry tows on barges, increase the length of leg sections removed from the top of the leg and carried on the platform.

Also, it was decided to determine, by experiment, whether thermal stress relief was beneficial to the fatigue life of the pinhole details. In addition, the experiments were designed to test the validity of the local strain model to pinhole life assessment.

Local Strain Model

Although the bulk of a structural component behaves elastically during cyclic loading, the material at the root of a notch in that component may be subjected to cyclic plasticity at all but the lowest loads. Cyclic plasticity and subsequent fatigue crack initiation occur when the maximum and minimum cyclic stresses at the notch root (algebraic sum of the load-induced local stress and the residual stress) exceed, respectively, the tensile and compressive yield strengths of the material. In such a case, the material at the notch root yields in both tension and compression during each load cycle, Fig. 2. The local strain model [2] essentially simulates the cyclic stress-

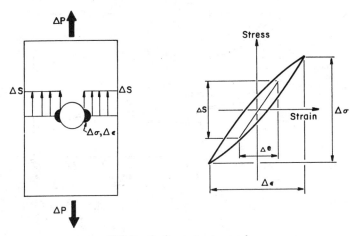

FIG 2—*Cyclic strain at a notch.*

strain behavior at the notch root through the combined use of: (*a*) the cyclic stress-strain relations from strain-controlled fatigue tests on unnotched specimens of the material at the notch, and (*b*) Neuber's rule. Neuber's rule relates the nominal stress range and notch severity in the notch to the cyclic stress range-strain range relation, and the cyclic strain range-life relation, of the unnotched material. The local strain model is generally used to estimate the life to development of a macroscopic crack (called crack initiation for brevity) in the notch root.

Local strain techniques are widely used in the ground vehicle industry [3–5] and have been successfully used to predict crack initiation in butt welds [6] and transverse fillet welds [7], and in tubular *K*-joints for offshore structures [8]. Application of the local strain model to crack initiation at weld discontinuities has been demonstrated in longitudinal fillet weldments containing porosity [9] and at the root of a partial-penetration butt weld [10]. The local strain approach, extended to include the effects of residual and mean stresses, modeled the insensitivity to mean stress for the as-welded case and the dependency of fatigue life on mean stress for stress-relieved weldments [11]. The latter modification to the local strain model was developed for application to the problem described in this paper and is outlined next.

Cyclic Strain Relationships

The cyclic strain-life relation (analogous to an *S-N* curve), Fig. 3, may be expressed by

$$\Delta\epsilon/2 = (\sigma_f'/E)(2N_i)^b + \epsilon_f'(2N_i)^c \tag{1}$$

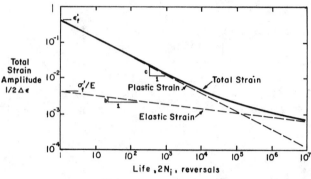

FIG 3—*Strain-life curve.*

where

$$\Delta\epsilon \ = \ \text{total strain range,}$$
$$\sigma_f' \ = \ \text{fatigue strength coefficient,}$$
$$E \ = \ \text{Young's modulus,}$$
$$N_i \ = \ \text{cycles to crack initiation,}$$
$$b \ = \ \text{fatigue strength exponent,}$$
$$\epsilon_f' \ = \ \text{fatigue ductility coefficient, and}$$
$$c \ = \ \text{fatigue ductility exponent.}$$

The cyclic strain range may be partitioned into its two components

$$\Delta\epsilon \ = \ \Delta\epsilon_e \ + \ \Delta\epsilon_p \tag{2}$$

where $\Delta\epsilon_e$ and $\Delta\epsilon_p$ are, respectively, the elastic and plastic strain ranges. The elastic and plastic strain ranges may be expressed by

$$\Delta\epsilon_e \ = \ \Delta\sigma/E \tag{3}$$

or

$$\Delta\epsilon_e \ = \ 2(\sigma_f'/E)(2N_i)^b \tag{4}$$

and

$$\Delta\epsilon_p \ = \ 2\epsilon_f'(\Delta\sigma/2\sigma_f')^{1/n'} \tag{5}$$

or

$$\Delta\epsilon_p \ = \ 2\epsilon_f'(2N_i)^c \tag{6}$$

where $\Delta\sigma$ and n' are, respectively, the stress range and cyclic strain-hardening exponent.

Neuber [12] analyzed a specific notch geometry and loading and derived a rule for nonlinear material behavior at the notch root. Neuber stated that the theoretical concentration factor is the geometric mean of the actual stress and strain concentration factors, that is, Neuber's rule

$$K_t = \sqrt{(K_\sigma K_\epsilon)} \tag{7}$$

The cyclic stresses and strains at a notch root in a thin part, that is, plane stress conditions [13], may be related to the cyclic stresses and strains away from the root by modifying Neuber's rule [14]

$$K_f = \sqrt{(K_\sigma K_\epsilon)} \tag{8}^2$$

with

$$K_\sigma = \Delta\sigma/\Delta S \quad \text{and} \quad K_\epsilon = \Delta\epsilon/\Delta e \tag{9}$$

where

$\Delta S, \Delta e$ = nominal stress and strain ranges, and
$\Delta\sigma, \Delta\epsilon$ = local (notch root) stress and strain ranges.

Equation 8 can be expressed as

$$K_f\sqrt{(\Delta S \Delta e E)} = \sqrt{(\Delta\sigma\Delta\epsilon E)} \tag{10}$$

The fatigue notch factor K_f is related to the stress concentration factor K_t, the notch root radius r and a material constant a by

$$K_f = 1 + (K_t - 1)/(1 + a/r) \tag{11}$$

The material constant a is, in turn, a function of the tensile strength Su [14]

$$a = 0.0254 \cdot (2079/S_u)^{1.8} \tag{12a}$$

with a and S_u expressed in, respectively, mm and MPa, or

$$a = 0.001 \cdot (300/S_u)^{1.8} \tag{12b}$$

with a and S_u expressed in, respectively, in. and ksi. When ΔS and Δe are elastic, Eq 10 becomes

$$K_f\Delta S = \sqrt{(\Delta\sigma\Delta\epsilon E)} \tag{13}$$

---

[2] It is generally accepted that the ratio of notched fatigue strength to unnotched fatigue strength for a given material is usually less than the stress concentration factor K_t. This ratio is defined as the fatigue notch factor K_f [13].

It has also been shown [14] that, for a given material, when

$$K_f \Delta S|_{\text{notched}} = \sqrt{(\Delta\sigma\Delta\epsilon E)}|_{\text{unnotched}} \qquad (14)$$

similar lives to crack initiation will be observed, or

$$N_i|_{\text{notched}} = N_i|_{\text{unnotched}} \qquad (15)$$

The Neuber relation or life prediction model for the material, $\sqrt{(\Delta\sigma\Delta\epsilon E)}$ versus $2N_i$, is plotted in one of two ways. If strain-controlled test data are available, the term $\sqrt{(\Delta\sigma\Delta\epsilon E)}$ is computed for each test with $\Delta\sigma$ and $\Delta\epsilon$ determined from the half-life hysteresis stress-strain loop and plotted versus the reversals to crack initiation $2N_i$, for example, Fig. 4. (It is generally accepted in the application of the local strain model that crack initiation and complete separation in strain-controlled tests of small unnotched specimens are synonymous.) If, on the other hand, the cyclic stress-strain and strain-life parameters are taken from publications, for example, Ref. 15, the life prediction model is computed as follows. First, an arbitrary life $2N_i$ is selected and the strain range $\Delta\epsilon$ is computed from Eq 1. Next the stress range $\Delta\sigma$ for the previously determined $\Delta\epsilon$ is computed by the iterative solution of the cyclic stress-strain curve

$$\Delta\epsilon/2 = \Delta\sigma/2E + (\Delta\sigma/2K')^{1/n'} \qquad (16)$$

where K' is the cyclic strength coefficient. Finally the term $\sqrt{(\Delta\sigma\Delta\epsilon E)}$ is formed and plotted versus the assumed value of $2N_i$, for example, Fig. 4.

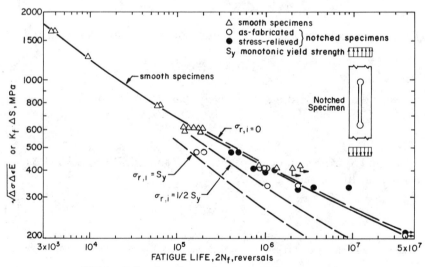

FIG 4—Full-scale and smooth specimen fatigue test results.

Nonzero Mean Stress

The foregoing analyses apply strictly to completely reversed local cyclic stresses. The stress-strain function of Smith et al [16] that includes the effect of mean stress led the writer [11] to incorporate mean stress in notched fatigue analysis as follows.

It has been shown [16] that similar fatigue lives result for two unnotched specimens at different mean stresses, σ_m, when

$$\Delta\sigma\Delta\epsilon|_{\sigma_m=0} = 2\sigma_{max}\Delta\epsilon|_{\sigma_m\neq0} \qquad (17)$$

With Neuber's rule, Eq 7, and

$$K_\sigma = \sigma_{max}/S_{max} \quad \text{and} \quad K_\epsilon = \Delta\epsilon E/\Delta S \qquad (18)$$

Equation 17 becomes, for $\sigma_m \neq 0$

$$\sqrt{(\Delta\sigma\Delta\epsilon E)} = K_f\sqrt{(2S_{max}\Delta S)} \qquad (19)$$

The maximum nominal stress Smax of Eq 19 is [11]

$$S_{max} = S_m + \Delta S/2 + \sigma_{res}/K_\sigma \qquad (20)$$

where S_m, and σ_{res} are, respectively, nominal mean stress, and notch root residual stress. The stress concentration factor K_σ is found from the iterative solution of

$$K_\sigma = K_f[1 + E/(K')^{1/n'} \cdot [K_\sigma(S_{max}) + \sigma_{res}]^{(1-n)/n}]^{1/2} \qquad (21)$$

For a nonzero mean stress, the product $K_f\sqrt{(2S_{max}\Delta S)}$ is formed and used to enter the cyclic strain-life relation $\sqrt{(\Delta\sigma\Delta\epsilon E)}$ versus N_i similar to Fig. 4 to determine life N_i.

Experiment

The experiment was designed to: (1) determine whether thermal stress relief influenced the fatigue resistance of full-scale, axially loaded specimens that simulated the pinhole detail, (2) test the ability of the local strain model to estimate the fatigue resistance of full-scale specimens and, thus, the pinhole, and (3) permit measurements of the notch residual stresses resulting from fabrication of the notch.

Materials

The full-scale specimens and smooth, unnotched specimens were fabricated from 51 mm (2 in.) thick ASTM A131 Grade C carbon steel plates

TABLE 1—*Chemical composition (% by weight).*

C	Mn	P	S	Si	Cu	N
0.14	0.64	0.004	0.020	0.19	0.04	0.0048

(normalized at 900°C (1650°F) for ½ h per 25 mm (1 in.) of thickness), and, where required, stress-relieved at 600°C (1100°F) for 1 h per 25 mm (1 in.) of thickness. The chemical composition is listed in Table 1, and the monotonic and cyclic properties are listed in Table 2. The cyclic fatigue parameters of Table 2, σ_f', b, ϵ_f', and c, were determined by least-squares fits of Eqs 4 and 6 to the strain-life data of the smooth, unnotched data. The cyclic stress-strain parameters of Table 2, n' and K', were estimated from a least-squares fit of Eq 16 to the stress range-strain range pairs from the half-life hysteresis loops of the smooth fatigue specimens.

Fatigue Specimens and Tests

The full-scale specimens, Fig. 5, were 190 mm (7.5 in.) wide, 38.1 mm (1.5 in.) thick,[3] and contained two 38.1 mm (1.5 in.) diameter holes drilled 368 mm (14.5 in.) on centers on the longitudinal center line of the specimen. The two holes were joined by a 12.7 mm (0.5 in.) wide slot gas-cut to introduce tensile residual stresses on the surfaces of the drilled holes. The elastic stress concentration factor of the notch, $K_t = 3.1$, (based on the gross-section nominal stress) was determined by both finite element and photoelastic analysis.

The smooth, unnotched specimens with a square cross section 8.9 × 8.9 mm (0.35 × 0.35 in.) and a gage length of 12.7 mm (0.5 in.) were subjected to constant-amplitude reversed cyclic strains in compliance with ASTM Standard Recommended Practice for Constant Amplitude, Low-Cycle Fatigue Testing (E 606-80).

The full-scale specimens were cycled under load control, Fig. 5, while the unnotched specimens were cycled under strain control in servohydraulic test systems. The capacities of the test systems were 2.7 MN (600 kip) and 156 KN (35 kip) for, respectively, the full-scale and smooth specimen tests. Both specimen configurations were tested in complete reversal under a triangular strain or load-time wave form at similar strain rates—0.0015/s.

Residual Stress Measurements

Residual stresses were measured at the notch roots in untested full-scale specimens using the same technique as previously used in the examination

---

[3] The 51-mm (2-in.)-thick specimen blanks were milled to 38.1 mm (1.5 in.) to insure flatness and eliminate bending in the fatigue tests.

TABLE 2—*Monotonic and cyclic mechanical properties.*

Monotonic yield strength (0.2% offset), S_y	239 MPa (34.7 ksi)
Monotonic tensile strength, S_u	415 MPa (60.2 ksi)
Reduction of area, RA	68.6%
Fatigue strength coefficient, σ_f'	1630 MPa (236 ksi)
Fatigue strength exponent, b	−0.160
Fatigue ductility coefficient, ϵ_f'	0.510
Fatigue ductility exponent, c	−0.512
Cyclic strength coefficient, K'	1050 MPa (153 ksi)
Cyclic strain-hardening exponent, n'	0.237

of the pinholes from leg sections. Electric resistance strain gages were installed in the drilled holes at the plate midthickness to monitor the circumferential strains after the material surrounding the gage was removed by sawing. Stabilized strain gage readings were obtained approximately 16 h after saw cutting. The residual stresses under the strain gages were then estimated by X-ray analysis.

In addition, the notches for all full-scale test specimens were instrumented with electric resistance strain gages, Fig. 5, to permit an approximate estimation of notch residual stresses during the first loading cycle.

Test Results

Measurement of the tensile residual stresses by sectioning strain gaged, untested notched specimens showed that the maximum surface residual stress varied from specimen to specimen but, in some cases, equaled the monotonic yield strength of the material. These observations were corroborated by strain gage response of the instrumented full-scale fatigue test specimens.

In Fig. 4, the results of the strain-cycle fatigue tests on the smooth specimens are compared with the test results for the full-scale notched specimens through the application of Neuber's rule. The smooth specimens are plotted using $\sqrt{\Delta\sigma\Delta\epsilon E}$ with $\Delta\sigma$ and $\Delta\epsilon$ being determined from the half-life hysteresis stress-strain loops, Fig. 2, and the reversals to end-of-test $2N_i$, that is, complete separation. The curve labeled "smooth specimens" in Fig. 4 represents the basic Neuber relation or life prediction model for the material and is also plotted as $\sqrt{(\Delta\sigma\Delta\epsilon E)}$ versus $2N_i$ where the radical term was computed as follows. First an arbitrary life $2N_i$ is selected, and the strain range is computed from the strain-life relation of Eq 1, with the cyclic parameters σ_f', b, ϵ_f', and c being determined from the strain-cycle fatigue tests on the smooth specimens (Table 2). Next, the stress range $\Delta\sigma$ for the previously determined $\Delta\epsilon$ is computed by the iterative solution of the cyclic stress range-strain range curve, Eq 16, with the cyclic strength coefficient and strain-hardening exponent, respectively, K', and n', from the smooth specimen tests. Finally, the term $\sqrt{(\Delta\sigma\Delta\epsilon E)}$ is formed and plotted versus the assumed value of $2N_i$.

FIG 5—*Full-scale fatigue tests.*

The full-scale specimens are plotted in Fig. 4 with $K_f = 3.05$ (from Eq 11 and 12) and ΔS and $2N_i$ from the full-scale fatigue tests. The life $2N_i$ was the number of reversals at which the crack in the notch root reached both faces of the specimen and, subsequently, continued to grow as a through-thickness crack.

All of the finite-life stress-relieved notched specimens are within a factor of 3, that is, ⅓ to 3 times, and seven are within a factor of 2 of the lives predicted by the Neuber curve representing the material (Fig. 4). The lives of seven as-fabricated notched specimens varied by factors of one third to one times the predicted lives. The greater variability demonstrated by the as-fabricated specimens is quite likely due to the presence of residual stresses that vary in magnitude from specimen to specimen. However, the inability to measure the residual stresses in the as-fabricated specimens means that local-strain analysis that includes the effect of residual stresses cannot be quantitatively applied to the individual as-fabricated notched specimens.

Neuber's rule for zero initial residual and nominal mean stress is expressed from Eq 13

$$K_f = \sqrt{(\Delta\sigma\Delta\epsilon E)} \tag{13}$$

For a nonzero initial residual stress σ_{res} and zero nominal mean stress, Neuber's rule from Eqs 9, 19, and 20 is

$$K_f\Delta S \cdot \sqrt{(1 + 2\sigma_{res}/K_\sigma\Delta S)} = \sqrt{(\Delta\sigma\Delta\epsilon E)} \tag{22}$$

If the value of the notch residual stresses in the as-fabricated specimens were known, these specimens would be plotted in Fig. 4 as $2N_i$ versus $K_f\Delta S \cdot \sqrt{(1 + 2\sigma_{res}/K_\sigma\Delta S)}$ rather than $K_f\Delta S$. Since, as expected, the residual stresses in the notch are tensile the term in the radical is greater than unity. Thus, the plots of the as-fabricated specimens would be shifted vertically upward and closer to the Neuber material curve in Fig. 4 if the values of the notch residual stresses were known.

The lives of the notched specimens were estimated for three conditions $\sigma_{res} = 0$, $0.5S_y$, and S_y (S_y = monotonic yield strength), and are shown as dashed curves in Fig. 4. It may be seen that the curve for $\sigma_{res} = S_y$ is a lower bound to the data of the as-fabricated notched specimens. Also, the curve for $\sigma_{res} = 0$ agrees reasonably well with the smooth specimen characterization of the material.

Discussion

In general, thermal stress-relief improved the fatigue resistance of the full-scale specimens although some as-fabricated specimens showed similar lives as the stress-relieved specimens. Measurement of the strains in drilled holes during testing showed that the tensile stresses in the as-fabricated specimens varied from zero to the tensile yield strength. Therefore, it may be concluded that thermal stress-relief reduced the variability in life by eliminating the short-life outliers.

The behavior of the full-scale specimens was adequately modelled by the local strain model incorporating the results of strain-controlled fatigue tests on unnotched specimens.

Elimination of Vortex Shedding

The effectiveness of spiral or helical strakes of circular cross section in reducing the response to vortex shedding of circular, cylindrical members has been demonstrated by Davenport [17]. Davenport studied the wind-induced vibrations in 102-mm (4-in.)-diameter antenna members and found that helical spoilers over 20 and 40% or more of the member's length reduced the lateral deflection by, respectively, 2 and 74%.

Arrays of helical strakes of rectangular cross section are generally called Scruton spoilers. Scruton [18] found that helical spoilers of rectangular cross section were more effective than those of circular cross section. The optimum array was three strakes with a height and pitch of, respectively, 0.09 and 5 times the diameter of the member. In addition, the excitation of circular, cylindrical bodies fitted with rectangular strakes was independent of Reynolds number thus increasing the reliability of model studies. When mounted on a flexible cantilever such as a smokestack (or jack-up leg), the spoilers needed only to be applied to the top one third of the member.

Easily installed and removed Scruton spoilers are now bolted to the top 30% of each leg of Bethlehem's jack-up rigs during transit and operation in shallow depths, Fig. 6. The Scruton spoilers consist of 3 helices of thin plates 120 deg apart with a pitch of 5 to 6 times the leg diameter. The height of the spoilers is equal to or greater than 0.07 times the column diameter.

FIG 6—*Scruton spoilers installed on jack-up rig legs.*

The implementation of Scruton spoilers has eliminated vortex shedding and adverse aerodynamic response of the legs. No further fatigue cracking has been observed at pinholes.

Conclusions

The local-strain approach adequately predicted the fatigue life of the notched components using the strain-controlled fatigue test results of unnotched material.

Thermal stress-relief reduced the variability in notched fatigue life by reducing the deleterious residual stresses that decrease fatigue life.

Since the introduction of stress relief of the legs and the addition of spoilers to eliminate vortex shedding (and reduce the concomitant fatigue loading), no further fatigue cracking has been observed at pinholes in mat-type jack-up drilling rigs.

Acknowledgments

The writer thanks Dr. B. D. Macdonald and Mr. A. J. Schweitzer for the performance of the A131 notched fatigue tests and Dr. G. A. Miller and Mr. T. R. Sharron for performing the A131 smooth specimen tests. Mr. Schweitzer is also acknowledged for his significant contributions in the computer programming of the local-strain analysis. Also acknowledged are Prof. J. P. den Hartog for suggesting the application of helical spoilers, and J. E. Steele, J. O. Whitley, Jr., and R. M. Whiddon for their contributions and discussions.

References

[1] "Innovative Mobilization of Two Jack-ups," *Ocean Industry*, Vol. 17, No. 5, May 1982, pp. 19–27.
[2] Reemsnyder, H. S., "Constant Amplitude Fatigue Life Assessment Models," Paper 820688, *SAE Transactions*, Society of Automotive Engineers, Warrendale, PA, Vol. 91, 1983, pp. 2337–2350.
[3] "Fatigue Design Handbook," *Advances in Engineering, Vol. 4*, J. A. Graham, Ed., Society of Automotive Engineers, Warrendale, PA, 1968.
[4] "Fatigue Under Complex Loading: Analyses and Experiments," *Advances in Engineering, Vol. 6*, R. M. Wetzel, Ed., Society of Automotive Engineers, Warrendale, PA, 1977.
[5] "Proceedings of the SAE Fatigue Conference, Dearborn, Michigan," P-109, Society of Automotive Engineers, Warrendale, PA, 1982.
[6] Mattos, R. J. and Lawrence, F. V., "Estimation of the Fatigue Crack Initiation Life in Welds Using Low Cycle Fatigue Concepts," SP-424, Society of Automotive Engineers, Warrendale, PA, 1977.
[7] Smith, K. N., El Haddad, M., and Martin, J. F., *Journal of Testing and Evaluation*, Vol. 5, July 1977, pp. 327–332.
[8] Asuta, T., Toma, S., Kurobane, Y., and Mitsui, Y., "Fatigue Design of an Offshore Structure," OTC Paper No. 2607, Eighth Annual Offshore Technology Conference, Houston, May 1976.
[9] van der Zanden, A. M., Robins, D. B., and Topper, T. H. in *Testing for Prediction of Material Performance in Structures and Components, ASTM STP 515*, American Society for Testing and Materials, Philadelphia, 1972, pp. 268–284.

[10] Lawrence, F. V., Mattos, R. J., Higashida, Y., and Burk, J. D. in *Fatigue Testing of Weldments, ASTM STP 648,* American Society for Testing and Materials, Philadelphia, 1978, pp. 134–158.

[11] Reemsnyder, H. S., "Evaluating the Effect of Residual Stresses on Notched Fatigue Resistance," *Materials, Experimentation, and Design in Fatigue-Proceedings of Fatigue '81,* Westbury Press, Guildford, England, 1981, pp. 273–295, distributed in the United States by Ann Arbor Science Publishers, Woburn, MA. Also abstracted in *Residual Stress Effects in Fatigue, ASTM STP 776,* American Society for Testing and Materials, Philadelphia, 1982, p. 32.

[12] Neuber, H., *Transactions ASME, Journal of Applied Mechanics,* Vol. 28, 1961, pp. 544–550.

[13] Reemsnyder, H. S., "Stress Analysis," Paper 820683, *SAE Transactions,* Society of Automotive Engineers, Warrendale, PA. Vol. 91, 1983, pp. 2327–2336.

[14] Topper, T. H., Wetzel, R. M., and Morrow, J., *Journal of Materials,* Vol. 4, March 1969, pp. 200–209.

[15] Technical Report on Fatigue Properties-SAEJ1099, *SAE Handbook, Part 1,* Society of Automotive Engineers, Warrendale, PA, 1981, pp. 4.57–4.65.

[16] Smith, K. N., Watson, P., and Topper, T. H., *Journal of Materials,* American Society for Testing and Materials, Philadelphia, PA, Vol. 5, Dec. 1970, pp. 767–778.

[17] Bretschneider, C. L. in *Handbook of Ocean and Underwater Engineering,* J. J. Myers, Ed., McGraw-Hill, New York, 1969, pp. 12-23–12-24.

[18] Scruton, C. in *Wind Effects on Buildings and Structures,* National Physical Laboratory, H. M. Stationery Office, London, 1965, p. 798–836.

Roy W. Hampton[1] *and Howard G. Nelson*[1]

Failure Analysis of a Large Wind Tunnel Compressor Blade

REFERENCE: Hampton, R. W. and Nelson, H. G., "**Failure Analysis of a Large Wind Tunnel Compressor Blade,**" *Case Histories Involving Fatigue and Fracture Mechanics, ASTM STP 918,* C. M. Hudson and T. P. Rich, Eds., American Society for Testing and Materials, Philadelphia, 1986, pp. 153–180.

ABSTRACT: A failure analysis was performed to establish the cause and prevent the recurrence of a 2014-T6 aluminum compressor blade failure in a large NASA-Ames Research Center wind tunnel. A metallurgical failure analysis showed that a 0.13 mm (0.005 in.) deep scratch in the shank of the blade had acted as an initiation site for a fatigue crack. The crack subsequently grew by Stage II fatigue across most of the blade's 22.9-cm (9-in.) base transition region before final fracture occurred by unstable crack growth. Extensive fractographic characterization of both the blade and laboratory specimens was performed using a scanning electron microscope. Surface morphology, including fatigue striations, was used to estimate the vibration stress levels, the time to grow the crack, and the magnitude of the local mean stress present.

A structural failure investigation was conducted to determine the resonance vibration condition which made the blade sensitive to a scratch. Studies of the blade natural frequencies were made to determine the mode responsible for the vibrations which grew the crack. The effect of the crack on modal frequency was measured and computed. Tunnel measurements were used to determine the vibration resonance tunnel speed, magnitude, and source. These data were used to compare with the crack growth data from the fractographic studies and to make recommendations to avoid future blade failures.

KEY WORDS: fractography, fatigue crack propagation, aluminum alloys, fatigue striations, compressor blades, wind tunnels

A failure analysis was performed to establish the cause of a compressor rotor blade failure in a large wind tunnel at NASA-Ames Research Center and to determine the required actions to prevent its recurrence. This facility is a closed return atmospheric wind tunnel with a 4.3-m (14-ft)-diameter slotted wall test section. It has a Mach number range of 0.25 to 1.2 which is controlled by varying the compressor rotor speed from 70 to 485 rpm. The three stage axial compressor has several possible rotor blade vibration sources from the aerodynamic effects of various stator and vane combinations.

[1] Research engineer and chief, respectively, Test Engineering and Analysis Branch, NASA-Ames Research Center, Moffett Field, CA 94035.

FIG. 1—*Fourteen-foot wind tunnel rotor blade.*

The rotor blade shape shown in Fig. 1 is the same for all three rotor stages. The blades are machined from aluminum 2014-T6 hand forgings and have a finished weight of 77 kg (170 lb) and a length of 1.7 m (67 in.). There is a 15.2-cm (6-in.)-diameter shank connecting the blade to a hub which is bolted to a steel rotor.

The blades are routinely removed from the tunnel for refinish sanding after about 500 h of usage. This refinishing smoothes the airfoil surface which is subject to impact damage from tunnel debris. Also, the refinishing greatly increases the fatigue life of the blades. The blades which failed had been refinished and operated in the compressor at various speeds for 267 h before the failure occurred. The crack which led to the blade failure initiated in a third stage rotor blade shank, and grew across most of the 22.9-cm (9-in.)-wide shank to airfoil transition region before an overload failed the remaining blade section. The blade failure caused damage to five other third stage rotor blades and required a shutdown of the facility for five months.

Metallurgical Blade Failure Analysis

The failed blade had been in service and refinished three times before the last installation, which attested to the overall blade quality. Also, hardness checks on the failed blade and other blades showed negligible differences. Since there were no indications of significant materials deficiencies associated with the blade failure, the metallurgical failure analysis consisted primarily of fractographic studies. The primary objectives of the fractographic and related laboratory investigations were (1) to establish the most probable cause for the initiation of the crack which eventually resulted in the blade failure, and (2) to define, as accurately as possible, the conditions which existed during the growth of the crack.

Macroscopic Observations

A light photograph of the fracture surface of the failed blade is shown in Fig. 2. The crack initiated at Point 1, which is shown in greater detail in Fig. 3. The markings on the crack surface in Fig. 2 suggest that it grew in Stage II fatigue (noncrystallographic crack path) from near the initiation Point 1 through Point 2 to Point 3 where an overload failure occurred. This fatigue crack growth can be separated into three regions as shown in Fig. 2.

Region I which contains the initiation site is at a different angle than the other regions and is foreshortened in Fig. 2. Regions I and II have dark areas which were determined to be oil residues and iron flakes from the tunnel environment. Markings around the initiation site in Region I are visible in Fig. 3 at Points 4 and 5. These markings are evident because of residue buildup during fatigue crack growth, and probably represent either a single tunnel run period or an event which yielded high contamination in the tunnel environment. Additionally, the fact that the residue is fairly evenly distributed along the fatigue markings throughout Region I suggests that the crack tip was highly stressed, yielding an open crack where residue could easily enter.

Region II, shown in Fig. 2, is in a different plane than Region I. It is relatively flat at the crack edges and has light colored fatigue markings. The residue in this region is located near the outside edges of the fracture,

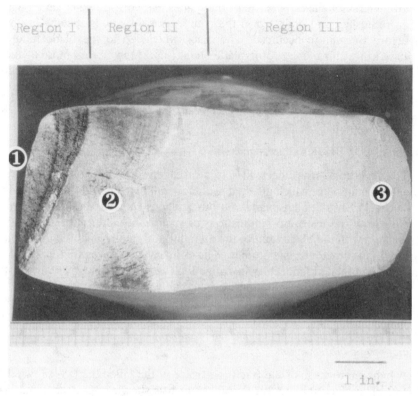

FIG. 2—*Light photograph of the fracture surface observed on the shank end of the failed blade. The fracture surface is made up of three regions of fracture.*

which suggests the crack surface was less accessible to the tunnel environment, and less highly stressed than in Region I. The area near Point 2 in Fig. 2 is an internal defect which acted to retard the growth of the crack in the center of the blade.

Region III of the crack is shown in Fig. 2 and contains shear lips at the crack edges. Here the fatigue markings are less distinct and greatly separated. The overall surface has a dull, matted appearance similar to the macroscopic appearance of the overload failure (beyond Point 3 in this figure). All of these observations strongly suggest that the crack growth in this region occurred as a result of a cyclic load under a very high static tensile load.

Microscopic Observations

Specimens were removed from the fracture surface and examined using the scanning electron microscope (SEM). (During SEM analysis, specimens were tilted 35 deg to the incident electron beam unless otherwise noted, and a 10 KV accelerating voltage was used.)

The initiation site (shown in Fig. 3) was examined with the SEM. A montage of the fracture surface around the initiation site is shown in Fig. 4. The site has a surface indentation or scratch indicated as Site 1 which is about 0.13 mm (0.005 in.) deep, and 2.5 cm (1 in.) long. There is no evidence of corrosion on this defect, which appears to have been present before the fatigue crack grew. The fatigue crack initiates just beyond the indentation, and radiates out in all directions.

Fatigue striations were found throughout most of Regions I and II of the crack. In particular, Site 3 in Fig. 4 is shown in greater detail in Fig. 5a. Site 8 in Fig. 5a is shown enlarged in Fig. 5b where fatigue striations are visible with a spacing of 0.8 μm (3 $\times$ 10^{-5} in.). Another example of striations is seen at Site 6 in Fig. 4 which is enlarged in Fig. 6a. Site 9 in Fig. 6a is enlarged in Fig. 6b, where striation spacings of 2 μm (8 $\times$ 10^{-5} in.) are found. These fracture surface sites all exhibit transgranular fracture,

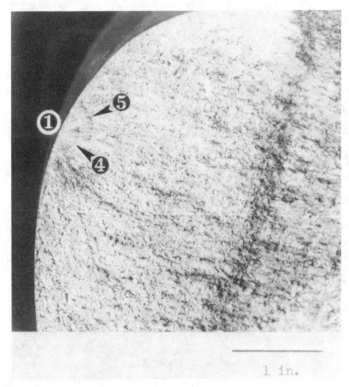

1 in.

FIG. 3—*Region I of fatigue fracture surface rotated to display true size, and showing the point of crack initiation, Point 1. Note the two cresent shaped markings, Points 4 and 5, about the point of initiation.*

FIG. 4—A SEM fractograph montage of the fracture surface about the point of crack initiation. Initiation occurred at Point 1 from a surface indentation in Area 2. Fatigue markings are evident in Areas 3 through 7.

2.5×10^{-3} in,

with classic fatigue striations as is often observed in Stage II fatigue of aluminum alloys.

The portion of the fracture surface from 1.3 to 5.6 mm (0.50 to 0.22 in.) away from the crack initiation point does not show distinct fatigue striations. Instead, the surface appears highly worked or deformed on a local level, and contains numerous cracks of different orientations as shown in Fig. 7. Studies described later showed this to be typical of fatigue crack growth where striation width is extremely small [less than 0.13 μm (5×10^{-6} in.)].

The crack surface from 5.6 mm (0.22 in.) away from the initiation point to the end of Region I shows striations with spacings of 0.74 to 0.81 μm (2.9×10^{-5} to 3.2×10^{-5} in.). In Region II, most of the fracture surface appears to contain coarse, uneven, and unresolvable fatigue markings. Very near to the transition from Regions II to III, resolvable striations begin to reappear. Unlike the striations observed in Region I, these striations appear in packets, or short segments with an average striation width of 0.16 μm (6.3×10^{-6} in.). The striation packets are separated by one or more deeper and broader striations, as seen in Fig. 8.

The fracture surface throughout Region III does not show any evidence of fatigue crack growth on a microscopic level. Region III (shown in Fig. 9) is similar to the area of final overload failure, which shows a typical ductile overload fracture surface.

Laboratory Studies

In an effort to better understand the fractographic observations made on the failed blade surface, and to quantify these observations where possible, a number of fatigue experiments were performed in our laboratory on the 2014-T6 aluminum alloy obtained from the failed blade.

The fracture surface characteristics associated with fatigue crack initiation at a notch under three extreme loading conditions were developed using compact tension test specimens. Two static load combinations were studied: one test with a large mean stress, and two conditions with a low minimum stress. Typical fracture appearances for these three cases are shown in Figs. 10 through 12, respectively.

Figure 10, for a large mean tensile stress and a low alternating stress at an *R*-ratio of 0.6, exhibits Stage I fatigue (crystallographic crack path) where the crack propagates on a given plane within a single grain, with its direction changing abruptly as the crack enters another grain. There is no evidence of fatigue striations even at high magnification.

Figure 11*a*, for a near zero minimum stress and a high alternating stress at an *R* of 0.05, shows a transgranular and very noncrystallographic fracture surface (Stage II fatigue where the crack propagates across the grains, but does not follow any particular plane). Fatigue striations are observed, as shown in Fig. 11*b*, even very near the initiation point.

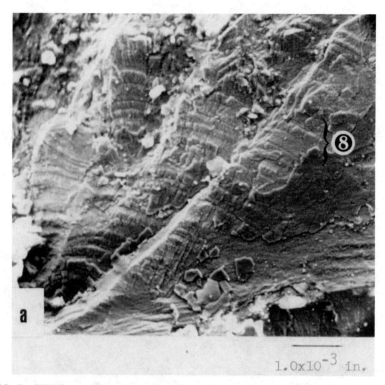

FIG. 5—*SEM fractograph of Area 3 in Fig. 4 showing distinct fatigue markings very near the point of crack initiation in* (a), *with striations of average width 0.8 μm (3 × 10⁻⁵ in.) in* (b).

Figure 12a, for a near zero minimum stress and a low alternating stress at an R of 0.05, shows a fracture surface which is intermediate between crystallographic (Fig. 10) and noncrystallographic (Fig. 11). In general, the crack appears to be following a given set of planes within each grain; however, the corners of the plane facets are smoothed and less abrupt. An indication of fine fatigue striations is also present, as seen in Fig. 12b.

The fractographic characteristics associated with fatigue crack growth well away from the point of initiation were also studied with the compact tension specimens cut from the failed blade. Two minimum load conditions were studied: (1) a high mean tensile stress with a low alternating stress, and (2) a low minimum tensile stress with varying alternating stresses.

The fracture surface resulting from a test under high mean load with low alternating stress is shown in Fig. 13. The fracture surface looks like an overload fracture surface. No fatigue markings or fatigue striations could be identified.

The low minimum load test showed different fracture surface characteristics, depending on the magnitude of the alternating load. At stress intensity

FIG. 5—*Continued.*

levels less than 11 MPa · m$^{1/2}$ (10 ksi · in$^{1/2}$), the surface looks highly worked as seen in Fig. 14. Fatigue striations are not observed. Instead, there are many cracks of different orientations approximating the location of the crack front. At median stress intensities from 11 to 22 MPa · m$^{1/2}$ (10 to 20 ksi · in$^{1/2}$), resolvable fatigue striations are visible, as shown in Fig. 15. At stress intensities above 22 MPa · m$^{1/2}$ (20 ksi · in$^{1/2}$), the fracture surface appears similar to what would be expected from an overload failure; however, it contains resolvable fatigue striations as shown in Fig. 16. The appearance of these striations differ from what is observed at lower alternating stress intensities in that the striations are very wavy and the overall striated surface appears highly deformed.

To quantify the meaning of the fatigue striations, the test specimen striation spacings at an R-ratio of 0.05 were correlated to the applied stress intensity as shown in Fig. 17. Also shown on this plot is average crack growth rate data previously measured for similar blade material samples at an R-ratio of 0.1 [1]. These data suggest a nearly one to one correlation of striation spacing to crack growth rate for the striation widths measured, and the R-ratio used.

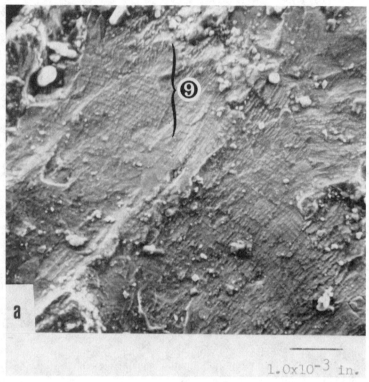

FIG. 6—*SEM fractograph* (a) *of Area 6 in Fig. 4, showing fatigue striations of average width 2 μm (8 × 10⁻⁵ in.) in* (b).

Interpretation of Blade Failure Surface

The controlling factors of the crack growth as determined from the macro- and microscopic observations and the laboratory studies are summarized next. The fatigue crack initiated at 0.13 mm (0.005 in.) deep surface scratch in the blade shank. There was no fractographic evidence of corrosion as- sisted crack initiation or growth. Both the crack initiation and subsequent growth appear to have occurred simply as the result of an applied static and cyclic load.

In order to determine the cause of the scratch in the blade, the blade installation procedures were reviewed. These studies suggested that the scratch probably was caused by the tools used to install the bolts holding the blade hub to the rotor. Similar, but less severe, scratches were found on the shanks of other blades when the compressor was disassembled after the blade failure.

To define the initial local mean stress acting on the scratch, the fracture surface data were reviewed. Near the initiation site, the crack surface is

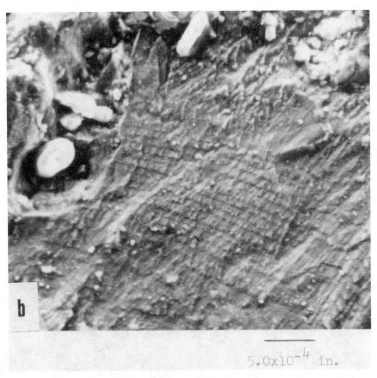

FIG. 6—*Continued.*

transgranular and noncrystallographic and shows fatigue markings (Fig. 5). A comparison to surface morphologies obtained in laboratory tests suggests that the initial crack growth occurred with a high alternating stress and a near zero minimum stress (Fig. 11). There are three possible mean stress sources at the crack initiation site: aerodynamic stresses, centrifugal stresses, and residual stresses. The aerodynamic drag bending stress is negligible, and the lift stresses are near zero at this location near the flexural neutral axis. The centrifugal stresses were computed to be only 3 ksi (21 MPa) at maximum speed. Therefore, since the total mean stress is low by fractographic evidence, the only remaining source of mean stress, residual stress, must also be low, and therefore did not greatly influence the blade failure.

The alternating stress which acted on the scratch is of interest, and may be estimated by the fatigue markings near the initiation site (Fig. 4). These markings are similar to classic fatigue striations which are often reported, and which were observed in laboratory tests with this material (Fig. 11). Table 1 is a summary of the striation spacing information obtained from the fracture surface. The stress-intensity factor based upon a one to one correspondence between striation spacing and crack growth (Fig. 17) is also listed in Table 1 and is used in the following correlations. (Any possible

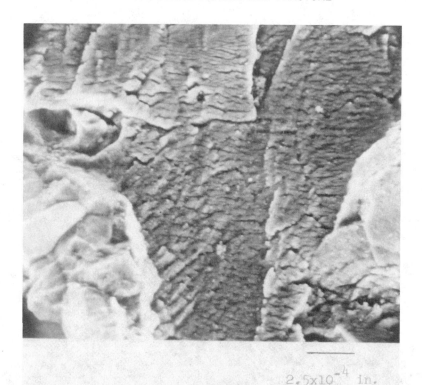

2.5x10^{-4} in.

FIG. 7—*SEM fractograph taken approximately 4.4 mm (0.175 in.) from the initiation site, showing a highly worked surface with cracks of different orientations, and no distinct fatigue striations.*

difference in *R*-ratio between blade crack growth and the laboratory data at an *R*-ratio of 0.1 is neglected since the blade crack growth conditions are not precisely known.) A calculation of the alternating stress, Δs, to produce this stress-intensity factor is given for the various cracks assumed, as noted in Table 1. Two solutions were used to compute the K_I due to bending (the modal action generating the crack). The two solutions were for a part-through-crack (PTC) [2] and an edge crack [3]. These two solutions bound the actual crack problem, which has a contoured shape and quickly changes from the long, shallow initial scratch to a semicircular crack (Fig. 3).

The estimated alternating stress at the crack initiation sites [shown in Table 1 for crack depths of 0.38 and 0.66 mm (0.015 and 0.026 in.)] appears to be too large for both types of crack solutions. A correction can be made for short cracks as shown by Topper and El Haddad [4], which adds a correction length, l_0, of 0.05 mm (0.002 in.) to the measured crack length. However, the increment to crack length does not appreciably change the predicted stress levels. These stresses are much larger than the material fatigue strength, and would have caused an earlier failure even without a

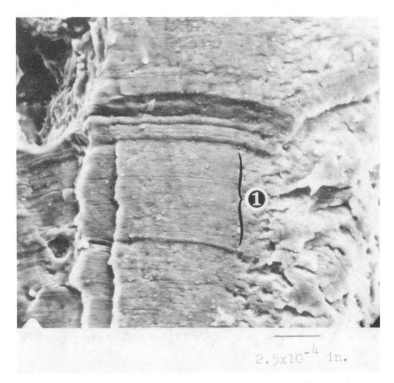

FIG. 8—*SEM fractograph showing a fine striation packet at about 10 cm (4 in.) from the crack initiation surface. The average striation width in this packet is 0.16 μm (6.3 × 10⁻⁶ in.).*

scratch. Therefore, it is concluded that these striations must not correspond to a single load cycle but instead are the result of a number of load cycles. This could be the result of a variation in the cyclic load form, magnitude, or frequency.

In order to obtain a lower bound for the magnitude of the alternating stress needed to initiate the crack from the scratch, the threshold stress-intensity factor for fatigue crack growth at a low R-ratio is needed. The threshold for crack propagation at an R-ratio of 0.1 was projected to be 2.7 MPa · m$^{1/2}$ (2.5 ksi · in.$^{1/2}$) from threshold data at higher R-ratios [1]. When these data are applied to the initial defect size of 0.13 mm (0.005 in.), plus the 0.05 mm (0.002 in.) short crack correction, the edge crack solution gives a value of 103 MPa (15 ksi). While the initial defect was a long scratch, which could be considered an edge crack, the black markings near the initiation site clearly show in Fig. 3 that the crack shape rapidly became semi-circular. Therefore, by assuming a semi-circular PTC and using the threshold stress intensity factor with the smallest site with visible striations [at a equal to 0.38 mm plus 0.05 mm equals 0.43 mm (0.017 in.)], an alternating stress of 110 MPa (16 ksi) is computed. This level of stress is

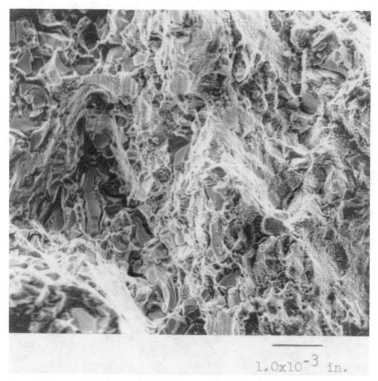

FIG. 9—*SEM fractograph of the typical fracture surface throughout the fatigue crack growth area of Region III. The fracture appearance is similar to overload failure and is free of any evidence of fatigue striations.*

supported by the calculation previously given and is similar to the striation data in Table 1 at crack depths of 1.3 to 5.7 mm (0.050 to 0.226 in.). Alternating stress predictions for the outer parts of the fracture surface are made as a function of the striation spacings shown in Table 1 and are summarized in Fig. 18.

The fracture surface in Region III (Fig. 9) is similar to that observed in the laboratory tests for a high mean tensile stress with a low alternating stress (Fig. 13). These high mean stresses would arise from the centrifugal loading acting on the very large crack present at this part of the crack growth. The crack growth in this region was probably very rapid.

An idea of the time involved in the crack propagation phase of the blade failure can be obtained from a quantitative interpretation of the average fatigue striations observed over the crack surface regions. The beginning of the first region did not give well correlated striation data with applied stress, so a lower resolvable bound for growth rate is used as shown in Table 2. The other average striation widths used are taken from Table 1. Assuming a loading frequency of 100 Hz, the time for crack propagation

1.0×10^{-3} in.

FIG. 10—*Typical fracture appearance near the point of crack initiation in a laboratory specimen tested under a high mean tensile stress superimposed upon a low alternating stress (R = 0.6). Fracture surface is very crystallographic with no evidence of fatigue striations.*

can be estimated, and is summarized in Table 2. It is seen that the actual crack propagation time is very short at this frequency. Of course, the time in the tunnel to grow the crack can be much longer, since many tunnel conditions will not produce alternating stresses large enough to grow the crack.

Structural Failure Investigations

Early in the failure analysis it was apparent that the blade was vibrating in a resonance condition to produce such sensitivity to scratches on the blade. Therefore, studies of the blade natural frequencies were made to determine the mode responsible for the vibrations. These studies included measured modal frequencies and shapes, finite-element computer modeling, and vibration measurements during the tunnel operation. Tunnel measurements were used to determine the vibration resonance tunnel speed, magnitude, and source. These data were used in a comparison with the observed crack growth and to make recommendations to avoid future failures.

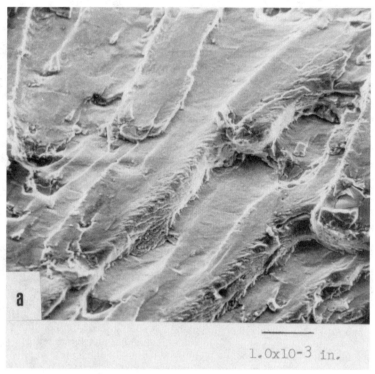

1.0×10^{-3} in.

FIG. 11—*Typical fracture appearance near the point of crack initiation in a laboratory specimen tested under a high alternating stress and a near zero minimum tensile stress (R = 0.05). Fracture appearance in (a) is transgranular and noncrystallographic, with fine fatigue striations as shown in (b).*

Blade Modal Tests

In an effort to identify the resonance conditions it can develop, a rotor blade was mounted in a test fixture with a fairly stiff base to simulate the mounting of the blade on the compressor rotor. The instrumented hammer technique [5] was used with a modal analyzer to determine the blade mode frequencies and shape. Several modes were found, including modes which were previously known and avoided in tunnel operations since they could produce resonance conditions in the compressor. Mode 3, at 100 Hz, was a previously unknown mode. (Later modal measurements with the blades installed in the more massive tunnel rotor gave a frequency of 110 Hz due to the different end conditions.) Studies of the modal deformation plots from the analyzer indicated that Modes 2 and 3 had a modal motion at the blade shank in a direction which would cause the observed crack growth direction. This was significant, because the crack growth direction was 90 deg away from the expected direction from aerodynamic loadings, and was an initial mystery.

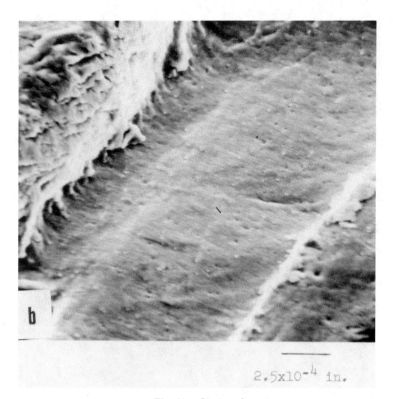

2.5x10⁻⁴ in.

Fig. 11—*Continued.*

To further demonstrate the possible involvement of these modes in the crack propagation, a tip damaged blade from the accident was mounted in the test stand for studies of the effects of a crack on modal frequency. To simulate the crack, a hacksaw was used to notch the blade shank at the location and orientation of the crack in the failed blade. The modal analyzer was used to monitor the blade frequencies as the cut was made deeper. The results showed that for a cut up to 5 cm (2 in.) deep (corresponding to the end of Region I of the crack, where it changed direction as shown in Fig. 2) there was no effect on the frequencies of modes except for Modes 2 and 3. The change in their frequency is shown in Table 3. This evidence shows that the effect of the crack is measurable as it gets about 5 cm (2 in.) deep, and also confirmed the observation that Modes 2 and 3 have the most action on the crack site.

It is interesting to note that the change in natural frequency of these modes with notch depth can be easily and accurately computed from the additional compliance of the blade due to a crack. In this procedure, the initial flexibility of the blade is modeled as a cantilever spring with a concentrated mass at the blade center of mass. Then the cracked blade is

FIG. 12—(a) *Typical fracture appearance near the point of crack initiation in a laboratory specimen tested under a low alternating stress and a near zero minimum tensile stress* (R = 0.05). *In* (b) *fine fatigue markings are visible.*

modeled as a rigid body from the mass to the crack. A virtual force at the mass center generates a moment which acts on the crack like a three-point bend specimen, and defines the equivalent spring due to the crack. Combining the two springs gives a very accurate prediction of the frequency shift due to crack length since it is referenced to the uncracked blade frequency. (For example, the data in Table 3 was modeled within 1% error.) While the change in frequency is sufficient to use as an inspection tool to detect a 5-cm (2-in.)-deep crack, the actual vibration time for such a crack to propagate to failure could be too short to use this as a practical inspection method.

Blade Finite Element Analyses

A NASTRAN finite element model (FEM) of the blade was generated to study the blade modal stresses. The computed mode shape for Mode 3 is shown as resultant motion vectors (the dotted lines) superimposed on the finite element mesh in Fig. 19. Studies of the stresses in these mode shapes showed that only Mode 3 produced significant stresses at the crack location,

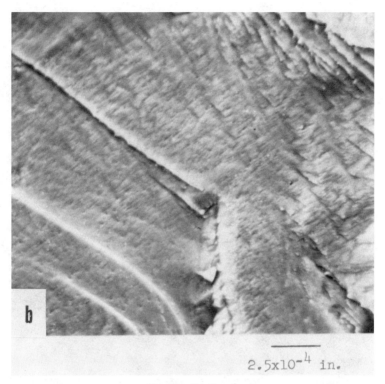

Fig. 12—*Continued.*

without requiring very large stresses elsewhere in the blade. (Large stresses in the airfoil section, for example, would have likely led to an earlier failure since the airfoil section is subject to impact damage from tunnel debris, and therefore has numerous possible initiation sites.) Thus, Mode 3 was determined to be the most likely mode responsible for the resonance condition which produced the vibration loading necessary to grow the crack in the blade.

The FEM model was also used to prepare for tunnel tests by selecting strain gage locations to monitor the peak strains occurring in each of the first five modes of the blade. (This included the crack initiation site for Mode 3.)

Tunnel Measurements

After the accident, a replacement set of blades was prepared and installed with special precautions to prevent any scratches in the blade shank areas. The stresses on the blades in the compressor operating environment were obtained, with special emphasis on defining the resonance points.

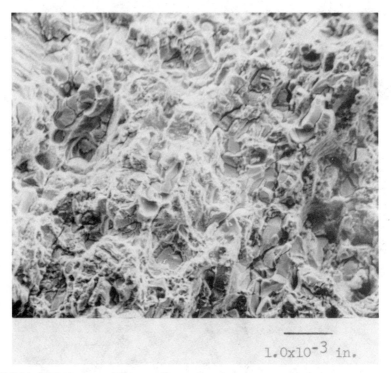

1.0x10^{-3} in.

FIG. 13—*SEM fractograph taken well away from the point of crack initiation in a laboratory specimen tested under a high mean tensile stress and a low alternating stress.*

These data established that the "new" Mode 3 was excited by the combined action of 33 interstage stator blades and 11 exit vanes to produce a beat excitation of 22 pulses per revolution. The resonance measured stress was 90 MPa (13 ksi) peak to peak and was highest on the third rotor stage. Data on the mode 3 resonance skirt showed the damping was 0.2% of critical dampening, which produces a very sharp resonance peak.

It is interesting to note that the decreasing stress with increasing crack length which was shown by the fractography data (Fig. 18) is probably due to increases in the modal dampening as a result of the crack plastic action. (The resonance frequency shift due to crack size changes would not have affected stress levels since the tunnel compressor operations required moving through the resonance condition wherever it was located.) An airfoil in a compressor usually has small modal structural dampening in comparison to the aerodynamic dampening. However, since the Mode 3 motion is almost parallel to the blade chord, the aerodynamic dampening is negligible, and the structural dampening becomes dominant. By ratioing the peak alternating stress levels found from the crack studies to the peak initial stress acting on the crack, and neglecting aerodynamic dampening, the modal

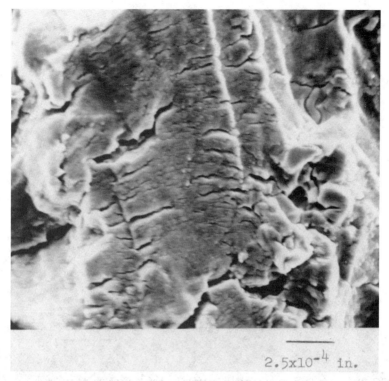

FIG. 14—*SEM fractograph taken well away from the initiation site on a laboratory specimen tested at a low alternating stress intensity [<11 MPa · m$^{1/2}$ (10 ksi · in.$^{1/2}$)]. Surface appears highly worked with many cracks with different orientations.*

dampening due to the crack can be estimated. The critical dampening ratio, and the ratio of the dampening to the uncracked modal dampening is shown in Table 4. Clearly, the crack effect dominates the dampening of this mode, and made it possible for the stresses to decrease as the crack grows. Without this action, the blade would have reached an overload condition and broken off at a much smaller crack size.

Computed Crack Propagation

With the measured Mode 3 vibration stress of 90 MPa (13 ksi) and the applicable crack sizes and stress data given in Table 1, it is possible to compute the number of cycles to grow the crack in the blade. The crack growth is computed from the threshold size for a Stage II fatigue crack, c_i, of 0.25 mm (0.010 in.) to a final computed crack size c_f, of 12.7 cm (5.03 in.) as shown in Table 5. Note that the effective crack depth was previously determined to be 0.18 mm (0.007 in.) which is slightly smaller than the 0.25

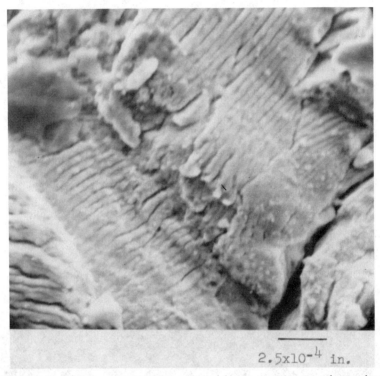

2.5×10^{-4} in.

FIG. 15—*Typical fatigue striations observed in a laboratory specimen tested at an alternating stress intensity between 11 and 22 MPa · $m^{1/2}$ (10 and 20 ksi · $in.^{1/2}$).*

mm (0.010 in.) used here. It is not clear whether the difference should be ascribed to errors in measuring the vibration stress, the threshold stress intensity, or the effective crack depth. However, it is felt that this agreement is close enough to support the position that the scratch was the feature which led to the growth of the crack under the applied stresses described.

Since the detailed data as to where the alternating stress changed as the crack grew are not available, an assumption is needed for the stress acting over the crack length increments from the fractography data. An assumption was made that the stress shown at the start of a crack growth increment acted for as much of the increment as possible without causing a brittle fracture. Since this procedure may use a larger alternating stress than was present, this calculation gives a lower bound of computed crack life. The program used to compute the crack growth [6] is for edge cracks under tension. Therefore, a reduced tension stress was used to give equivalent stress-intensity factors for the actual stress acting in bending. As seen in Table 5, the crack propagation is estimated to take 141 min at 110 Hz, or 9.3×10^5 cycles. This is in reasonable agreement with the fractographic estimate shown in Table 2.

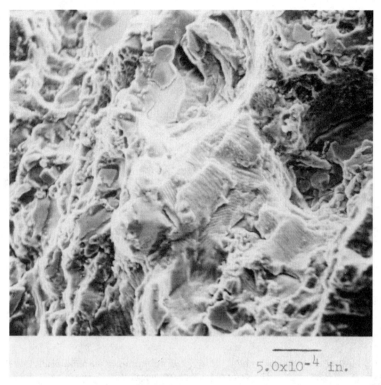

FIG. 16—*SEM fractograph of a laboratory specimen tested at a high alternating stress intensity* [>22 MPa · m$^{1/2}$ (20 ksi · in.$^{1/2}$)]. *The fracture is similar to overload failure except it contains many areas which exhibit distinct fatigue striations.*

To determine if this number of cycles could be accounted for in the 267 h of operation before the accident, the operating log was reviewed. No time was recorded at the actual resonance condition. However, the compressor passed through the Mode 3 resonance region 280 times. Since no warnings were in place, the operator may have moved gradually through this region. An average of 30 s in passage time would account for the computed cycles, which is likely to have occurred.

Recommendations

The tunnel measured stress data were used to develop recommended restricted operation ranges (placards) so as to minimize exposure to blade resonance points during normal compressor operations. The tunnel data showed that the steady-state resonance peak stresses were reached during the optimal passage of the compressor through the placard region, but that only 220 cycles were accumulated per passage. It was estimated this would give approximately 1.5 × 10^5 cycles at or just below the peak stress level

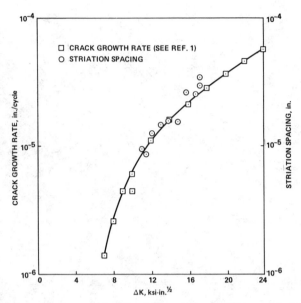

FIG. 17—*Laboratory data correlating striation width and fatigue crack growth rate as a function of alternating stress intensity in forged 2014-T6 aluminum.*

TABLE 1—*Summary of fatigue striations observed on the fracture of the failed blade, correlated ΔK_I, and projected alternating stress.*

Crack Depth, in.[f]	Observed Striation, in.	Correlated ΔK_I,[a] ksi in.[g]	Alternating Stress	
			Edge Crack,[b] ΔS (ksi)[h]	PTC,[c] ΔS (ksi)
0.015	3.0×10^{-5}	18.5	76.1	129.0
0.026	8.0×10^{-5}	>24.0	>75.0	>127.0
0.050	$<5.0 \times 10^{-6d}$	<9.4	<21.3	<36.0
0.175	$<5.0 \times 10^{-6d}$	<9.4	<11.4	<19.6
0.226	1.0×10^{-5}	11.6	12.6	21.5
1.334	2.9×10^{-5}	18.3	8.7	
2.176	3.2×10^{-5}	19.1	6.9	
TRANSITION FROM REGION I TO REGION II				
2.250	$<5.0 \times 10^{-6d}$	<9.4	<3.4	
4.000	$>6.3 \times 10^{-6}$	>10.0	>2.1	
TRANSITION FROM REGION II TO REGION III				
4.100	$>6.0 \times 10^{-5e}$	>24.0	>5.0	

[a] Assumes striation width equals crack growth (Fig. 17).
[b] Edge crack solution [3].
[c] Part-through-crack solution [2].
[d] Assumes striation width $<5.0 \times 10^{-6}$ in. is unresolvable.
[e] Assumes striation width $>6.0 \times 10^{-5}$ in. is masked in the fine structure of the fracture.
[f] 1 in. = 2.54 cm.
[g] 1 ksi $\sqrt{\text{in.}}$ = 1.099 MPa $\sqrt{\text{m}}$.
[h] 1 ksi = 6.89 MPa.

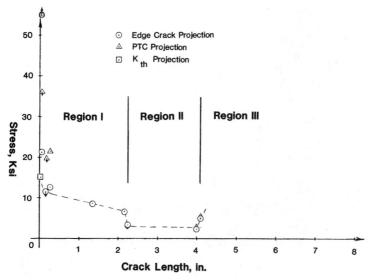

FIG. 18—*Estimated gross alternating stress in the unfailed blade ligament based on the fractography data as the fatigue crack propagates across the blade shank.*

during the usage period btween refinishing of the blades. The stress for fatigue failure at 10 times this life, 1.5×10^6 cycles, is given as 262 MPa (38 ksi) for smooth specimens at an *R*-ratio of 0 by MIL-HDBK-5D [7]. This would yield an adequate fatigue margin if there are no surface scratches on the blade shank (which can act as crack initiation sites).

To prevent the recurrence of the surface damage, several actions were recommended. First, since corrosion was noted at other locations on the blade (although it did not appear to contribute to the blade failure), a protective surface coating (Alodine 1200) was prescribed for the blades. Second, the installation bolting procedure was modified to reduce the possibility of scratches in the blade shanks, and a molded urethane fiber glass reinforced protective boot was made up to protect the blades. Finally, since the blades are especially vulnerable to defects and cracks in the blade shank

TABLE 2—*Summary of the estimated time involved in the crack propagation phase of the blade failure as determined from fractographic analysis.*

Region	Crack Depth, in.[b]	Average Striations	Cycles per in.[b]	Time,[a] min.
I	0.005 to 0.22	$<5 \times 10^{-6}$	$>4.3 \times 10^{+4}$	>7
I	0.22 to 2.2	2.3×10^{-5}	$8.6 \times 10^{+4}$	13
II	2.2 to 4.0	$<5 \times 10^{-6}$	$>3.6 \times 10^{+5}$	>60
III	4.0 to 8.0	$>6 \times 10^{-5}$	$<6.7 \times 10^{+4}$	<11

[a] Time is based on a loading frequency of 100 Hz.

[b] 1 in. = 2.54 cm.

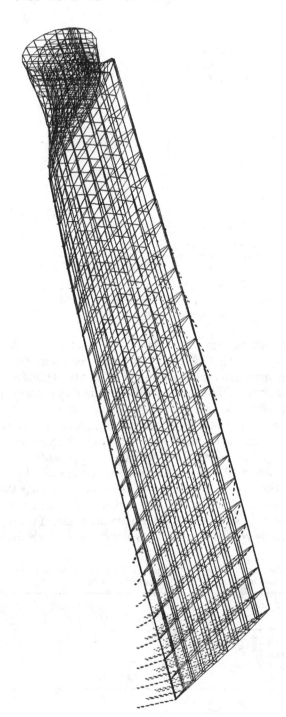

FIG. 19—NASTRAN model of fourteen-foot wind tunnel rotor blade showing dashed Mode 3 motion vectors superimposed on the finite element mesh.

TABLE 3—*Effect of saw cut simulation of cracked blade on blade natural frequency.*

Notch Depth, in.[a]	Measured Frequency, Hz	
	Mode 2	Mode 3
0.0	80.00	100.26
0.5	80.00	100.00
1.0	79.50	99.50
1.5	78.75	98.26
2.0	78.00	97.00

[a] 1 in. = 2.54 cm.

area, a high sensitivity fluorescent penetrant inspection was specified for the blade shanks after each 500 h usage period.

Summary

Fractographic analysis of the fracture surface of the failed blade was able to determine the crack initiation feature, the stresses acting on the cracked blade during the crack growth, and the time involved in the active crack growth. These data were useful in determining the factors which led to blade failure. The alternating stresses were predicted by the combination of threshold stress-intensity factor data and the striation measurement data.

Investigations of the blade modal frequencies and mode shapes were able to identify the mode responsible for the vibration stresses and the effect of the crack on modal frequency. The studies of the vibration conditions in the compressor validated the modal studies and identified the vibration excitation source. The measured stress levels and the computed crack life were found to be in reasonable agreement with the fractography studies. Also, the data were used to show that the observed crack growth was controlled by both the tunnel resonance and the dampening from the crack itself.

TABLE 4—*Change in critical modal dampening ratio with increase in crack size.*

Crack Depth, in.[a]	Vibration Stress, ΔS (ksi)[b]	Critical Dampening Ratio	Ratio of Increased Dampening
0.000	13.0	0.0020[c]	. . .
0.226	13.0	0.0020	1.0
1.334	8.7	0.0030	1.5
2.176	6.9	0.0038	1.9
2.250	<3.4	>0.0076	>3.8
4.000	2.1	0.0124	6.2

[a] 1 in. = 2.54 cm.
[b] 1 ksi = 6.89 MPa.
[c] From tunnel resonance vibration measurement of uncracked rotor blade.

TABLE 5—*Computed crack growth based on edge crack relationship* [6].

Crack Interval		Alternating Stress		ksi $\sqrt{\text{in.}}$[f]		
Initial, Ci (in.)[d]	Final, Cf (in.)[c]	Bending, ksi[e]	Tension,[a] ksi	Initial, ΔK_1	Final, ΔK_1	Cycles Blocks[b]
0.010	0.670	13.0	13.0	2.6	22.1	96.6
0.670	1.690	8.6	7.1	12.1	22.0	7.1
1.690	2.187	6.9	5.0	15.5	19.4	3.0
2.187	3.857	3.4	2.4	9.3	19.2	22.0
3.857	4.10	2.1	1.2	9.6	10.7	6.0
4.10	5.03	2.9	1.6	14.3	22.2	6.6
					Total blocks:	141.3

[a] The alternating tension stress is an equivalent stress to give the same ΔK_1 as the bending stress shown.
[b] One block is 6600 cycles, or 1 min at 110 Hz.
[c] Cf is determined when Kmax equals K_{lc} (22 ksi $\sqrt{\text{in.}}$), or else when the end of the range of crack size from Table 1 is reached.
[d] 1 in. = 2.54 cm.
[e] 1 ksi = 6.89 MPa.
[f] 1 ksi $\sqrt{\text{in.}}$ = 1.099 MPa $\sqrt{\text{m}}$.

The blade modal data were used with the tunnel vibration data to recommend placard regions in the compressor operating envelope. This action, combined with procedures to eliminate the damage to the blade shanks during installation into the compressor, were used to prevent a recurrence of this type of compressor blade failure.

References

[1] "Fracture and Fatigue Crack Propagation Rates of a Population of Aluminum Compressor Rotor Blades," Del West Associates, Inc., Woodland Hills, CA, 31 March, 1978 (under contract to NASA-Ames Research Center).
[2] Newman, J. C., Jr. and Raju, I. S., *Engineering Fracture Mechanics Journal,* Vol. 15, No. 1–2, 1981, pp. 185–192.
[3] Tada, H., Paris, P., and Irwin, G., *The Stress Analysis of Cracks Handbook,* Del Research Corporation, St. Louis, MO, 1973.
[4] Topper, T. H. and El Haddad, M. H., *Fatigue Thresholds Fundamentals and Engineering Applications, Volume II, Proceedings,* 1–3 June 1982, J. Backlund, A. F. Blom, and C. J. Beavers, Eds., Engineering Materials Advisory Services, Ltd., United Kingdom Conference sponsored by Aeronautics Research Institute of Sweden, pp. 777–798.
[5] Ramsey, K. A., *Sound and Vibration,* April 1976, pp. 18–30.
[6] Creager, M., "MSFC Crack Growth Analysis Computer Program Version II," August 1976 (under contract to NASA-Marshall Space Flight Center).
[7] "Metallic Materials and Elements for Aerospace Vehicle Structures," *Military Standardization Handbook,* Vol. 1, MIL-HDBK-5D, Air Force-11, Department of Defense, Washington, DC.

Russell C. Cipolla,[1] Jeffrey L. Grover,[1] and Roger H. Richman[1]

Analysis of a Compressor-Wheel Failure

REFERENCE: Cipolla, R. C., Grover, J. L., and Richman, R. H., "**Analysis of a Compressor-Wheel Failure,**" *Case Histories Involving Fatigue and Fracture Mechanics, ASTM STP 918,* C. M. Hudson and T. P. Rich, Eds., American Society for Testing and Materials, Philadelphia, 1986, pp. 181–210.

ABSTRACT: Pieces of a burst compressor wheel from a large combustion turbine revealed severe segregation of alloying constituents and impurities, high densities of sulfide inclusions, and intergranular, oxide filled microcracks. Analysis of oxidation kinetics for AISI 4140 steel indicates, however, that the service environment could not have caused the observed scale thicknesses in the microcracks. One- and two-dimensional stress analyses confirm modest cyclic stresses and maximum stresses in the range of 531 to 683 MPa (77 to 99 ksi), not the 827 MPa (120 ksi) level implied by crack growth from 0.76 mm (0.03 in.) flaws. To account for the observed cyclic life of about 1500 cycles, the initial defects would have to be 8.6 mm (0.34 in.) to 16.8 mm (0.66 in.) long, according to fracture mechanics analyses performed by an influence function method. On the basis of all the results, it is concluded that small cracks were already in the wheel when it went into service.

KEY WORDS: fatigue (materials), fracture toughness, fatigue crack initiation, fracture mechanics, crack propagation, steel-AISI 4140

Fourteen minutes after a normal cold start, the last-stage wheel in the compressor of a large combustion turbine burst into three major pieces. The turbine, of approximately the 15 MW (20 000 hp) class, had 1431 previous starts and had accumulated 5965 operating hours. The compressor was comprised of 16 stages, each constructed of a forged, individually rabbetted disk. Separation between compressor and turbine was by a tubular member, known as a distance piece, that was bolted and rabbetted to the first-stage turbine disk and to the 16th stage compressor disk. A balance journal was interference fit into a female rabbet near the bore of the last compressor wheel. The 0.965 m (38 in.) diameter, last-stage wheel is considerably more massive than the preceding 15 stages, being 82.6 mm (3.25 in.) thick over much of its cross section and 101.6 mm (4 in.) thick between diameters 0.65 m (25.5 in.) and 0.85 m (33.5 in.), which is the location of the bolt circle for the distance piece.

[1] Principal engineer, senior engineer, and principal engineer, respectively, Aptech Engineering Services, Inc., Palo Alto, CA 94303.

After a failure investigation, the manufacturer recommended that last-stage wheels be replaced in 141 similar turbines. Since most of the affected machines were in electricity generating service, the cost and availability impact of the recommended replacement program on the electric utilities was large. Therefore, a review of the manufacturer's investigation and conclusions was undertaken at the request of the Electric Power Research Institute (EPRI).

Review of the Manufacturer's Findings

Detailed examination had revealed that the steel (AISI 4140) of which the wheel was fabricated had a higher sulfur content than is now considered acceptable and that the sulfides were heavily segregated in streaks. It was stated by the manufacturer that cracks could initiate in segregated regions at stresses lower than those induced by thermal and centrifugal loads during normal startups. It was further stated that calculations showed that fracture could occur in about 1500 cycles by propagation of an initial crack only 0.76 mm (0.03 in.) long in material of 44 MPa $\sqrt{m}$ (40 ksi $\sqrt{in.}$) fracture toughness.

Initial fracture-mechanics calculations were made to estimate the stress levels and cyclic stress ranges that could cause crack growth from 0.76 mm (0.03 in.) to critical size in 1500 cycles, as implied by the manufacturer. By assuming a reasonable crack-growth law [1] and two reasonable critical crack sizes of 2.5 mm (0.1 in.) and 6.35 mm (0.25 in.), it was found that either the stress ranges were higher than could be sustained by material of low fracture toughness, or the initial defect size was much larger than 0.76 mm (0.03 in.). Therefore, a decision was made to proceed with an independent analysis.

Examination of the Failed Wheel

Macroscopic Observations

Since the three large pieces of the wheel were cut up to facilitate handling in the manufacturer's investigation, the main groupings were fit together in more or less their original positions. Smaller pieces that had been cut from regions of interest (at or near the main crack surfaces) were given more detailed reconstruction. From the patterns on the main cracks, it was agreed that the most probable fracture origin is at a rabbet corner 25.4 mm (1 in.) from the central bore surface, as concluded by the manufacturer. Two views of that region are shown in Fig. 1, from the two mating fracture surfaces. An equivalent reconstruction of a region at a bolt hole (Fig. 2) shows the presence of a large "old" crack that had obviously been present before the failure event.

FIG. 1a—*Reconstruction of fracture in region of bore and rabbet. The top piece* (arrow), *containing the most probable origin, had been encapsulated in plastic and then broken out. Residual mounting material is still adherent on this piece.*

Metallography

Metallographic examination of various cross sections of the wheel confirmed the frequency and morphology of manganese sulfide (MnS) inclusions, and the segregation of alloying constituents and impurities, reported by the manufacturer. Figure 3 illustrates those features, as well as some microcracks associated with the MnS inclusions. Note, however, that the microcracks are not parallel to the long axes of the inclusions as would be characteristic of link-up to form larger cracks but across the inclusions and, it turns out, parallel to the main crack.

Throughout the metallographic examinations, it was observed that the microcracks often occurred at inclusions near the main fractures. With increasing distance from a main crack, however, the frequency of microcracks at MnS inclusions diminished. At distances greater than about 2.5 mm (0.1 in.) from a main crack, microcracks at inclusions were no longer seen. This suggests that most of the microcracks are a consequence of the fast fracture, not a cause.

FIG. 1b—*Fracture surface that mates to the surface in Fig. 1a.*

FIG. 2—*Reconstruction showing the "old" crack at a bolt hole.*

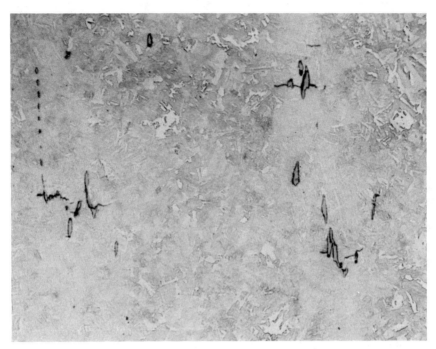

FIG. 3—*Metallographic section near main fracture, showing segregation, MnS inclusions, and some microcracks at inclusions (×150).*

Another class of rather longer microcracks was observed around bolt holes and rabbets. At high magnification, these cracks are seen to be filled with oxide (Fig. 4). Note the intergranular branches, also oxide filled, on the larger portion of the crack. Many such cracks were observed in the wheel pieces [2], including an old crack of substantial size at the bore rabbet (Fig. 5).

Fractography

Although many pieces of the broken wheel were examined by visible light and scanning electron microscopy, we concentrate here on a portion of the fracture that is believed to contain the most probable origin.

A piece about 15 mm (0.6 in.) on an edge, containing the crack origin, was cut from the right-hand portion of the reconstruction shown in Fig. 1b. The origin area is primarily intergranular fracture, as shown in Fig. 6, which was taken at about 3 mm (0.12 in.) from the rabbet corner. In contrast, Fig. 7 was taken at a distance of about 4 mm (0.16 in.) from the origin corner; it shows a mixed mode of ductile fracture with some intergranular cracking. Thus, at this point, the crack was probably still subcritical. Additional evidence can be seen in Fig. 8, which shows what appear to be

FIG. 4—*Segment of an intergranular crack near a bolt hole; the crack is filled with oxide* (×1000).

FIG. 5—*The backside of the bottom left-hand piece of Fig. 1a showing an old crack* (arrow).

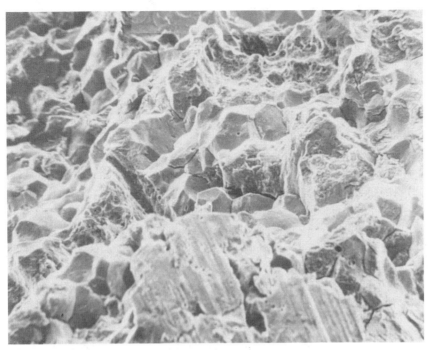

FIG. 6—*Scanning electron fractograph at 3 mm (0.12 in.) from the rabbet corner. Intergranular and secondary cracking are evident, along with some residual oxide and ductile tearing* (×500).

crack arrest lines (striations) at a distance of about 15 mm (0.6 in.) from the origin corner. Cleavage fracture was not observed anywhere on the 15 mm (0.6 in.) piece of Figs. 6 through 8. From an examination of adjacent pieces, along with allowances for the material lost by cutting, the best estimate of where the fracture becomes predominantly cleavage is about 31 mm (1.22 in.) from the most probable origin.

On the basis of these results, it would appear that the critical crack size for fast fracture is substantially larger than assumed by the manufacturer.

Age of Cracks

In view of the possibility that cracks could have been in the wheel for a long time (the old cracks), it was thought worthwhile to develop estimates of the time-temperature histories of the crack surfaces from the scale thicknesses.

Several cracks similar to the one in Fig. 4 were found to be still oxide filled at widths of 1×10^{-2} mm (0.4×10^{-3} in.) (that is, 0.5×10^{-2} mm (0.2×10^{-3} in.) oxide growth from each face). Since information about oxidation kinetics at low temperatures could not be found for steel, two

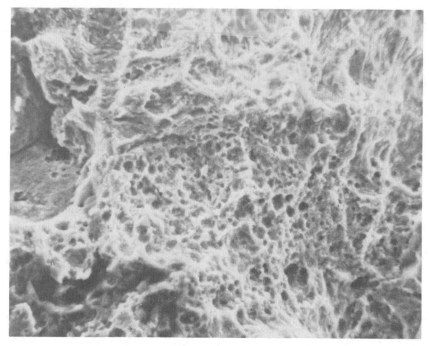

FIG. 7—*Scanning electron fractograph taken at about 4 mm (0.16 in.) from the origin corner, showing intergranular cracking, ductile tearing, and some residual oxide (×2000).*

methods were invoked to estimate the required time at 288°C (550°F), the temperature of the wheel at steady-state operation, to account for the oxide thickness. In the first, rate constants measured by Pinder [3] were extrapolated to 288°C (550°F) [4], and time was calculated from the well-known parabolic rate law

$$X^2 = Kt \tag{1}$$

where

   X = oxide thickness,
   K = parabolic rate constant, and
   t = exposure time.

The calculated time is 92 600 h, whereas the actual time of operation was slightly less than 6000 h.

To confirm this result, the oxidation data for AISI 4140 steel at 454°C (850°F), 538°C (1000°F), and 621°C (1150°F) tabulated in Ref 5 were analyzed by the methods of linear regression [6]. An excellent correlation was obtained [2]. The equation that describes the oxidation (for thickness in

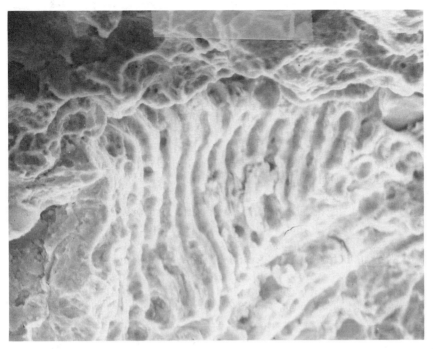

FIG. 8—*Scanning electron fractograph taken at about 15 mm (0.6 in.) from the origin corner, showing what appear to be crack arrest lines (×1000).*

inches, time in hours, and temperatures in degrees Fahrenheit) is

$$\log(\text{oxide penetration}) = -8.4306 + 0.4313 \log t + 0.00396T \quad (2)$$

An oxide scale thickness of 0.5×10^{-2} mm (0.2×10^{-3} in.) corresponds to a penetration of 2.2×10^{-3} mm (0.086×10^{-3} in.); substitution into Eq 2 yields an exposure time of 116 900 h. Again, it is seen that the operating environment could not have produced the observed oxide scales; therefore, the integranular, oxide filled cracks were probably not initiated by service stresses.

One-Dimensional Stress and Fracture Mechanics Analyses

Introduction

Early in the failure investigation, preliminary analyses were conducted to determine by simple models the disk stress due to the initial overspin condition and subsequent operational stresses brought about by centrifugal and thermal transient loads. One-dimensional idealizations of the geometry and loading provided guidance for the failure investigation and also offered insight into the failure conditions.

Stress Analysis

An unfailed disk provided by EPRI was measured with micrometer and dial calipers to an accuracy of 0.025 mm (0.001 in.). The wheel is about 0.97 m (38 in.) in diameter, up to 102 mm (4 in.) thick near the outer circumference, with a 51-mm (2-in.)-diameter bore hole. To determine stresses induced by rotation, a finite difference analysis [7] was performed for a one-dimensional axisymmetric representation of the wheel. No stress concentration effects were considered; however, the effects of bolt holes on the stress field were modelled by decreasing the disk thickness in the regions of bolt holes. Although bolt masses were included in the model, blade and nut masses were neglected.

For a wheel spinning at 5100 rpm, a peak hoop stress of 476 MPa (69.1 ksi) occurs at the bore. The hoop stress attenuates rapidly as a function of radial position, such that at the rabbet location the hoop stress is 285 MPa (41.4 ksi). Since the wheel was subjected to a 60% overspin at 116°C (240°F) during manufacture to produce beneficial residual stresses in the bore region, the stresses were computed elastically for 8170 rpm, and the amount of plastic deformation was estimated by a simple, unaxial model for contained plasticity based on Neuber's stress-strain concentration theory [8]

$$K_t^2 = K_t^\sigma K_t^\epsilon \qquad (3)$$

where

K_t^σ = stress concentration factor, and
K_t^ϵ = strain concentration factor.

Assuming that the material behaves according to a Ramberg-Osgood stress-strain rule in terms of elastic modulus, E, and fitting parameters, σ_0 and n

$$\epsilon = \frac{\sigma}{E} + \sigma_0 \left(\frac{\sigma}{E}\right)^n \qquad (4)$$

the stress-strain behavior at the spin temperature of 116°C (240°F) is defined by $n = 10.62$ and $\sigma_0 = 0.495$ in Eq 4. These values are determined from the yield strength of 844 MPa (122.4 ksi) and ultimate strength of 1020 MPa (147.9 ksi) at 121°C (250°F), reported by the manufacturer.

The elastic and elastic-plastic stresses for the 8170 rpm overspeed are shown in Fig. 9. Also shown is the residual stress field, determined by subtracting the elastic-plastic from the elastic stresses and the operational stress field. Note that the compressive residual stress attenuates very rapidly and is nearly zero at the rabbet location [$r = 51$ mm (2 in.)]. Peak steady-state stress occurs approximately 13 mm (0.5 in.) from the bore surface with a value of 317 MPa (46 ksi). At the rabbet location, the mechanical stress is 290 MPa (42 ksi).

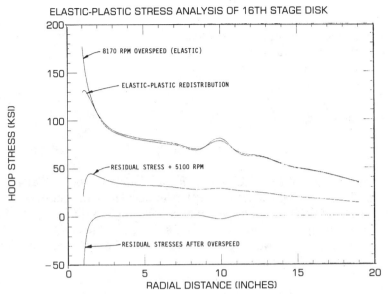

FIG. 9—*Effect of residual stress from overspeed condition in mechanical operating stress. 6.895 MPa = 1 ksi; 25.4 mm = 1 in.*

Thermal Stress Analysis

The governing equation for a finite difference model of the transient temperature distribution is

$$\frac{\partial^2 T}{\partial r} + \frac{1}{r}\frac{\partial T}{\partial r} = \frac{1}{\alpha_t}\frac{\partial T}{\partial t} \tag{5}$$

where

t = time,
r = radial coordinate,
T = temperature distribution $T(r,t)$, and
α_t = thermal diffusivity.

In the solution of Eq 5, the disk rim was allowed to heat up by convection, but the bore was assumed to be insulated. The boundary conditions are given by

$$\frac{\partial T}{\partial r} = \frac{h}{k}(T_f - T_w) \text{ at } r = r_0 \tag{6}$$

$$\frac{\partial T}{\partial r} = 0 \text{ at } r = r_i \tag{7}$$

where

h = surface heat transfer coefficient,
k = thermal conductivity,
T_f = fluid temperature, and
T_w = wall temperature.

The initial condition is

$$T(r,o) = T_0 \qquad (8)$$

In the solution of Eqs 5 through 8, h was assumed to be high at 408 800 KJ/(h m^2 °C) (20 000 BTU/(h ft^2 °F)). The maximum air temperature was taken as 288°C (550°F), and the initial condition was set at 16°C (60°F).

Once the temperature profile was established, the combined stresses (mechanical and thermal) were determined by the finite difference solution of the following expressions

$$\text{Equilibrium: } \frac{d}{dr}\,(rb\sigma_r) - b\sigma_\theta + \rho\omega^2 r^2 b = 0 \qquad (9)$$

$$\text{Compatibility: } \frac{d}{dr}\left(\frac{\sigma_\theta}{E}\right) - \frac{d}{dr}\left(\frac{v\sigma_r}{E}\right) + \frac{d}{dr}\,(\alpha\Delta T)$$
$$- \frac{(1 + v)(\sigma_r - \sigma_\theta)}{Er} = 0 \qquad (10)$$

where

r = radius,
b = thickness variation as a function of r,
ρ = disk density,
ω = angular velocity,
ΔT = temperature increment above the zero thermal stress state,
α = coefficient of thermal expansion,
E = modulus of elasticity, and
v = Poisson's ratio.

Table 1 summarizes the thermal stresses.

Fracture Mechanics Aspects

On the basis of the results of the one-dimensional analysis, calculations were made to assess the conditions that would lead to disk failure from a defect at the rabbet corner. Analysis details are summarized as follows:

1. The flaw was treated as a quarter-circular corner crack.
2. Applied stress level is as computed in the one-dimensional model.

TABLE 1—*Summary of maximum thermal stresses.*

Location	Transient Time, t (min)	Temperature, T (r,t), °C(°F)	Stress, σ_θ, MPa (ksi)
Bore [r = 25.4 mm (1 in.)]	26.0	40 (104)	352 (51.0)
Rabbet [r = 51 mm (2 in.)]	26.0	40 (104)	192 (27.8)
Bolt hole [r = 274 mm (10.8 in.)]	5.5	27 (80)	79.3 (11.5)[a]

[a] This value is a nominal stress level, since it excludes the stress concentration of the bolt hole.

3. The stress attenuates from the bore (Fig. 9).
4. Residual stress benefit at the rabbet is negligible.
5. The critical crack size was not constrained to be less than 6.35 mm (0.25 in.), as suggested by the manufacturer.

The stress-intensity factor as a function of crack depth was computed numerically by the influence function method [9], at 14 min into the startup. Then the number of machine starts required to grow a crack from the rabbet corner to critical size was computed. Crack growth per cycle was determined from the relations in Ref *1*. By subtracting 1500 cycles from the cyclic life results, the size of the initial flaw as a function of fracture toughness was also determined. These results are shown in Fig. 10. At any reasonable toughness level, a large defect, say 5 mm (0.2 in.) or larger, would

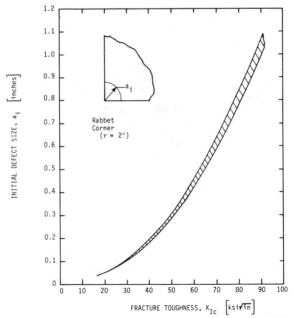

FIG. 10—*Initial defect size to cause failure in 1500 cycles (startups). 1.1 MPa $\sqrt{m}$ = 1 ksi $\sqrt{in.}$; 25.4 mm = 1 in.*

be required to predict the failure. Even at the 44 MPa $\sqrt{m}$ (40 ksi $\sqrt{in}$.) level, a defect of 4.3 mm (0.17 in.) is necessary. This preliminary assessment suggests that the disk should not have failed in 1500 cycles, even with a low toughness, if the initial flaw were of the order of 0.76 mm (0.03 in.).

Refined Stress and Failure Analysis

Introduction

A two-dimensional stress analysis of the 16th stage disk was performed to check the results of the one-dimensional analysis. In reviews of the one-dimensional analysis, the manufacturer indicated the following items as being important contributors to the stress in the 16th stage disk:

1. The interference fit between the 16th stage and other disks, the balance journal, and the distance piece.
2. The ΔT between the distance piece and the 16th stage disk.
3. The blade loadings.
4. The bolt loads and the fact that the 15th stage does not carry the mass of the bolt.

In this section, an analysis model was developed which incorporates these conditions in order to duplicate as closely as possible the manufacturer's stress analysis.

Model Geometry

A two-dimensional axisymmetric finite element model was developed for the ANSYS program. ANSYS is a general purpose finite element program capable of full three-dimensional analyses for static and dynamic stress and heat transfer analyses [10]. The two-dimensional model of the compressor stage included the 14th, 15th, and 16th stage disks, the distance piece, balance journal, and inner and outer attachment bolts and nuts. A geometry plot of the model is shown in Fig. 11.

The model consists of 695 nodes connected by 593 axisymmetric solid elements. The balance journal and the 16th stage disk are joined by ANSYS spring elements to allow for the modelling of the interference fit between the mating surfaces. The dimensions of the 14th and 15th stage disks, the balance journal, and the distance piece were first estimated by scaling them from a machine schematic drawing contained in a promotional brochure. The dimensions of the 14th and 15th stage disks were further refined by using a one-dimensional finite difference analysis to optimize the disk profile to obtain a uniform stress in the disk. All the dimensions were later verified as being reasonably accurate.

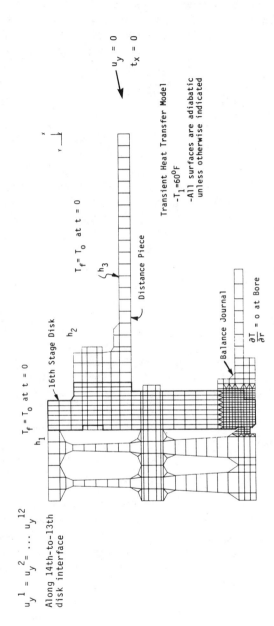

FIG. 11—*Axisymmetric finite element model of 14th, 15th, and 16th stage disks, balance journal, and distance piece showing model boundary conditions.* °C = 5/9 (°F −32).

The elastic properties assumed for the material are given in Table 2. Since the results of the one-dimensional analysis showed that the overspin condition did not influence the region near the inner rabbet, an elastic-plastic analysis of the 16th stage disk was not performed. For the transient heat transfer analysis, the thermal properties were assumed to vary with temperature. Table 2 summarizes the thermal properties at room temperature and at 200°C (392°F).

Mechanical Loading Conditions

Two types of mechanical loads were investigated: (1) centrifugal body forces due to rotation, and (2) interference fit between the 16th stage and mating pieces. In the analysis of the centrifugal forces, the effect of the blades was considered by estimating the weight of each blade and calculating the resulting blade load due to rotation at 5100 rpm. These blade loads were converted to a surface pressure, which was outwardly applied to the rims of the 14th, 15th, and 16th stage disks. The bolt masses were modelled by adjusting the densities of various regions of the model. As the 15th stage bolt hole is oversize, the weight of the bolt is shared by the 14th and 16th stages. Thus, the density in the bolt hole region of the 14th and 16th stage disks was increased, and the density in the 15th stage bolt hole was decreased.

The bolt heads and nuts were modelled with solid elements whose density was corrected by ratioing the gross area of the region to the net area of the nuts. The modulus of elasticity of the nuts was decreased so as not to increase the hoop stiffness of the disk. Similarly, the modulus of elasticity in the bolt hole regions was decreased to account for the reduced net area of the disk in the bolt hole region. The technique for scaling the modulus of elasticity is similar to the technique described in the one-dimensional analysis to account for the bolt holes.

The interference fit between the 16th stage and the balance journal was modelled by isothermally expanding the balance journal in the model. The diametral growth interference between the 16th stage and the balance journal was assumed to be 0.15 to 0.20 mm (0.006 to 0.008 in.) and both values [that is, 0.15 and 0.20 mm (0.006 and 0.008 in.)] were analyzed. Duplicate nodes were defined for the 16th stage-balance journal interference, and these nodes were connected by spring elements. The interference fit between the 15th and 16th stages [0.005 to 0.10 mm (0.002 to 0.004 in.)] was determined to be less than that used to install the balance journal, and it is expected to offset or reduce the stress at the rabbet location. For this reason, the 15th and 16th stage interference was not considered in the model.

The centrifugal forces (ω = 5100 rpm) were combined in the same load step as the interference fit case, and the conditions that would lead to lift-off between the 16th stage and the balance journal were examined. If lift-

TABLE 2—*Summary of material properties.*

	At Room Temperature	At 200°C (392°F)
Modulus of elasticity	207×10^3 MPa (30×10^3 ksi)	...
Poisson's ratio	0.3	...
Density	0.00785 kg/cm³ (0.283 lb/in.³)	
Coefficient of thermal expansion	12.20×10^{-6} mm/mm °C (6.78×10^{-6} in./in. °F)	12.71×10^{-6} mm/mm °C (7.06×10^{-6} in./in. °F)
Specific heat	0.451 kJ/kg °C (0.108 BTU/lb °F)	0.472 kJ/kg °C (0.113 BTU/lb °F)
Thermal conductivity	154.82 kJ/m h °C (24.85 BTU/ft h °F)	152.27 kJ/m h °C (24.44 BTU/ft h °F)

off occurs, then the springs indicating the separation (that is, tensile spring force) would be removed from the model, and the analysis would be performed again. The effect of complete separation was analyzed by effectively removing the balance journal from the model.

In the stress boundary conditions shown in Fig. 11, the end of the distance piece was treated as a line of symmetry which is restrained in the axial direction. The end surface of the 14th stage disk (that is, the interference between the 13th and 14th stages) was allowed to move axially and radially; however, the axial plane was restricted to move as a plane by coupling the normal displacements. This boundary condition will approximate the effect of other stages on that portion of the model.

Thermal Loading Conditions

The thermal loading and boundary conditions assumed in the thermal analysis for a machine startup transient are shown in Fig. 11. In estimating the operating conditions during startup, the compressor was assumed to have an 8:1 compression ratio and an efficiency of approximately 90%. The mass flow rate of air through the compressor was taken as 321 364 kg/h (707 000 lb/h). From these conditions, the heat flow rate was calculated to be 7.63×10^6 KJ/h (7.22×10^6 Btu/h) and the air temperature at the 16th stage would be about 288°C (550°F). Response to these assumptions by the manufacturer indicated an air temperature at the 16th stage of 321°C (610°F), not 288°C (550°F). Although the higher temperature would require a compression ratio of about 10.4:1 and at least 17 stages, a value of 321°C (610°F) for T_f was used in the thermal analysis.[2]

In the design of the machine, the turbine-side wheels are cooled by passing compressor air through the annulus between the distance piece and the outer shroud directly to the turbine. In this analysis, the air was assumed to come from the 16th stage at a temperature of 321°C (610°F) and at a flow rate of 16 360 kg/h (36,000 lb/h) along the distance piece. In similar fashion to the one-dimensional analysis, the air temperature in the compressor was assumed to change in a step fashion with time, from an initial temperature of 16°C (60°F).

The transfer of heat to the rotating components takes place at the wheel rims and along the outer surface of the distance piece. The boundary conditions along all other surfaces were assumed adiabatic. In estimating the film coefficients for the heat transfer surfaces, two sets of assumptions were invoked: one set which would result in a very conservative or "worst case" thermal response, and a second set which would be a "best estimate." It

---

[2] After this analysis was performed, the manufacturer agreed that 321°C (610°F) is too high for the machine under consideration; however, the use of this high value in the analysis will make the stress results more conservative (that is, higher stress magnitude).

was determined that the mode of heat transfer at the rim would be controlled by water condensation from the heated air to the initially cold wheel. An upperbound film coefficient (h) would result under the assumption that all the moisture in the air, under the conditions that pertain along the 14th, 15th, and 16th stages, condenses on the rim of the 16th wheel. This assumption yields a heat transfer coefficient of 7767 kJ/h m² °C (380 BTU/h ft² °F). Assumption of this value for h along the three wheel rims and along the distance piece surfaces provides for a conservative assessment. Hence, the worst case analysis under the boundary conditions shown in Fig. 11, assumes

$$h_1 = h_2 = h_3 = 7767 \text{ kJ/h m}^2 \text{ °C (380 Btu/h ft}^2 \text{ °F)} \qquad (11)$$

In contrast, a more realistic view was taken to arrive at best estimate values for the film coefficients. By assuming that condensation along the rim surfaces is uniform over the three stages and that the mode of heat transfer through the seal and along the distance piece is hot gas convection, the following best estimate values are computed

$$
\begin{aligned}
h_1 &= 2289 \text{ kJ/h m}^2 \text{ °C (112 Btu/h ft}^2 \text{ °F)} \\
h_2 &= 368 \text{ kJ/h m}^2 \text{ °C (18 Btu/h ft}^2 \text{ °F)} \qquad (12) \\
h_3 &= 61.3 \text{ kJ/h m}^2 \text{ °C (3 Btu/h ft}^2 \text{ °F)}
\end{aligned}
$$

In determining the h-values given by Eqs 11 and 12, the inlet conditions to the compressor were specified as air at 16°C (60°F) and 101.3 kPa (14.7 psi), and the air entering the 14th stage contains an amount of water vapor equal to saturation conditions at the local pressure and 16°C (60°F). Thermal stress analyses were performed for temperature distributions derived from the worst case and best estimate thermal transient responses.

The hoop stress at the rabbet region was determined for four load cases: centrifugal loads only for ω = 5100 rpm, interference fit only for δ = 0.20 mm (0.008 in.) or 8 mil diametral, centrifugal plus 8 mil interference, and centrifugal plus 6 mil interference. The hoop stress at the rabbet corner for each case is summarized in Table 3. Also shown in Table 3 for comparison are the stress levels calculated for the bore region. It should be noted that the load case where the interference fit and centrifugal loads are combined is not the sum of each load case treated separately. Clearly, the reason for this behavior is that connection between the disk and the balance journal is similar to a preloaded joint like a bolted flange and that the centrifugal load is reducing the preload and the compressive stress in the balance journal while it is raising the stress in the disk specifically at the rabbet. The net effect is that the stress at the rabbet does not significantly change as the disk is brought up to speed. For the case of 0.15 mm (6 mil) interference

TABLE 3—*Stresses due to interference fit and centrifugal force.*

Load Case	Stresses At Bore, MPa (ksi)	Stresses At Rabbet, MPa (ksi)
Centrifugal only	559.2 (81.1)	364.7 (52.9)
Interference fit of 0.20 mm (8 mil) only	143.4 (20.8)	133.1 (19.3)
Centrifugal plus 0.20 mm (8 mil) interference	578.5 (83.9)	382.0 (55.4)
Centrifugal plus 0.15 mm (6 mil) interference	510.9 (74.1)	364.7 (52.9)

fit, the model predicts wheel lift-off from the balance journal. In this case, the stress in the wheel is the same as for the centrifugal only problem.

Transient Response and Thermal Stresses

Worst Case Thermal Analysis—In the worst case assessment for thermal stress, the film coefficient of Eq 11 was assumed. The temperature distribution in the 16th stage disk was solved by ANSYS, and the time at which the maximum stress occurs in the bore region was determined by inspecting

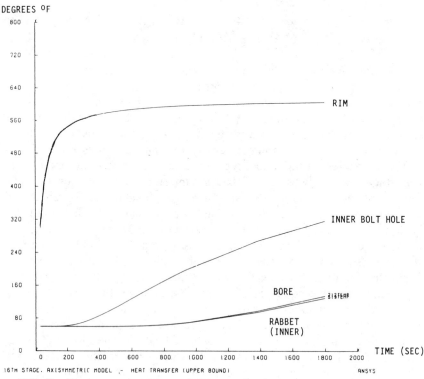

FIG. 12—*Temperature-time history of 16th stage disk (worst case).* $°C = 5/9 (°F - 32)$.

the temperature difference between the rim and the bore. The transient thermal response of the 16th stage disk for the bore, rabbet, bolt hole, and rim locations is shown in Fig. 12. The maximum temperature difference between the rim and the bore occur at approximately 13 min. Although the failure time (14 min) for the wheel occurs after the time to peak stress from this analysis, there is not a significant difference between these two times as far as the rim-to-bore ΔT or for the thermal stress at the bore.

The peak thermal stresses were calculated for the temperature distribution at 13 and 14 min. These stresses are later added to the centrifugal plus interference fit stresses for comparison with the one-dimensional analysis results. The actual stresses will be less than this summation because the thermal displacements will also act to reduce the interference fit. The stresses at the bore and inner rabbet locations are summarized in Table 4. For the worst case condition, the peak stress level occurs one minute before the stated accident time; the stress values are, however, almost equal. The stress level at the bore is significantly higher than at the inner rabbet, much as for the mechanical stresses.

A displaced shape plot for the maximum stress time of 13 min is shown in Fig. 13. Note that as the distance piece expands thermally, it presses against the 16th stage disk, thus causing the disk to deflect. The deflections shown are exaggerated, but the deflection of the rim of the 16th stage disk is enough to contact the 15th stage disk, despite the initial gap between the disks at the rim. Most of the deformation is confined to the region radially outward from the bolt holes, and it is felt that contact between disks at the rim would not significantly change the stresses in the rabbet region.

Best Estimate Thermal Analysis—A best estimate analysis was also performed in essentially the same way as the worst case analysis but with the film coefficients of Eq 12. The time after startup at which peak stress occurs is approximately 30 min, which is after the accident time. Stress levels at the bore surface and rabbet corner are listed in Table 4. The peak stress calculated for the best estimate case is approximately one half (54%) of

TABLE 4—*Thermal stresses.*

	Stresses At Bore, MPa (ksi)	Stresses At Rabbet, MPa (ksi)
WORST CASE		
Peak stress[a] (t = 13 min)	490.2 (71.1)	299.2 (43.4)
Time of accident (t = 14 min)	493.7 (71.6)	297.2 (43.1)
BEST ESTIMATE		
Peak stress[a] (t = 30 min)	295.1 (42.8)	160.7 (23.3)
Time of accident (t = 14 min)	243.4 (35.3)	151.0 (21.9)

[a] As determined at rabbet.

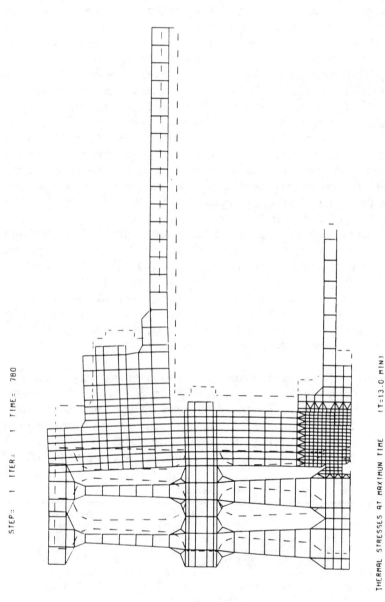

FIG. 13—*Displaced geometry plot at t = 13 min, from worst case thermal analysis.*

that of the worst case. This illustrates the sensitivity of the thermal stress to the assumption for heat transfer coefficient. It is believed that the actual thermal behavior of the wheel falls between the two cases analyzed.

Comparison of the One-Dimensional and Two-Dimensional Model Results

As a check on the accuracy of the one-dimensional finite analysis, a comparison was made between the one-dimensional analysis and the results of the two-dimensional finite difference analysis. Figure 14 is a comparison of the hoop stress due to centrifugal force from the one-dimensional analysis with the corresponding stress from the midthickness of the disk in the two-dimensional analysis. Away from corners and other geometric discontinuities, the difference between the two analyses is approximately 10%. Some of the difference can be attributed to the addition of blade and bolt loads which were not included in the one-dimensional model.

In Table 5, the results from each load case are summarized as well as the stress totals including maximum stress, minimum stress, mean stress, and stress range during the startup cycle. In comparing the one-dimensional with the two-dimensional, the major differences in the computed stress are due to the fact that the interference fit-up was included in the two-dimensional model while it was not considered in the early one-dimensional study,

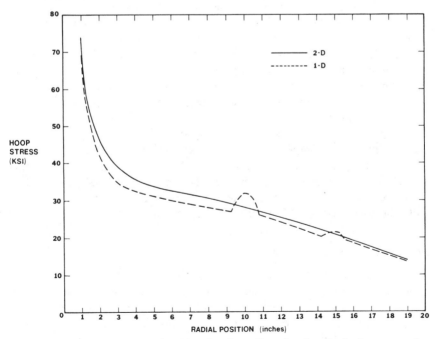

FIG. 14—*Comparison of one-dimensional and two-dimensional results for hoop stress due to centrifugal force. 6.895 MPa = 1 ksi; 25.4 mm = 1 in.*

TABLE 5—*Comparison of one-dimensional and two-dimensional stress results at the inner rabbet location.*

	One-Dimensional Model, MPa (ksi)	Two-Dimensional Model, MPa (ksi)	
		Worst Case	Best Estimate
Centrifugal only	285.5 (41.4)	364.7 (52.9)	364.7 (52.9)
Interference fit only [0.20 mm (8 mil)]	...[a]	133.1 (19.3)	133.1 (19.3)
Centrifugal plus 0.20 mm (8 mil) interference fit	285.5 (41.4)	382.0 (55.4)	382.0 (55.4)
Thermal stress at 14 min	182.0 (26.4)	297.2 (43.1)	151.0 (21.9)
Thermal stress at peak	210.3 (30.5) (26 min)	299.2 (43.4) (13 min)	160.7 (23.3) (30 min)
SUMMARY [δ = 0.20 mm (8 mil)]			
Stress at accident (t = 14 min)	467.5 (67.8)	679.2 (98.5)	533.0 (77.3)
Maximum stress, σ_{max}	495.8 (71.9)	681.2 (98.8)	542.6 (78.7)
Minimum stress, σ_{min}	0	133.1 (19.3)	133.1 (19.3)
Alternating stress, $\Delta\sigma$	495.8 (71.9)	548.2 (79.5)	409.6 (59.4)
Mean stress, σ_{mean}	248.2 (36.0)	407.5 (59.1)	337.9 (49.0)
$\sigma_{min}/\sigma_{max}$	0	0.195	0.245

[a] Not considered.

and that the worst case thermal analysis increased the maximum stress level. The effect of including the fit-up stress results in two interesting observations. The maximum stress increased somewhat, and a mean stress is introduced into the disk. However, because the interference fit stress elevates the minimum stress in the cycle to 133 MPa (19.3 ksi), the cyclic stress at the rabbet is only 11% higher under worst case conditions and actually is 17% less under best estimate conditions when compared to the one-dimensional analysis results.

The effect of having a higher maximum stress will be to cause a smaller critical flaw to be computed. Since the expressions for da/dN are upperbound fits to specimen data for $0 < R < 0.7$, these expressions include the effect of mean stress on crack growth. Therefore, it is expected that the two-dimensional stress results for the worst case and best estimate conditions will bracket the fatigue life for the wheel, as computed previously.

Fracture and Fatigue Results

Critical Flaw Size—The fracture and fatigue assessment was revised to consider the refined stress results. The method used to compute stress-intensity factors and fatigue life was the same as described previously, except that a quarter elliptical corner crack model was used, and the hoop stress variation around the inner rabbet was represented by a bivariate stress distribution. The magnitude of the stress field was defined by the stress output at the integration points within the elements from the two-dimensional finite element analysis.

The maximum stress-intensity factor was calculated for each of the crack tips by BIGIF [9], and these results are shown in Fig. 15. The advantage of using a two-degree-of-freedom crack model is that the change in crack shape is allowed in the fatigue analysis, and the effect of the gradient stress field on K can be investigated. This is especially important when there is a rapid decrease in stress in one direction when compared to the stress decrease in the other direction, which is the case for the stresses due to the interference fit.

The stress-intensity factor is highest at the rabbet side of the crack, as shown in Fig. 15. If a fracture toughness of 88 MPa $\sqrt{m}$ (80 ksi $\sqrt{in.}$) is assumed, then the critical flaw size would be between 12.2 and 13.5 mm (0.48 and 0.53 in.) for the worst case analysis and at least 19.2 mm (0.78 in.) with the results from the best estimate analysis. If a value of K_{Ic} as low as 44 MPa $\sqrt{m}$ (40 ksi $\sqrt{in.}$) is assumed, then the critical flaw size is computed to be approximately 2.8 mm (0.11 in.) and 4.3 mm (0.17 in.) for worst case and best estimate conditions, respectively. These critical flaw sizes are later used as terminal flaw sizes in the fatigue life analyses.

Fatigue Analysis—The earlier fatigue analysis was executed again with the refined stress distributions and a more representative flaw model. The computation of fatigue life indicates that significant life remains when the initial flaw size is assumed to be 0.76 mm (0.03 in.), which is the initial defect size suggested in the manufacturer's report. The results of the fatigue analysis are shown in Fig. 16, where the cyclic life to failure from a 0.76 mm (30 mil) crack is plotted against fracture toughness level. Also shown in Fig. 16 are the effects of the thermal analysis assumptions (that is, worst case versus best estimate) on the computed life as well as the sensitivity of the results to the interference fit between the wheel and balance journal.

For a fracture toughness of 88 MPa $\sqrt{m}$ (80 ksi $\sqrt{in.}$) and an interference fit of 0.20 mm (8 mil), the computed life is 18 000 or 38 000 cycles, depending on the thermal analysis assumptions. Clearly, this life is much greater than the 1500 cycles which the machine experienced before failure. In order to achieve wheel failure from a 0.76 mm (30 mil) flaw within 1500 cycles and with stress levels computed with a two-dimensional analysis, the fracture toughness would have to be less than 26 MPa $\sqrt{m}$ (24 ksi $\sqrt{in.}$). Changing the initial interference fit value from 0.20 mm (8 mil) to 0.15 mm (6 mil) does not significantly decrease the computed life. Even for the extreme case of no interference fit, the cyclic life is still much greater than the 1500 machine starts. All results shown in Fig. 16 were produced from the faster of the two crack growth relations given by Barsom [1].

By subtracting the number of machine cycles from the terminal life values, the initial flaw size required to cause wheel failure was determined. The

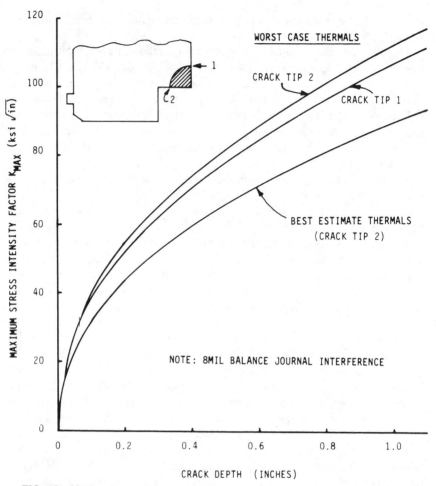

FIG. 15—*Maximum stress intensity factor for a crack at the rabbet corner during startup. 1.1 MPa $\sqrt{m}$ = 1 ksi $\sqrt{in.}$; 25.4 mm = 1 in.*

initial flaw size to achieve failure in 1500 cycles as a function of fracture toughness is shown in Fig. 17. For a fracture toughness of 88 MPa $\sqrt{m}$ (80 ksi $\sqrt{in.}$), the required initial flaw size is between 8.6 mm (0.34 in.) and 16.8 mm (0.66 in.), depending on thermal analysis assumptions. On the basis of this analysis, the initial flaw size required to cause the failure is much larger than the 0.76 mm (0.03 in.) flaw which the manufacturer indicated was the initiating defect.

Conclusions and Implications

Both the one-dimensional and two-dimensional analyses performed here indicate that the maximum stress and cyclic range are not sufficiently high

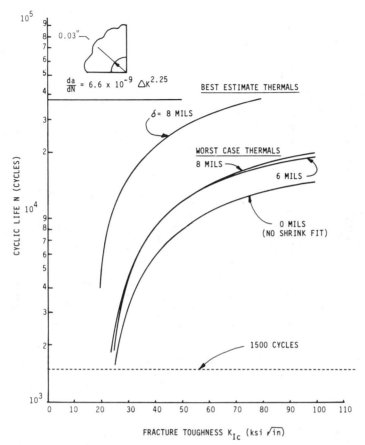

FIG. 16—*Computed fatigue life as a function of fracture toughness and interference fit for a 0.76 mm (0.03 in.) initial flaw. 1.1 MPa $\sqrt{m}$ = 1 ksi $\sqrt{in}$.; 0.025 mm = 1 mil.*

to explain the disk failure in 1500 cycles when a 0.76 mm (0.03 in.) crack is assumed. The fracture toughness of the disk material was measured to be as high as 90.2 MPa $\sqrt{m}$ (82 ksi $\sqrt{in}$.) at 22°C (71°F); however, a fracture toughness would have to be about 24 MPa $\sqrt{m}$ (22 ksi $\sqrt{in}$.) to give failure in 1500 cycles with the stress levels computed herein. The cyclic stress here computed is between 409 MPa (59.4 ksi) and 548 MPa (79.5 ksi), depending on the thermal analysis assumptions, whereas, the stress range suggested by the manufacturer must be approximately 861 MPa (125 ksi) to predict the failure under the condition which the manufacturer reports to exist. This stress level cannot be supported by the analyses performed herein. These analyses therefore suggest that a larger defect was present early in the disk life. The fracture mechanics calculations indicate a flaw of 8.6 to 16.8 mm (0.34 to 0.66 in.) would be required to explain the failure originating at the inner rabbet for the stresses here computed, with K_{Ic} = 88

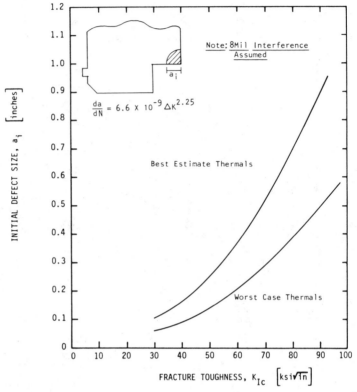

FIG. 17—*Initial defect size required to cause failure in 1500 startup cycles. 1.1 MPa $\sqrt{m}$ = 1 ksi $\sqrt{in.}$; 25.4 mm = 1 in.*

MPa $\sqrt{m}$ (80 ksi $\sqrt{in.}$) and an interference fit of 0.20 mm (8 mil) between the wheel and the balance journal.

Since the properties of the wheel should not have degraded in the service environment of only 288°C (550°F), it must be concluded that the burst was preceded by growth of a preexisting flaw to critical size (that is, the wheel had subcritical cracks when it went into service).

The small, oxide filled, intergranular cracks in regions around rabbets and bolt holes were the obvious candidates for preexisting flaws. They appeared to have all the characteristics of quench cracks [11]. Therefore, an additional experiment was performed. A piece about 25 mm (1 in.) on a side and about 76 mm (3 in.) long was cut from one of the wheel fragments. After it was austenitized for 1 h at 843°C (1550°F), it was quenched into water, which caused it to split almost in two. There is no implication here that the wheel itself was quenched in water; the intention in this experiment was only to ensure the generation of quench cracks, and many small, oxide free, intergranular cracks were in fact produced in addition to the large split. The piece was then tempered for 12 h at 482°C (900°F) to simulate the final part of the heat treatment of the wheel. Oxide thickness in the small cracks after tempering measured 0.43×10^{-2} mm (0.17×10^{-3} in.). It will be appreciated that this oxide thickness, plus whatever oxide growth could occur in service at 288°C (550°F), could account for the observed scale on the small cracks in the wheel fragments if those cracks are quench cracks.

At the time of this investigation, the occurrence of quench cracks could not be explained because machining after heat treatment removes a layer of metal thicker than is typically affected by quench cracks. Although it was known that the burst wheel and some others in the population had been heat treated a second time owing to a specification change that was instituted while the wheels were in the manufacturing process, it was believed that both heat treatments had preceded any machining. However, it was later discovered that a group of forgings, including this wheel, had been rough machined and center bored *before* the second heat treatment. Thus, there is good reason to suspect quench cracks as the preexisting defects.

The primary question posed at the outset of this investigation was: To what degree is the failed wheel the same as or different from the other wheels in the "at risk" fleet? If all the wheels are essentially the same as the wheel that burst, the replacement recommendation was correct. However, it is very unlikely that they are all the same. Only those wheels that were heat treated a second time after rough machining should be retired.

Acknowledgments

The authors are grateful to T. W. Rettig and R. L. Cargill for assistance throughout the project. This work was performed for the Electric Power Research Institute under Contract EPRI RP 1802-2.

References

[1] Barsom, J. M., "Fatigue Behavior of Pressure-Vessel Steels," Welding Research Council Bulletin 194, May 1974.

[2] "Analysis of Sixteenth Stage Compressor Wheel Failure," *Investigation and Research of Specific Combustion-Turbine and Combined-Cycle Field Problems,* EPRI AP-2888, Electric Power Research Institute, Palo Alto, CA, Aptech Engineering Services, Inc., March 1983.

[3] Pinder, L. W., "Oxide Kinetics and Scale Morphology of Mild Steel and Low Chromium Alloy Steels in $N_2/20\%$ O_2 Between 500 and 850°C," Report SSD/MID/R58/77, Central Electricity Generating Board, Nottingham, England, 1977.

[4] Richman, R. H. In *Proceedings of ISTFA 82,* International Society For Testing and Failure Analysis, Torrance, CA, 1982, pp. 219–225.

[5] *Metals Handbook, Vol. 1, Properties and Selection of Metals,* Eighth Edition, American Society For Metals, Metals Park, OH, 1961, p. 265.

[6] Kleinbaum, D. G. and Kupper, L. L., *Applied Regression Analysis and Other Multivariable Methods,* Duxbury Press, North Scituate, MA, 1978.

[7] Manson, S. S., "Determination of Elastic Stresses in Gas Turbine Disks," Report 871, National Advisory Committee For Aeronautics, Cleveland, OH, Feb. 1947.

[8] Neuber, H., *Journal of Applied Mechanics,* Dec. 1961, p. 544.

[9] Cipolla, R. C., Besuner, P. M., and Peters, D. C., "BIGIF: Fracture Mechanics Code for Structures, User's Manual," EPRI NP-838, Electric Power Research Institute, Palo Alto, CA, Aug. 1978.

[10] DeSalvo, G. J. and Swanson, J. A., "ANSYS Engineering Analysis System User's Manual, Vol. 1," Revision 3, Swanson Analysis Systems, Houston, PA, July 1979.

[11] *Metals Handbook, Vol. 10, Failure Analysis and Prevention,* 8th ed., American Society For Metals, Metals Park, OH, 1975, pp. 298–299.

Allen Selz[1] *and David B. Peterson*[1]

Preventing Fracture By Inspection and Analysis

REFERENCE: Selz, A. and Peterson, D. B., **"Preventing Fracture By Inspection and Analysis,"** *Case Histories Involving Fatigue and Fracture Mechanics, ASTM STP 918,* C. M. Hudson and T. P. Rich, Eds., American Society for Testing and Materials, Philadelphia, 1986, pp. 211–225.

ABSTRACT: A problem in operating pressure components under cyclic loading conditions is establishing an in-service inspection frequency to assure continued operability without threat of catastrophic failure. For some extremely high pressure components, this has been done on a rational basis by (1) establishing the regions of highest stress, usually thread roots or other local stress raisers in a high stress field, (2) establishing the minimum-size reliably observable crack-like defect, using proven inspection means, such as magnetic particle or eddy current testing, and (3) determining the number of design cycles this minimum observable crack needs to grow to critical size, using Paris law or analogous crack propagation relations, and then dividing the resulting number of cycles by a factor of at least three. This factor of three is based on retaining cyclic margin to failure, even if a given defect is below perceptible size at initial inspection and is missed during the next inspection. A factor of four would accommodate two consecutive failures to observe a defect. Methods have been also developed which permit modification of the inspection interval if the operating logs show the imposition of a distribution of operating cycles other than the design cycles upon the component.

The methods described were originally developed and applied to predicting safe life and establishing inspection intervals for a hypervelocity wind tunnel. They have since been applied to commercial pressure vessels and for recertification of pressure vessels for the National Aeronautics and Space Administration.

KEY WORDS: crack propagation, fatigue, fracture, high pressure, inspection, isostatic press, pressure vessel

Pressure vessels and other components which operate at very high pressures—104 MPa (15 000 psi) and more—must be guarded against failure due to fatigue and fracture, particularly in the region of the closure. Several examples of these regions in such vessels are shown in Figs. 1 through 4.

[1] Vice presidents, O'Donnell & Associates, Inc., Pittsburgh, PA 15236.

FIG. 1—*Gas heater vessel.*

The design of these devices to withstand fatigue failure is difficult because high-strength steels must be used, in which yield strength and ultimate tensile strength are close to one another, and the common stress limit of one third ultimate tensile strength is usually exceeded. We have analyzed several failed vessels to determine the cause of failure and to see if we could analytically verify the cycles-to-failure sustained by these vessels. Using the fatigue design rules given in the American Society of Mechanical Engineers Boiler and Pressure Vessel Code, Section VIII, Division 2, we were able in most cases to predict that a failure would occur in approximately the number of cycles in which one actually took place. More generally, records of failure demonstrate that over 80% of structural failures in pressure vessels are cyclic in origin.

The predominance of fatigue as a cause of vessel failure in general, and in high pressure vessels in particular, means that one cannot simply place such a vessel into service and forget it. The vessel must be inspected periodically to make sure that failure due to fatigue and fracture is not imminent. Of course, the problem becomes one of where, how, and how often to inspect the vessel to head off a fracture failure.

FIG. 2—*Heater vessel plug.*

Crack Initiation and Growth

A little background on how a component fails due to fatigue is appropriate. The sequence of events leading to fatigue-initiated failure can be somewhat arbitrarily divided into four phases, as follows:

1. Cyclic straining of the metal in regions of existing surface irregularities, internal inclusions, and other volumetric strain risers initiates microcracks.

2. A flaw grows to a size at which it can first be observed in an engineering sense.

3. Continued cycling leads to continued crack propagation.

4. Critical flaw size is reached, and the component fails.

These phases are shown schematically in Fig. 5, and an apportionment of the fatigue curve between crack initiation and crack propagation is shown schematically in Fig. 6.

Approach

We have developed a rational approach to the determination of an inspection cycle to prevent fracture. This approach utilizes the predictability

FIG. 3—*Vessel, plug, and discharge nozzle.*

of the cyclic life of a component between the initial observation of a flaw, and the time the flaw has propagated to a critical size. First, stress analysis is used to establish the regions of high stress in a component, because observable flaws will develop first in a component in and around the regions of highest stress. Then a conservative evaluation is made of the minimum flaw which can be observed in a particular location, using a particular inspection method. Next, K_{Ic}, the critical stress-intensity factor for the material of interest, is used to determine the critical flaw or crack size at which fast fracture occurs. Then, using known data for da/dN, the crack extension per cycle, the number of cycles required for the crack to increase from its minimum observable size to the critical size, is found. This number is then divided by an appropriate safety factor to guard against missing a defect during a specific examination, to establish an acceptable number of cycles between examinations for flaws.

Stress Analysis

The stress analysis of high pressure vessels in general is a complex problem. It is made more difficult because no formal stress criteria exist in current U.S. codes and standards. However, stress analysis is used herein only for

FIG. 4—*Close-up of interrupted threads.*

determining the level of stress as a function of pressure, so the fracture
analysis previously outlined can be performed.

A large body of experience exists for evaluating the stresses in threaded
closures where the materials of construction have been well characterized.
A key factor is that the friction forces between mating threads have been
found to dissipate as threaded closures are subjected to repeated pressur-
izations. Thus, the effects of such friction forces are quite small after the
first few cycles of operation. For conservatism, however, we analyze thread
behavior with and without friction, and the condition which causes the worst
stresses is assumed to exist.

For the purpose of analysis by finite element methods, models were
developed for each vessel. Overall closure models were used to calculate
the thread loads in the threaded end closures. A computer-drawn scale plot
of the overall model for one representative vessel is shown in Fig. 7. The
scale is shown on the axes of the figure. A single detailed model of one
thread was then prepared and used to calculate the maximum stress in the
thread root due to the maximum thread load on each particular threaded
end closure. A computer-drawn scale plot of this model is shown in Fig. 8.
Most of the computer analyses were performed using the digital computer

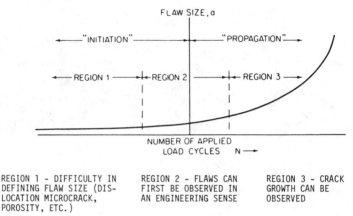

| REGION 1 - DIFFICULTY IN DEFINING FLAW SIZE (DIS-LOCATION MICROCRACK, POROSITY, ETC.) | REGION 2 - FLAWS CAN FIRST BE OBSERVED IN AN ENGINEERING SENSE | REGION 3 - CRACK GROWTH CAN BE OBSERVED |

FIG. 5—*Schematic flaw growth curve.*

progam ANSYS. The other analyses were performed using special computer programs or hand calculations or both.

The overall model of each closure was used in conjunction with the ANSYS computer program to calculate the maximum loads and stresses in the closure due to internal operating pressure loading. These models contain many axisymmetrical isoparametric (STIF42) elements. The cylindrical portion was modeled to a distance of approximately $\pi\lambda$ beyond the nonuniform portions of the closure, where λ is the attenuation length.

The nut and cylinder threads were modeled individually. The male and female threads were coupled by specifying thread interfaces to have the same displacements in the direction perpendicular to the thread tooth surface. This allowed the load to be transmitted from one thread tooth to the other with no sliding friction between mating threads.

An internal pressure of 104 MPa (15 000 psi) was applied to the inside surfaces of the overall model. The load on each of the individual threads was then obtained from the computer results. The resulting maximum thread load was then converted into an equivalent uniform load and applied to the tooth surface on the detailed thread model. In addition, displacements from the overall model in the vicinity of the thread where the maximum thread load occurs were applied as boundary conditions on the edges of the detailed model.

The detailed thread model shown in Fig. 8 is the model which was used to calculate the maximum stresses in the thread root for the closure. A computer-drawn stress-contour plot for these threads is shown in Fig. 9.

Fracture Mechanics Evaluation

The material from which this vessel and closure were made, is modified 4340 steel, commonly known as "gun-steel." The ultimate strength of the

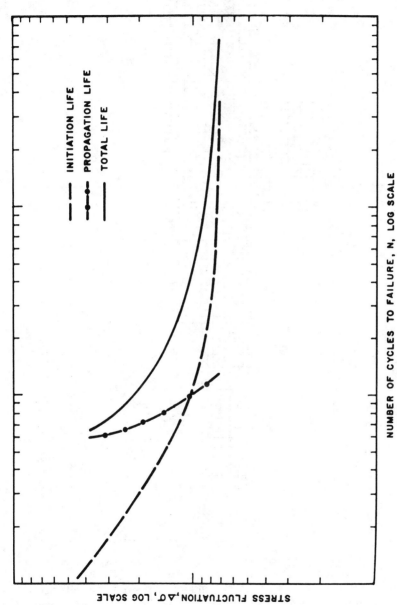

FIG. 6—*Schematic S–N curve divided into "initiation" and "propagation" components.*

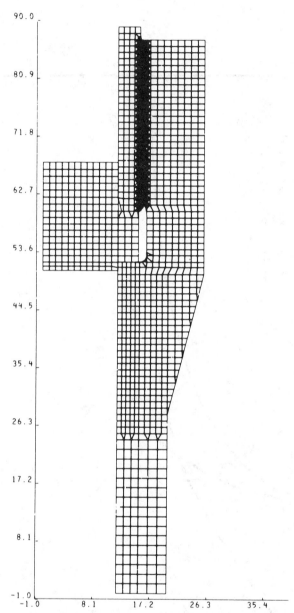

FIG. 7—*Overall finite element model of buttress threads.*

vessel material is 1000 MPa (145 ksi) and its yield strength is 897 MPa (130 ksi). Therefore, although ASTM Specification for Alloy Steel Forgings for High-Strength Pressure Component Application (A 723-80) was not yet issued when this vessel was built, its tensile properties fall about halfway between those of Class 2 and Class 3 material of that specification.

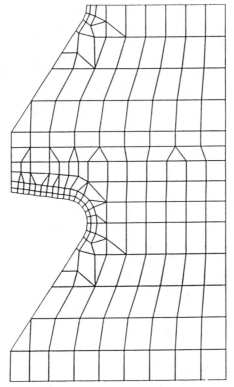

FIG. 8—*Localized finite element model of buttress thread.*

This material is of reasonably low yield strength for modified 4340. For this material, K_{Ic} was found to exceed 3475 MPa $\sqrt{mm}$ (100 ksi $\sqrt{in.}$); and this value was used in the fracture mechanics analysis.

Figure 10 shows numerous data on the crack propagation rate, da/dN, for various 4340 steels. It is seen that they follow the well known Paris law, $da/dN = C_0 \Delta k^n$ closely; and for this material, $da/dN = 0.66 \times 10^{-8} (\Delta k)^{2.25}$.

For a thick-walled region containing a shallow surface defect ($a/l \cong 0$) oriented normally to the maximum surface stress, the critical crack depth, a_{cr} is

$$a_{cr} \cong \frac{K_{Ic}^2}{1.25 \pi \sigma^2}$$ (1)

where

a_{cr} = critical crack depth, mm (in.),
$K_{I}c$ = fracture toughness, MPa$\sqrt{mm}$ (ksi$\sqrt{in.}$), and
σ = maximum surface stress, MPa (ksi).

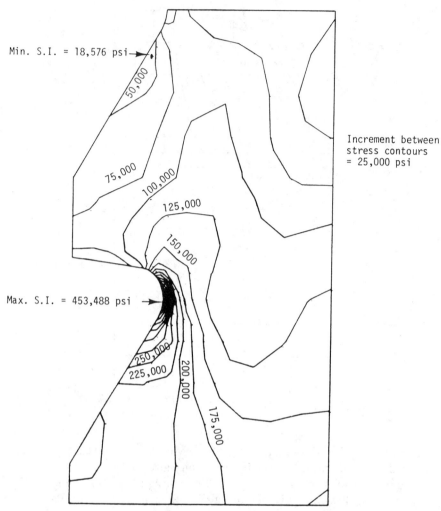

FIG. 9—*Computer plot of stress contours for buttress thread.*

Now the number of cycles required to grow a flaw from an initial size to critical flaw size (failure), N, is found by integration to be

$$N = \frac{2}{(n - 2)C_0 M^{n/2} \Delta\sigma^n} \left(\frac{1}{a_i^{(n-2)/2}} - \frac{1}{a_{cr}^{(n-2)/2}} \right) \qquad (2)$$

where

N = number of cycles to failure,
a_i = initial crack depth, mm (in.),
n = slope of da/dN versus log ΔK curve,

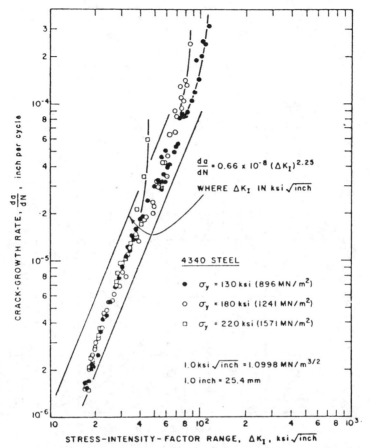

FIG. 10—*Fatigue crack propagation in 4340 steel of various yield strengths.*

a_{cr} = critical crack depth, mm (in.),

C_0 = empirical intercept constant for ΔK in MPa$\sqrt{\text{mm}}$ (psi$\sqrt{\text{in.}}$),

$\Delta\sigma$ = applied cyclic stress range, MPa (psi), and

M = 1.25π.

For K_{Ic} = 3475 MPa$\sqrt{\text{mm}}$ (100 ksi$\sqrt{\text{in.}}$), and for the stresses calculated for the representative region calculated above, a_{cr} = 0.37 mm (0.0145 in.). Figure 11 shows, for this critical flaw size, the number of full pressure cycles required to increase the crack depth from any initial value to this critical size.

The critical flaw size is so small relative to vessel wall thickness that small crack behavior governs: that is, the effect of the crack depth upon average stress in the section is negligible, and the effect of pressure within the crack is also negligible.

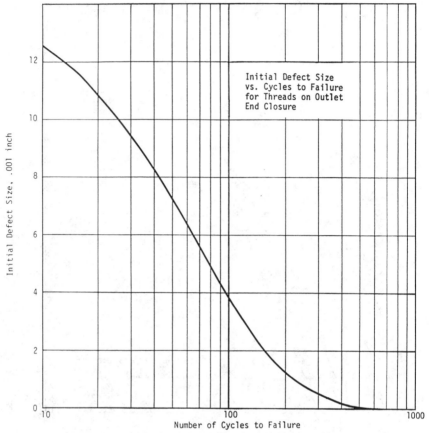

FIG. 11—*Fracture mechanics evaluation of gas storage vessel threads.*

Minimum Observable Defect

The number of cycles during which crack propagation proceeds before unstable crack propagation is estimated from the crack propagation function, da/dN. When estimating a "safe" interval, an initial flaw or crack size (a_0) must be assigned from which propagation starts at rate da/dN. The assumed or measured initial size (a_0) must equal or exceed the minimum flaw size reliably detectable by nondestructive evaluation (NDE). Larger flaws than this minimum will result in premature failures due to their faster growth rates. Thus, the prediction of the time to reach critical flaw size depends heavily on the ability to detect small flaws. Therefore, the establishment of the lower limit of flaw detectability with demonstrated reliability (probability of detection at a given confidence level) is a *critical* factor in the characterization of materials and structures. It is also an important element in determining the frequency of nondestructive inspection.

TABLE 1—*Liquid penetrant and magnetic particle flaw sensitivity levels.*

Type	Sensitivity	
	Width	Depth
Zyglo ZL-15 high sensitivity water-washable liquid penetrant or other Group 1 or Group 6 penetrant per NAVSHIPS 250–1500	10 to 20 μm (0.0004 to 0.0008 in.)	200 μm (0.008 in.)
Magnetic particle with AC device or DE Parker-Probe	10 to 20 μm (0.0004 to 0.0008 in.)	100 μm (0.004 in.)

The minimum observable defect size depends on the nondestructive evaluation technique that is used, the material, and the geometry. There are three potential ways of examining surfaces for crack-like defects in notches and similar areas of stress concentration at which they usually occur. These are eddy current inspection, magnetic particle examination (MT), and liquid penetrant examination (PT). Of these, eddy current examination is the most difficult to use. It has the highest sensitivity, but because it is difficult to use, especially in regions such as the roots of threads, it is not as practical or reliable as MT or PT. Therefore, MT and PT methods are commonly used to inspect surfaces, with the sensitivities shown in Table 1.

The detection of surface defects is highly dependent on the width of the flaw. Therefore, the flaw depth that can be detected is really a function of the flaw width. Thus, the most significant parameter in Table 1 is the width measurement.

The sensitivity levels given in Table 1 represent the best detection capabilities that can be expected using the most sensitive examination methods carried out by the best inspection personnel. These methods can be employed, but without the necessary high probability of detecting the small flaw sizes specified in Table 1.

Other methods of examination, such as ultrasonic examination and radiography, can be also used depending on the material and geometry to be inspected. For example, they are of no use on the buttress thread forms considered herein. The minimum observable defect size which these methods can detect must be established if they are employed to perform periodic inspections.

In Ref. 3, data from test programs performed at several major aerospace corporations have been statistically evaluated to establish flaw sizes with a 90% probability of detection and a 95% confidence level. Inspectors certified to Level II of MIL-STD-410, working to normal aerospace industry and military specifications, have been determined capable of detecting (with the 90/95 statistics) the "standard NDE" flaw size given in Table 2 (from Ref 3).

The flaw sizes shown in Table 2 are considered to be practical values for standard NDE capability for fracture analysis. After verifying that a quantitative NDE method is employed to inspect the component, the fracture analyst can safely assume that no flaw larger that the appropriate "standard flaw" will be present in the component.

A further evaluation of the flaw detection statistics reveals that some inspectors consistently detect smaller defects than others and that an improved level of flaw detection can be established. Test specimens carefully fabricated with known flaws are used to evaluate capabilities of the most qualified inspectors. An inspector who demonstrates the capability to detect smaller flaws while using a part-specific inspection procedure has been determined capable of detecting (again with 90/95 statistics) the "special NDE" flaw sizes given in Table 2 (again from Ref 3).

The inspection capabilities given in the "Standard NDE" column of Table 2 are about what can be expected with normal commercial inspection techniques and training. Thus, the data given in Tables 1 and 2 represent the range of inspection capabilities that can be expected.

Initial analysis of the thread considered herein showed that the readily detectable flaw size from Table 2 was larger than the critical flaw size. Therefore, a combination of MT and expertly done eddy current testing at the thread roots was applied. For magnetic particle examination alone, Table 1 shows a minimum detectable depth of 100 μm (0.004 in.). For conservatism, this minimum detectable depth was doubled, and an initial flaw depth of 200 μm (0.008 in.) was assumed.

Inspection Interval

From Fig. 11, a flaw will grow from 8 mil to the critical size of 14.5 mil in about 42 cycles. If a flaw exists during one examination, and is just below the limit of detection, it could grow to cause failure in 42 cycles. If such a flaw exists, and is missed in an examination established to be performed

TABLE 2—NDE capabilities.

Inspection Method	Flaw Type	Standard NDE	Special NDE
Penetrant or magnetic particle	surface flaw (depth × length)	0.19 × 0.38 cm (0.075 × 0.150 in.) or equivalent area	0.063 × 0.127 cm (0.025 × 0.050 in.) or equivalent area
Ultrasonics	embedded flaw (diameter)	2% thickness	
Radiographic	surface or embedded flaw (depth × length)	2% thickness	

every 21 cycles, the vessel could again fail at 42 cycles. Therefore, to insure against the presence of a flaw just below detectable size, and then the failure to see it on one subsequent examination, the inspection interval must be less than one half the number of cycles to increase the crack from detectable size to failure. We have typically used a factor of one third. Therefore, reinspection was recommended every 14 cycles.

Modified Inspection Interval

It is convenient to the user to maximize the interval between inspections. A way in which to do this is to account for the actual pressure loading imposed instead of basing the interval on design pressure loading. This can be done by rewriting the expression for da/dN as

$$\frac{da}{dN} = C_0[\Delta\sigma a^{1/2}M^{1/2}]^n/[1 - R]^{0.5} \tag{3}$$

where

$R = P_{min}/P_{max}$ and
$\Delta\sigma = \sigma_{design} (\Delta P/P_{design})$.

Thus $\Delta\sigma$ is reduced when a peak pressure lower than design is applied, or when pressure does not go totally to zero, but the damaging effect of mean stress is accounted for in the factor $[1 - R]$.

Following this procedure, the user can account for the actual damage imposed during operation, and when sufficient operation has accumulated so analysis shows that one third of the growth between minimum perceptible size and size for failure has occurred, the vessel is shut down for inspection.

References

[1] Imhof, E. J. and Barsom, J. M. in *Progress in Flaw Growth and Fracture Toughness Testing, ASTM STP 536*, American Society for Testing and Materials, Philadelphia, 1973, pp. 182–205.
[2] Wessel, E. T. and Mager, T. R., "Fracture Mechanics Technology as Applied to Thick-Walled Nuclear Pressure Vessels," Proceedings, Conference on Practical Application of Fracture Mechanics to Pressure Vessel Technology, Institution of Mechanical Engineers, 1971.
[3] Ehret, R. M., *Transactions*, American Society of Mechanical Engineers, *Journal of Engineering Materials and Technology*, Jan. 1980, Vol. 102, pp. 40–44.

James B. Chang[1]

Fatigue Crack Growth Predictions of Welded Aircraft Structures Containing Flaws in the Residual Stress Field

REFERENCE: Chang, J. B., **"Fatigue Crack Growth Predictions of Welded Aircraft Structures Containing Flaws in the Residual Stress Field,"** *Case Histories Involving Fatigue and Fracture Mechanics,* ASTM STP 918, C. M. Hudson and T. P. Rich, Eds., American Society for Testing and Materials, Philadelphia, 1986, pp. 226–242.

ABSTRACT: Welded structures often contain flaws or crack-like defects. The need to develop a reliable methodology to predict the fatigue crack growth behavior of such flaws, whether embedded in the body or exposed to the surface of the welded structure, is obvious. This paper describes the use of the analytical prediction methodology implemented in a computer code, EFFGRO III, for performing the fatigue crack life prediction of an embedded flaw contained in a welded aircraft structure subjected to a fully reversed flight spectrum loading. Effects of residual stresses to the crack growth were accounted for in the analytical prediction. Results of the analytical prediction were compared to the test data. Good correlations were shown.

KEY WORDS: fatigue (materials), crack growth, embedded flaw, residual stress, spectrum loading, 9Ni-4Co-0.2C steel

In the research, development, test and evaluation (RDT&E) phase of an advanced aircraft, several full-scale design verification test (DVT) articles which represented major portions of the aircraft were built, and fatigue tests were conducted. The objectives of the full-scale fatigue tests were to:

1. Demonstrate that the fatigue life of the test article was greater than the design service life requirement when subjected to the design service load/environment spectra.
2. Locate critical areas of the airframe which were not identified by analysis or component testing.
3. Provide a basis for establishing special inspection and modification requirements for the fleet airplanes.

All the full-scale fatigue tests were conducted in the test laboratory at

---

[1] Adjunct professor, Northrop University, Inglewood, CA.

room temperature, ambient air environment conditions. The test loadings were flight-by-flight spectrum loads which were developed from the analytical design spectrum for each test article. One such DVT article identified as DVT-2 aft fuselage/empennage specimen, as shown in Fig. 1, was tested under the flight-by-flight empennage spectrum loads with 469.2 MPa (68 ksi) maximum spectrum stress. Most of the load cycles were fully reversed. Figure 2 shows the peaks and valleys of a typical flight of this spectrum. After the application of 991 flights of the fatigue spectrum loadings, a 12.15-cm (4.785-in.) crack was discovered on the exterior surface of the horizontal stabilizer spindle support fitting right-hand side plate. A sketch of the horizontal spindle support fitting and the cracked region is illustrated in Fig. 3. The fitting was a welded assembly of HP 9Ni-4Co-0.2C steel forging, heat-treated to 1380 MPa (200 ksi) ultimate tensile strength. Because of the

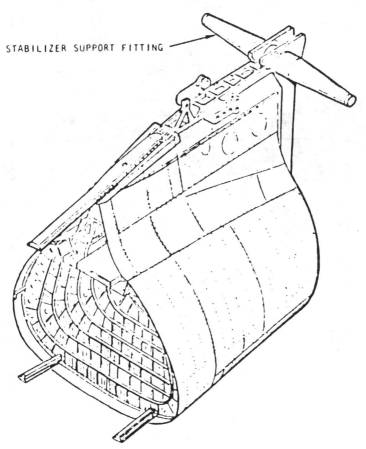

STABILIZER SUPPORT FITTING

FIG. 1—*DVT-2 aft fuselage/empennage test specimen.*

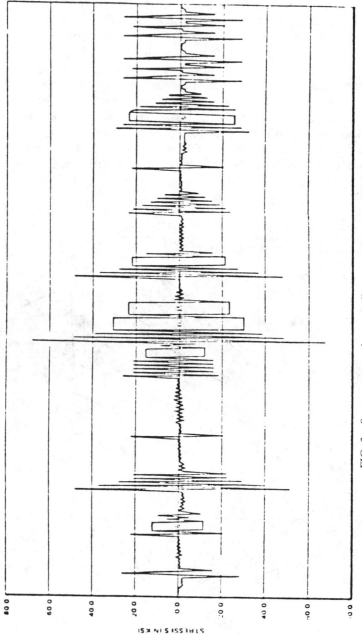

FIG. 2—Stress spectrum shart, spindle support fitting, one flight.

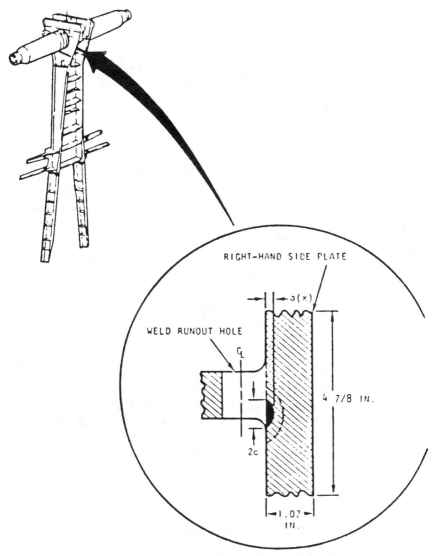

FIG. 3—*Spindle support fitting and its cracked region.*

large thickness of the part, more than 100 weld passes were required to make the joint. The weld joint was in a double U-groove configuration. Partial stress-relief operation was performed after the welding.

The cracked region was sectioned off from the test article. Fractographic analysis was then conducted. The crack was determined to originate from the inner surface of the weld runout hole. The initial flaw was identified as an intergranular anomaly, elliptical in shape. The initial crack depth was 0.051 cm (0.02 in.). A fracture mechanics correlation study was conducted by assuming: (1) no residual stresses, (2) residual stress stayed in constant, and (3) residual stress changed. Results of the correlations are shown in Figs. 4, 5, and 6. It was concluded that the tensile residual stresses introduced through the welding process caused the "unexpected" fast crack growth in the test article, although a stress-relief operation at 510°C (950°F) was performed after the welding of the test article. The detailed description of the correlation study was documented in Ref 1.

The test article was repair-welded and put back into the test fixture, and the fatigue testing was continued. After the 1946 additional flights were applied to the test article, a 1.78-cm (0.7-in.)-long crack was detected on the surface of the left-hand side plate of the same spindle fitting during a routine visual inspection. From the ultrasonic signals measured in the cracked region, it was determined that the crack originated from the weld defect embedded in the weld joint. Fatigue crack growth analysis was performed in order to predict the remaining crack life and the residual strength of the damaged test article, such that a decision could be made if the costly full-scale fatigue test should be continued. This paper describes the analytical crack growth prediction methodology, fatigue crack growth rate data, estimations of the residual stress distributions, and the stress-intensity factor solutions for an embedded flaw used in the analysis.

Analysis Methodology

The fatigue crack growth analysis was conducted, employing a modified version of an existing in-house crack growth computer code—the EFFGRO program [2]—which was developed at Rockwell and identified as EFFGRO III. This crack growth computer code is essentially a special integration routine where an initial crack size, a_i, is given and the cyclic growth rate, da/dN, is calculated to yield the relationship between "a" and "N" for a structure containing cracks subjected to a given spectrum loading. A brief description of the methodology which was implemented into the modified EFFGRO is presented in the following paragraphs.

Baseline Fatigue Crack Growth Rate Equations

EFFGRO calculates crack growth in cyclic loaded structures, based on linear elastic fracture mechanics (LEFM) principles on a cycle-by-cycle basis. For tension-tension loadings, the Walker equation [3] is used in EFFGRO to model the fatigue crack growth rates on a cycle-by-cycle basis. The Walker equation can be expressed as follows:
For $\Delta K > \Delta K_{th}$

$$da/dN = C[\Delta K/(1 - R)^{1-m}]^n$$
$$0 \le R < R_{cut}{}^+, R = R \qquad (1)$$
$$0 \le R \ge R_{cut}{}^+, R = R_{cut}{}^+$$

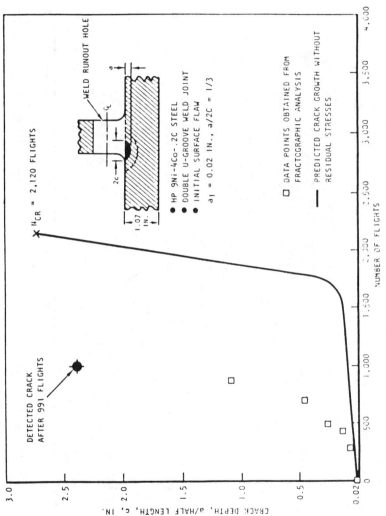

FIG. 4—*Test data/predicted crack growth correlation, no residual stresses assumed in analysis.*

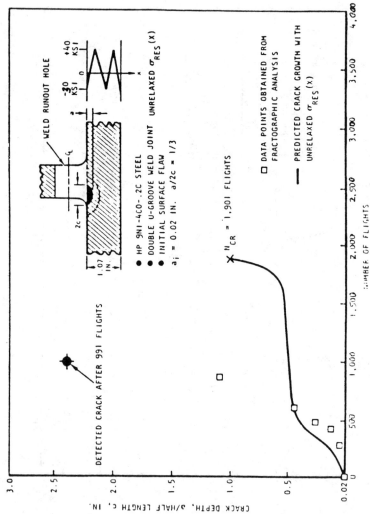

FIG. 5—*Test data/predicted crack growth correlation, unrelaxed residual stresses assumed in analysis.*

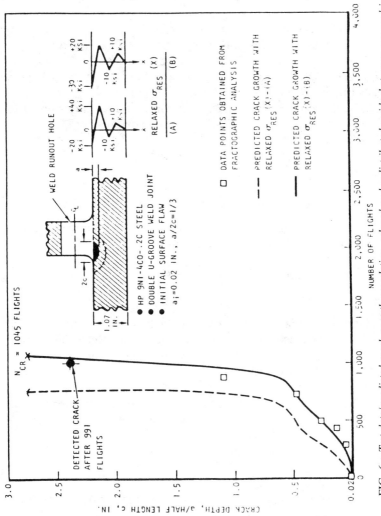

FIG. 6—*Test data/predicted crack growth correlation, relaxed and redistributed residual stresses assumed in analysis.*

For $\Delta K \leq \Delta K_{th}$

$$da/dN = 0$$

where

C and n = materials fatigue crack growth rate constants,
R = stress ratio,
m = stress ratio collapsing factor, and
ΔK = crack tip stress intensity factor range.

The R_{cut}^+ is the cutoff value of the stress ratio above which no further stress-ratio layering is shown in a da/dN versus ΔK plot.

The ΔK_{th} is the threshold of ΔK. The values of ΔK_{th} are assumed to be varied with respect to stress ratios in the following manner

$$\Delta K_{th} = (1 - R)\,\Delta K_{tho} \tag{2}$$

where

ΔK_{tho} = threshold value of ΔK obtained from the $R = 0$ test.

For tension-compression load cycles, the crack growth rate equation proposed by Chang is used [4]

$$da/dN = C[(1 - R)^q\,K_{max}]^n,\ R < 0 \tag{3}$$

where

C and n = material's fatigue crack growth rate constants identical to those used in the tension-tension load cycles, and
K_{max} = stress-intensity factor corresponding to the maximum stress of a load cycle.

The exponent q in Eq 3 is the "acceleration index" for negative stress ratios. For a specific negative stress ratio, q can be determined by

$$q = [\ln(r)/\ln(1 - R)]/n,\ R < 0 \tag{4}$$

where

r = ratio of the crack growth rate at a specific negative stress ratio to its $R = 0$ counterpart measured from tests, and
n = baseline fatigue crack growth rate constant determined from the $R = 0$ crack growth data.

For a given negative stress ratio, there is a corresponding q. However, to simplify the calculation procedure (and eliminating a large test program), an average q is used in EFFGRO III for a range of negative stress ratios.

Load Interaction Model

For tension-tension load cycles, EFFGRO III employs the Vroman retardation model [2] to account for the overload retardation effects on crack growth rates. The Vroman model utilizes the effective stress-intensity factor range concept which reduces the value of ΔK by the following relationship when an overload cycle is existing

$$\Delta K_{\text{eff}} = \Delta K - 0.333[(a_{ol} + R_{yol} - a)/R_{yol})K_{\text{max}_{ol}} - K_{\text{max}}] \qquad (5)$$

where

a_{ol} and $k_{\text{max}_{ol}}$ are the crack size and stress-intensity factor value corresponding to the previously applied tensile overload, and
R_{yol} is the radius of the plastic zone produced by the tensile overload.

It is determined from

$$R_{yol} = \frac{1}{\alpha\pi}(K_{\text{max}_{ol}}/F_{ty})^2 \qquad (6)$$

where

F_{ty} = material yield strength, and
α = plane stress/plane strain coefficients.

For the plane stress condition, $\alpha = 2$. For the plane strain condition, $\alpha = 6$.

For variable amplitude loadings which contain tensile overload cycles, EFFGRO III calculates the fatigue crack growth rate in terms of ΔK_{eff} as follows

$$\begin{aligned} da/dN &= C[\Delta K_{\text{eff}}/(1 - \overline{R})^{1-m}]^n \\ 0 &\le R < R_{\text{cut}}^+, \overline{R} = R \\ 0 &\le R \ge R_{\text{cut}}^+, \overline{R} = R_{\text{cut}}^+ \end{aligned} \qquad (7)$$

Since the numerical value of ΔK_{eff} is always less than ΔK when the overload is existing, the corresponding value of the fatigue crack growth rate is smaller than its constant-amplitude loading counterpart; this results in crack growth retardation.

Effects of Residual Stresses

For a welded structure containing residual stresses, σ_{res}, under a remotely applied tensile stress, σ_∞, the resultant stress-intensity factor concept proposed by Elber [5] was adopted to account for the residual stress effects to the crack growth. The resultant K-factor can be expressed as

$$K^* = \beta\sigma_\infty\sqrt{\pi a} + \sigma_{res}\sqrt{\pi a} \qquad (8)$$

or

$$K^* = (\beta\sigma_\infty + \sigma_{res})\sqrt{\pi a} \qquad (9)$$

where

β = geometrical correction function on a stress-intensity factor for a crack with a size a.

The maximum and minimum resultant K-factors during a loads cycle are then

$$K^*_{max} = (\beta\sigma_{\infty max} + \sigma_{res})\sqrt{\pi a}$$
$$K^*_{min} = (\beta\sigma_{\infty min} + \sigma_{res})\sqrt{\pi a} \qquad (10)$$

The resultant stress-intensity factor range and stress ratio are calculated based on the resultant K-factor as

$$\Delta K^* = K^*_{max} - K^*_{min} = \beta(\sigma_{\infty max} - \sigma_{\infty min})\sqrt{\pi a} = \Delta K_\infty$$
$$R^* = K^*_{min}/K^*_{max} = \frac{(\beta\sigma_{\infty min} + \sigma_{res})}{(\beta\sigma_{\infty max} + \sigma_{res})} \qquad (11)$$

The fatigue crack growth rate equation is expressed in terms of these resultant parameters as follows

$$da/dN = C[\Delta K/(1 - R^*)^{1-m}]^n, \; R^* \geq 0$$
$$da/dN = C[(1 - R^*)^q K^*_{max}]^n, \; R^* < 0 \qquad (12)$$

Hence, if the magnitude and distribution of the residual stress is known, its effect to the fatigue crack growth rate can be accounted for using Eq 12. Note that if the residual stress is in tension, $R^* > R$; the result is faster growth rate. On the other hand, if the residual stress is in compression, $R^* < R$, which results in slower growth rate. The effect of the compressive residual stress introduced through the welding process is equivalent to the compressive residual stress introduced by tensile overload in many retardation models such as the Willenborg retardation model [6]. The same effect has been discovered for crack growth at a cold worked fastener hole [7,8].

Stress-Intensity Factor Solutions

EFFGRO III consists of a CRACK-LIBRARY module which is a collection of many subroutines, each containing the stress-intensity factor equation for a specific crack configuration such as the surface flaw, a corner crack at a fastener hole, etc. A crack code system is used to execute each subroutine. For an embedded crack, the opening mode stress-intensity factor implemented into the CRACK-LIBRARY is

$$K_I = M_E \sqrt{\frac{\pi a}{Q}} \tag{13}$$

where M_E is the magnification factor for an embedded flaw in a finite thickness plate which is a function of eccentricity and the crack depth.

The quantity "Q" is a combined factor or the elliptical shape normalizing factor, Φ, and the ratio of the applied stress to the material yield strength, σ_{ty}, as

$$Q = \Phi^2 - 0.212 \, (\sigma/\sigma_{ty})^2 \tag{14}$$

Values of Φ are the complete elliptical integral of the second kind, which can be expressed as

$$\Phi = \int_0^{\pi/2} (\sqrt{1 - (1 - 4\xi^2) \sin^2\theta}) \, d\theta, \; \xi = a/2c \tag{15}$$

After the embedded flaw breaks through one of the free surfaces, it becomes a part-through crack (surface flaw). The Mode I stress-intensity factor for a surface flaw used in EFFGRO is expressed as

$$K_I = M_f \, (a/c) \, M_b \, (a/t, \, a/c) \, M_w \left(\frac{c}{w}\right) \sqrt{\frac{\pi a}{Q}} \tag{16}$$

where M_f and M_b are the front and back face correction factors, respectively. For various a/c and a/t values, M_f and M_b are shown in Tables 1 and 2. They are derived from the results reported by Kobayashi and Moss [9] and Shah and Kobayashi [10].

In EFFGRO III, the width correction function M_w used in the stress-intensity factor solution is the Tada's modified secant function which takes the following form [11]

$$M_w = [1 - 0.025 \, (2c/w)^2 + 0.06 \, (2c/w)^4 \sqrt{\sec\left(\frac{\pi c}{w}\right)} \tag{17}$$

TABLE 1—*Surface crack front face correction factor.*

a/c	0	0.1	0.2	0.4	0.6	0.8	1.0
M_f	1.12	1.11	1.1	1.08	1.06	1.04	1.03

Analytical Crack Growth Prediction

The analytical crack growth prediction was performed by running EFF-GRO III with the following inputs:

Structure and Crack Geometries

Thickness (t): 2.718 cm (1.07 in.)
Width (w): 21.59 cm (8.5 in.)
Initial embedded flaw size: a_i = 0.051 cm (0.02 in.)
$2c_i$ = 0.203 cm (0.08 in.)

Material's Crack Growth Constants and Fracture Properties

$C = 5.376 \times 10^{-19}$ (psi unit)
$n = 2.99$
$m = 0.4, q = 0.3$
$R_{cut}^+ = 0.75, R_{cut}^- = -0.3$
$\Delta K_{th} = 6.05$ MPa $\sqrt{m}$ (5 500 psi $\sqrt{in.}$)
$K_{Ic} = 181.5$ MPa $\sqrt{m}$ (165 000 psi $\sqrt{in.}$)
$K_c = 253$ MPa $\sqrt{m}$ (230 000 psi $\sqrt{in.}$)
$\sigma_{ty} = 1242$ MPa (180 000 psi)

Residual Stress Profile

a, cm (in.)	σ_{res}, MPa (psi)
0　　(0)	−207 (−30 000)
0.635 (0.25)	+138 (+20 000)
1.27　(0.50)	−69　(−10 000)
1.905 (0.75)	+69　(+10 000)
2.718 (1.07)	0　(0)

The material crack growth rate constants used in the analysis are shown in Fig. 7. They were derived from the baseline constant amplitude test data generated from a previous experimental program [12]. The residual stress distribution across the thickness of the uncracked sideplate containing the double U-groove weld joint was best estimated from the X-ray diffraction measurements. The relaxation of the residual stresses during the crack growth under cyclic loadings were considered in the analyses, as shown in Fig. 6.

TABLE 2—*Surface crack back face correction factor.*

a/c \ a/t	0.0	0.2	0.3	0.4	0.5	0.6	0.7	0.8	0.9
0.1	1.0	1.02	1.04	1.05	1.08	1.13	1.20	1.36	1.76
0.2	1.0	1.0	1.02	1.03	1.06	1.10	1.16	1.27	1.53
0.4	1.0	1.0	1.02	1.01	1.04	1.07	1.13	1.20	1.37
0.6	1.0	1.0	1.0	1.01	1.02	1.05	1.09	1.16	1.28
0.8	1.0	1.0	1.0	1.01	1.02	1.04	1.07	1.13	1.24
1.0	1.0	1.0	1.0	1.0	1.01	1.02	1.05	1.10	1.19

This phenomenon has been studied by many investigators [13,14]. Radhakrishnan et al established a functional relationship of the decay of the residual stress to the applied number of fatigue load cycles. In the current work, it was assumed that the tensile residual stress near the surface decreased 50%. It was further assumed that when the crack grew through the thickness of the sideplate, the residual stress vanished.

The test flight spectrum as shown in Fig. 2 was used in the analytical prediction. The fatigue spectrum applied on the test article was the horizontal stabilizer spindle support fitting spectrum in a flight-by-flight format. Each flight contains 138 load steps. The total cycles in one flight were 239. The maximum spectrum stress was 469.2 MPa (68 000 psi) which applied once for every 100 flights. Most of the load steps in this spindle support fitting spectrum are fully reversed, that is, $R = -1$.

The growth behavior of the embedded flaw with an initial size, $a_i = 0.051$ cm (0.02 in.), and an aspect ratio, $a/2c = 0.25$, was predicted by running EFFGRO III using the previously listed crack growth rate constants and parameters. Figure 8 shows the prediction which indicated that under the applied fully reversed spectrum loading, an embedded flaw contained in the weld-joint with an initial size, $a_i = 0.051$ cm (0.02 in.), $c_i = 0.102$ cm (0.04 in.), would grow to a surface crack, $c = 1.778$ cm (0.7 in.), after 2250 flights; and the test specimen would fail at 2275 flights. The 1.718 cm (0.7-in.) half length surface crack was discovered after the application of 2937 flights of spectrum loading. In comparison to the analytical prediction, a good correlation was shown (within 25%). It demonstrated that the residual stress profile established in the previous correlation study on the surface crack growth in the right-hand side-plate was adequate.

The predicted growth indicated that not much life remained in the cracked test article (25 flights), so it was decided to continue the test until the surface crack grew through the outer face of the sideplate. After the test was restarted, the crack growth was carefully monitored. The crack was detected to finally break through the outer face just after the application of 30 flights of the spectrum loading. Compared to the analytical prediction of 25 flights, the prediction was considered to be very good.

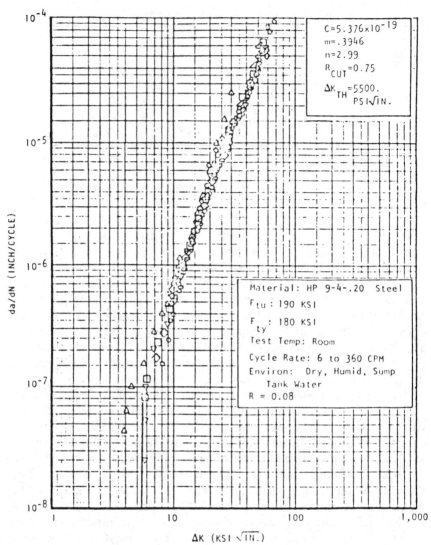

FIG. 7—*Material properties, HP 9Ni-4Co-0.2C steel.*

Concluding Remarks

A case study of employing state-of-the-art linear elastic fracture mechanics technology to predict the fatigue crack growth behavior of an embedded flaw contained in the welded aircraft structure has been presented. From the result of this study, the following conclusions were drawn:

1. Use of linear elastic fracture mechanics as the analytical tool for predicting the growth behavior of an embedded flaw contained in the welded aircraft structure is adequate.

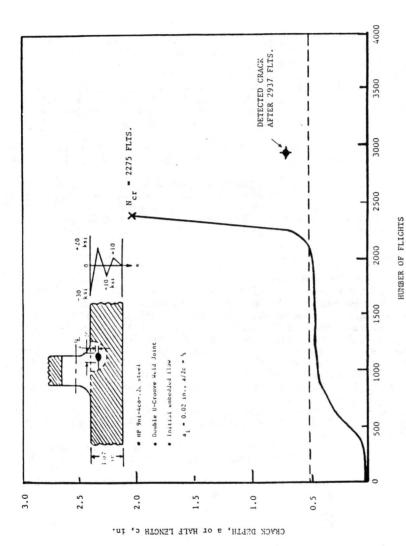

FIG. 8—*Predicted crack growth for an embedded crack.*

2. Residual stresses play a very important role for a flaw or a crack growing in the welded structure. The residual stress effect to the crack growth behavior should be accounted for in the analysis.

References

[1] Chang, J. B., "Fracture Mechanics Correlation Study on Cracked Aircraft Structures," AMMRC-MS-77-5, *Case Studies in Fracture Mechanics,* T. P. Rich and D. J. Cartwright, Eds., Army Materials and Mechanics Research Center, Watertown, MA, 1977.

[2] Szamossi, M., "Crack Propagation Analysis by Vroman's Model, Computer Analysis Summary, Program EFFGRO," NA-72-94, Rockwell International, 1972.

[3] Walker, K., "The Effect of Stress Ratio During Crack Propagation and Fatigue for 2024-T3 and 7075-T6 Aluminum," *Effects of Environment and Complex Load History on Fatigue Life, ASTM STP 462,* American Society for Testing and Materials, Philadelphia, 1970.

[4] Chang, J. B., "Development of Fatigue Crack Growth Model for Flight Spectra Containing Compressive Load Cycles," NA-76-858, Rockwell International, B-1 Division, 1977.

[5] Elber, W., "Effect of Shot-Peening Residual Stresses on the Fracture and Crack-Growth Properties of D6AC Steel," *Fracture Toughness and Slow-Stable Cracking, ASTM STP 559,* American Society for Testing and Materials, Philadelphia, 1974.

[6] Willenborg, J., Engle, R. M., and Wood, H. A., "A Crack Growth Retardation Model Using an Effective Stress Concept," AFFDL-7R-71-1, Air Force Flight Dynamics Laboratory, Wright-Patterson Air Force Base, OH, 1971.

[7] Speakman, E. R., "Fatigue Life Improvement Through Coining Method," presented at the 72nd Annual Meeting, ASTM, Atlantic City, NJ, 22–27 June 1969.

[8] Chang, J. B., "Analytical Prediction of Fatigue Crack Growth at Coldworked Fastener Holes," AIAA paper No. 75-805; presented at AIAA/ASME/SAE 16th Structures, Structural Dynamics, and Materials Conference; Denver, CO, 27–29 May 1975.

[9] Kobayashi, A. S., and Moss, W. L., "Stress Intensity Magnification Factor for Surface-Flawed Tension Plate and Notched Round Tension Bar," *Proceedings of the Second International Conference on Fracture,* Brighton, England, 1969.

[10] Shah, R. C., and Kobayashi, A. S., "On Surface Flaw Problem," *The Surface Crack: Physical Problem and Computational Solutions,* American Society of Mechanical Engineers, New York, 1972.

[11] Tada, H., *The Stress Analysis of Cracks Handbook,* Del Research Corporation, Hellertown, PA, 1973.

[12] Ferguson, R. R., and Berryman, R. C., "Fracture Mechanics Evaluation of B-1 Materials," AFML-TR-76-137, Air Force Materials Laboratory, Wright-Patterson Air Force Base, OH, 1976.

[13] Morrow, JoDean, Ross, A. S., and Sinclair, G. M., "Relaxation of Residual Stresses Due to Fatigue Loading," SAE Transaction, Society of Automotive Engineers, Vol. 68, 1960.

[14] Radhakrishnan, V. M. and Prasad, C. R., "Relaxation of Residual Stress With Fatigue Loading," Engineering Fracture Mechanics, Vol. 8, 1976.

Daniel L. Rich,[1] *R. E. Pinckert,*[1] *and T. F. Christian, Jr.*[2]

Fatigue and Fracture Mechanics Analysis of Compression Loaded Aircraft Structure

REFERENCE: Rich, D. L., Pinckert, R. E., and Christian, T. F., Jr., **"Fatigue and Fracture Mechanics Analysis of Compression Loaded Aircraft Structure,"** *Case Histories Involving Fatigue and Fracture Mechanics, ASTM STP 918,* C. M. Hudson and T. P. Rich, Eds., American Society for Testing and Materials, Philadelphia, 1986, pp. 243–258.

ABSTRACT: During depot maintenance of a U.S. Air Force F-15 fighter aircraft, cracks were discovered in the wing upper spar cap at the inboard end of the spar. Since the upper spar cap is a compression member during normal flight, these cracks were an unanticipated phenomenon which required extensive fatigue and fracture mechanics analyses. The tensile stresses resulting from maneuvers and landing were not considered sufficient to cause cracking. Investigation indicated that the local stresses around holes in the compression-loaded spar cap exceeded compression yield, and, upon unloading, residual tensile stresses were induced at the holes. To determine this cause and the exact nature of service loadings, analytical and experimental studies were performed. A large, detailed finite element model was constructed to determine the local loads in the area of fastener holes. A comprehensive program of element specimens subjected to spectrum loading was also conducted to simulate the in-service cracking and determine the effectiveness of various repair configurations. Crack initiation and crack growth characteristics were obtained from the tests.

The number of flight hours to crack initiation at the side of the critical fastener hole was predicted. Crack growth analyses were made using a contact stress model computer program, modified to take into account the residual stresses adjacent to the fastener hole. The results of this crack growth analysis are being used to guide Air Force planning for in-service modifications of aircraft.

KEY WORDS: crack propagation, fatigue (materials), design, stress analysis, plastic properties, mathematical prediction, residual stress

The past decade has been a period of great effort and achievement in the field of aircraft fatigue and fracture analysis. In the mid 1970s the U.S. Air Force introduced the philosophy of durability and damage tolerance analysis (DADTA) based on fracture mechanics principles. This analysis

[1] Unit chief and branch chief, Technology, respectively, McDonnell Aircraft Company, McDonnell Douglas Corporation, St. Louis, MO 63166.

[2] Aerospace engineer, Warner Robins Air Logistics Center, United States Air Force, Robins AFB, GA 31098.

was used to predict crack growth and subsequent inspection intervals for both in-service and new aircraft as part of its aircraft structural integrity program (ASIP) [1,2]. Since that time numerous baseline DADTA assessments of U.S. Air Force aircraft have been conducted by various airframe manufacturers with guidance from Air Force steering groups comprised of Aeronautical Systems Division, Flight Dynamics Laboratory, and Air Logistics Center personnel. All of these baseline assessments, such as the one reported in Ref 3, focused on developing crack growth curves and making inspection recommendations for the various critical areas of the airframe being studied. The determination of the critical areas to be evaluated is an exhaustive process which requires identification of high design stress areas, local geometric stress concentrations such as radii and joints, fatigue test article results, and in-service crack findings.

It is virtually axiomatic that fatigue cracking occurs in a tensile stress field; hence, the DADTA sieving of critical areas has naturally concentrated on the tension-designed components of an airframe such as the lower wing skin. There are, however, unusual instances where the unexpected can occur such as this paper will discuss. Several years ago, cracks were found during depot maintenance in the upper spar cap of a U.S. Air Force F-15 fighter aircraft at the very inboard end of the intermediate spar as shown in Fig. 1. All of the cracks radiated from either 7.94 or 3.18-mm-diameter fastener holes located in the spar cap flange where the net section was reduced by the machining of a wing fuel seal groove. The 7.94-mm fasteners attached the aluminum upper skins to the titanium spar and 3.18-mm fasteners attached a plate nut gang channel to the inside of the spar cap. The cracks originated at approximately 45 deg to the spar axis and propagated toward the edge of the seal groove. The upper spar cap is loaded primarily in compression during normal flight due to up-bending on the wing. Since the maximum tensile loads are only 26% of the maximum compressive loads, the structure had been thought to be resistant to fatigue cracking. The appearance of the cracking in aircraft with 1000 to 1500 service hours was an unanticipated phenomenon. Since all of the compression load in the upper flange is reacted through the upper lug which is attached to the spar web and not directly to the flange, the internal stresses concentrate at the inboard end of the flange. This concentration of stress coupled with the reduced section at the seal groove caused the stresses at the fastener holes to exceed compressive yield, and, consequently, residual tension stresses occurred after unloading.

This paper will delineate the extensive analytical and experimental study which was necessary to determine the exact nature of this cracking phenomenon. A large, detailed finite element model was developed to determine the local stresses at the fastener holes in the seal groove as well as the overall stresses in the flange. The complementary experimental study was comprised of a comprehensive test program of element specimens tested

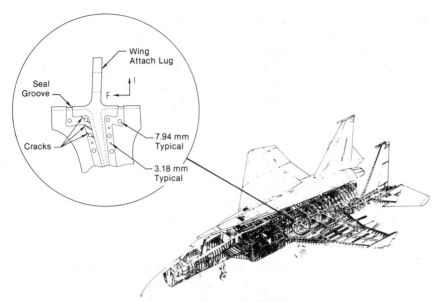

FIG. 1—*Cracking location in wing upper spar cap.*

under spectrum loading to simulate the observed in-service cracking. This experimental investigation produced crack initiation and growth behavior for the as-built spar configuration as well as potential repair modifications.

This paper will also discuss the predictions of the flight hours during the crack initiation and crack growth phases. These analytical predictions are based on actual recorded average flight loads and notch stresses and strains at critical fastener holes. A contact stress retardation model with an additional residual stress intensity term to account for the effect of the residual stresses adjacent to the fastener hole is utilized for the crack growth predictions. The analysis predicts the crack will grow rapidly in the seal groove due to a residual tension stress field, but then the growth rate will slow appreciably as the crack propagates into the full thickness of the flange where the residual stress level is reduced. This analysis procedure also verified the efficiency of repairs for operational aircraft in which cracks are found and the improvement to new production aircraft by relocating the seal groove and increasing the thickness of the spar cap.

Service Loading

The first step of any fatigue investigation is defining the service loading which is applied to the structure. It was fortunate that the F-15 had extensive data recorded from direct data measurements of wing loading for typical F-15 usage. Data were collected for 4469 flight hours using signal data recorders. These devices, which are installed in 20% of the fleet aircraft, continuously record 22 flight parameters on a tape cassette at rates from 1 to 30 times a second, depending on the parameter. The data are processed into a format usable for fatigue spectrum development. These data allowed calculation of a typical wing loading spectrum for the F-15 as summarized in Fig. 2. The maximum up- and down-bending loads in 1000 service h are 115 and -30% of limit load respectively. Up-bending on the wing produces compressive loads in the upper spar cap in the critical area. Bending moment for the wing at limit load is 0.884 MN m at the wing root where the wing attaches to the fuselage.

Finite Element Analysis

The original finite element analysis for the wing, performed 15 years ago, utilized too coarse a grid to determine local stresses around holes. That early model had indicated gross stress levels of only -345 MPa in the wing spar.

A new finite element model was thus constructed with greater detail to more accurately define the local loads around the fastener holes. The finite element analysis was conducted using the NASTRAN computer program.

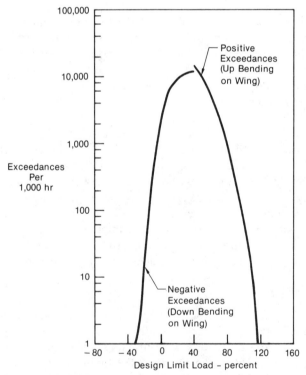

FIG. 2—*Fatigue spectrum cumulative exceedances.*

Six model substructures were developed using quadrilateral and triangular membrane elements. These substructures simulated the upper spar cap, lower spar cap, spar web and upper lug, the closure rib, and the forward and aft upper skins as shown in Fig. 3. The two dimensional substructures were connected together to provide a three dimensional arrangement. The spar cap substructure is shown in Fig. 4 with location of the seal groove and fasteners noted. Shear load transfer between the skin and spar cap was provided using rod elements. Rod stiffnesses were selected to provide fastener flexibility using the relationships of Barrois [4]. Membrane thicknesses identical to the airplane configuration were used. In the vicinity of the hole the thickness at the seal groove is 2.29 mm, and the basic flange thickness is 3.81 mm.

Stresses were predicted using the model for two loading conditions: wing up-bending and wing down-bending. These conditions produce compression and tension loads, respectively, in the upper spar cap. External loads introduced through the upper skins, spar web, spar lug, and closure beam, and the spar cap had been determined from the earlier finite element analysis of the entire wing structure. Both the fastener bearing stress and the through stress levels were found to peak at the innermost hole in the seal groove adjacent to the web. Fastener bearing stresses of 1540 and 441 MPa and through stresses of −903 and 262 MPa for the 115% up-bending and −30% down-bending conditions, respectively, were predicted. The predicted stress levels decreased rapidly outboard of the critical fastener hole.

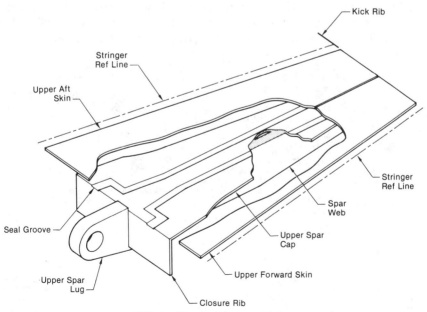

FIG. 3—Sketch of area modeled.

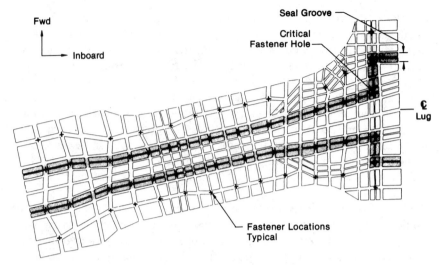

FIG. 4—*Spar cap substructure.*

Cause of Cracking in Compression Flange

Elastic stresses adjacent to the fastener hole were calculated by superimposing three separate stresses, determined from the finite element analysis, as shown in Fig. 5. These stresses are the fastener bearing stress f_{br}, the through stress f_t, and transverse stress f_a. Using stress concentration factors from the literature [5–7], the sum of these stresses times the elastic stress concentration factors were calculated to be -2137 and 1400 MPa, for the 115% up-bending and -30% down-bending load conditions, re-

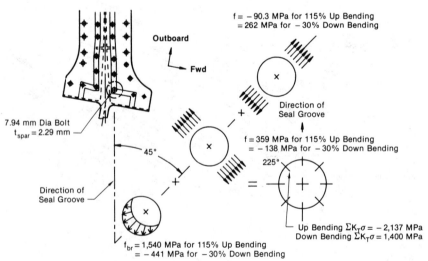

FIG. 5—*Calculation of elastic notch stresses.*

spectively. The stress concentration for a compression bearing load as given in Ref 8 was reduced by 50% to agree with previous finite element analysis work of the authors. The −2137 MPa elastic stress at the hole wall corresponding to the 115% up bending load is considerably above the −930 MPa compressive yield stress of the 6A1-4V mill annealed titanium from which the spar cap is made; hence, extensive compressive yielding is expected for that condition.

Upon unloading, a high residual tensile stress results as shown in Fig. 6. Subsequent compressive loading produces a stress range from the positive residual value to a lower stress value depending on the load applied. A typical hysteresis loop for a −5% ground load to a +95% air load back to a −5% ground load is shown in Fig. 6. Every load cycle thus has a high positive peak stress, and becomes effective for initiating and propagating a fatigue crack.

Crack Initiation Analysis

The crack initiation life for the upper spar cap was developed using the crack initiation model of Rich and Impellizzeri [8]. This procedure predicts stresses and strains at the edge of the stress concentration using the Neuber notch analysis and calculates damages based on closed hysteresis loops. Through correlation with many previous element test results, the model has been shown to accurately predict the initiation of a 0.25-mm crack.

The crack initiation life as a function of $\Sigma K_T \sigma$ is shown in Fig. 7 for the wing spar upper cap. A crack initiation life of 720 h for the critical hole

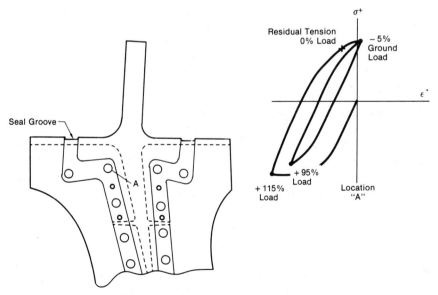

FIG. 6—*Stress strain cycles at fastener hole.*

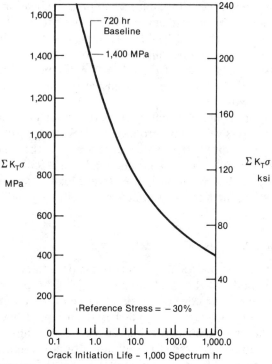

FIG. 7—*Crack initiation life curve for wing upper spar cap.*

location in the upper spar cap is predicted for a nominal notch stress of 1400 MPa. The 720-h crack initiation life agrees favorably with the observed in-service crack initiation lives of approximately 800 to 1000 flight h for the F-15 fleet.

Crack Growth Analysis

Spectrum crack growth analyses were conducted using the Contact Stress Model computer program of Dill and Saff [9]. The contact stress model accounts for crack growth retardation under spectrum loading based on contact stresses left in the wake of the moving crack. An additional residual stress intensity term was added to account for residual tensile stresses adjacent to the fastener hole similar to investigations by Impellizzeri and Rich [8,10]. The effective stress intensity, K_{eff}, is given by the following equation

$$K_{eff} = K_{applied} + K_{residual}$$

$K_{applied}$ is the stress intensity determined using the slice synthesis method of Fujimoto [11] for the stress gradient shown in Fig. 8. This stress gradient

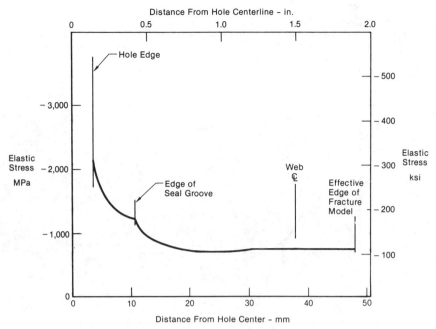

FIG. 8—*Stress gradient for 115% up-bending load on wing.*

had been determined from the finite element analysis and the elastic notch analysis discussed previously. The peak stress at the hole wall is -2137 MPa, and the stress decreases in magnitude rapidly to the edge of the seal groove where it is reduced even further to approximately -655 MPa as a result of the increased spar cap thickness.

The term $K_{residual}$ represents the residual stress intensity determined using the residual stress gradient shown in Fig. 9 which exists after application and removal of the 115% load. The tension residual stress at the hole wall is 1070 MPa and decreases to zero at a distance of 1.8 mm from the hole wall. Figure 10 shows that the residual stress intensity factor peaks at 35 MPa $\sqrt{m}$ at approximately 0.6 mm and decreases to zero at 7 mm.

The predicted crack growth, as shown in Fig. 11, has three distinct growth stages. The first stage exhibits very rapid crack growth. The cracks on each side of the hole propagate quickly from 0.25 mm corner cracks into through-the-thickness cracks in the region of high positive residual stress intensity. In this region, every cycle in the fatigue spectrum causes the crack to grow a finite amount, because the positive residual stress intensity is so large. Once the crack reaches the edge of the seal groove the residual stress intensity factor has decayed. The crack is predicted to grow in this second stage only during the few positive, wing down-bending, load cycles in the spectrum. The predicted crack growth rate is also slower in the second stage

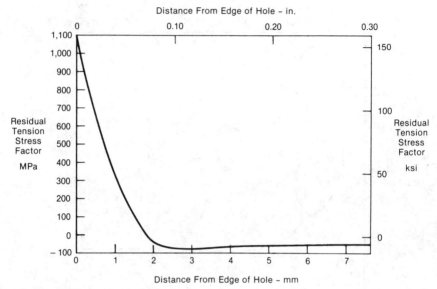

FIG. 9—*Residual stress gradient resulting from compression overload.*

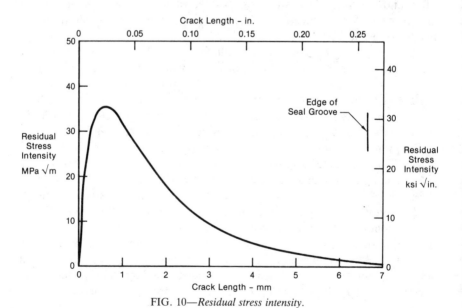

FIG. 10—*Residual stress intensity.*

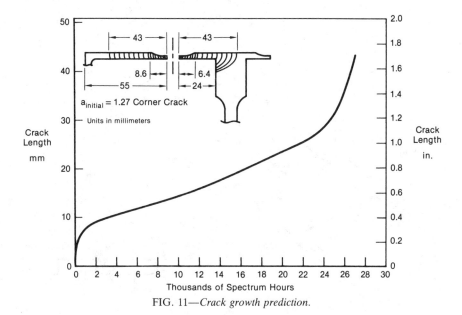

FIG. 11—*Crack growth prediction.*

since the stress is decreasing due to the stress gradient effect and due to the thicker part of the spar cap. The third stage of predicted crack growth occurs when the crack is in the vicinity of the spar web where there is a slight influence of the web. The total crack growth life is predicted to be 27 000 flight hours. The total life prediction of 27 720 h is the sum of the crack initiation and crack growth lives. Although catastrophic failure of the wing structure is not a problem with such a long crack growth life, there is a potential for fuel leakage which could impair the use of the wing as a fuel tank and release potentially hazardous fuel vapors. This consideration is discussed in a later section on redesign and repair.

Element Test Program

A comprehensive test program of element specimens subjected to spectrum loading was conducted to simulate the in-service cracking and to determine the effectiveness of various repair configurations. The test specimen, shown in Fig. 12, consisted of two, 102-mm-wide sheets of 6A1-4V titanium. One sheet was 406 mm long and was loaded at both ends of the specimen. The second sheet was 235 mm long loaded at one end of the specimen and attached to the other sheet with a fastener located in the center of the specimen. The fastener was located in a machined seal groove along with a 3.18-mm-diameter hole and rivet. The rivet was through only one sheet and did not transfer load from one sheet to the other. The specimens were loaded with a spectrum simulating the upper spar cap load environment. The baseline tests simulated the original aircraft configura-

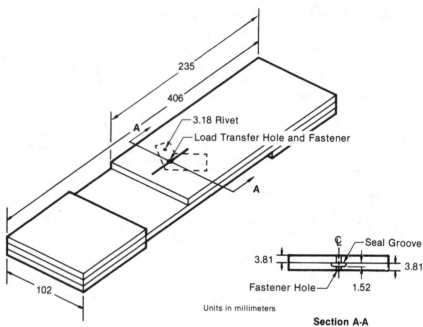

FIG. 12—*Spar cap seal groove element specimen configuration.*

tion. Subsequent tests simulated various repair configurations, many of which did not improve the crack initiation life.

The baseline configuration specimens were found cracked after 1000 to 3000 simulated flight hours of testing. The cracks in the specimens were similar to those found in F-15 airplanes with approximately the same number of flight hours. The results of the baseline test are shown in Fig. 13 along with a prediction of crack growth. The prediction was similar to that for the spar cap except for the difference in specimen geometry. Even though cracks in the specimens initiated quickly, all baseline specimens lasted 32 000 h without failure.

Repair and Redesign

Although the total predicted life for the wing spars is several times the F-15 service life of 8000 h, repair and redesign as shown in Fig. 14 are necessary to ensure that fuel leakage does not occur. Redesign for future production airplanes was easily defined, as shown in Fig. 14. The seal groove was relocated to the second fastener row where the stress levels are lower in magnitude, the spar cap and web were made thicker, and larger diameter fasteners were installed. This change was evaluated by test and analysis. The basis for the analysis was a revised finite element model. The test and analysis showed that the crack initiation time for the redesign was greater than 32 000 h.

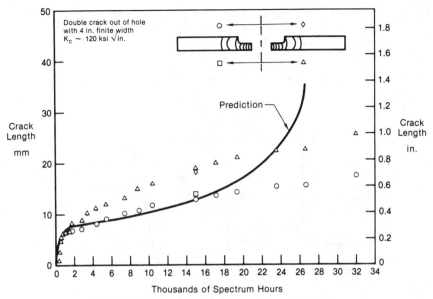

FIG. 13—*Element test data with crack growth prediction.*

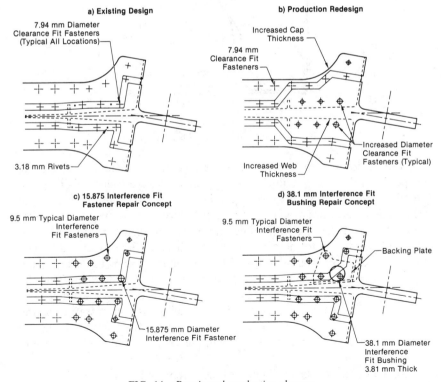

FIG. 14—*Repair and production changes.*

Limit	Symbol	Baseline (hr)	15.98 I.F. Fastener (hr)	38.10 I.F. Bushing (hr)
Safety	□	27,000	25,000	16,000
Functional impairment	O	4,500	9,500	11,700
Economic	△	1,000	N/A	N/A

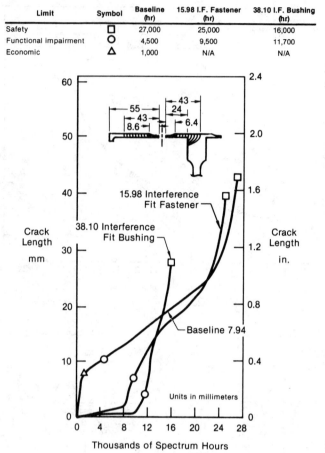

FIG. 15—*Operational limits for wing spar.*

Several different repair concepts were evaluated by element test using the specimen defined previously. Many concepts were tried and discarded as inadequate. The following concepts, however, were found to reduce the early cracking and are being used to repair aircraft. The use of a 15.875-mm-diameter interference fit fastener at the critical corner location as shown in Fig. 14 was found to increase the crack initiation time to 22 000 h. The 15.875-mm fastener was large enough to effectively span the groove and could transfer compressive stress across the groove without magnifying the stress as much as the 7.94-mm-diameter baseline clearance fit fastener. The interference fit also causes a reduction in the stress concentration at the hole wall. This concept is being used to repair aircraft when the cracks are small enough to be removed by machining the hole for the 15.875-mm fastener.

For longer cracks, a large diameter hole is required to clean up the crack. For these holes, special 38.1-mm-diameter steel interference fit bushings are used for repair as shown in Fig. 14. As with the 15.875-mm fastener, these bushings also span the seal groove and were found by the element tests to increase the crack initiation time to between 20 000 and 32 000 h. In addition, a seal groove was machined in the steel bushings to bypass the critical fastener hole.

Interference fit 9.5-mm-diameter fasteners are used in both of the previously mentioned repair concepts to provide required life at the other fastener holes adjacent to the critical hole. This diameter of fastener at the corner hole had been shown by test to be inadequate because it did not bridge the seal groove. The stresses at the outboard holes were low enough so that the smaller diameter interference fit fasteners provided sufficient stress concentration reduction to provide 24 000 h of life.

Three operational limits were developed for all baseline, repair, and redesign configurations to aid the Air Force in force planning. The safety limit is equal to the crack growth life for an initial crack to grow to critical size when catastrophic failure of the spar would occur. The functional impairment limit is the time for an initial crack to grow to a size large enough for fuel leakage. The economic repair limit is the time for the initial crack to grow to a size that would be too large for repair by the 38.1-mm-diameter bushings and would require a more costly repair.

The operational limits for the baseline, the 15.875-mm interference fit fastener and the 38.1-mm-diameter bushing configurations are shown in Fig. 15. The safety limits are all equal to or in excess of two 8000-h lifetimes. The safety limit is longer for the baseline 7.94-mm fasteners than for any of the repairs. The repairs all have larger diameter holes which influence the crack growth for a greater distance than the baseline. However, when there are no cracks, these repair configurations increase the crack initiation time and make the functional impairment limit longer.

Conclusions

Compression loaded structures should receive attention with regard to fatigue and fracture mechanics analyses as do tension loaded structures. State of the art fatigue and fracture analyses give accurate predictions of in-service crack initiation and crack growth provided residual stresses are included. The analyses were useful for the F-15 aircraft in defining limits when aircraft are likely to have cracks and when they must be repaired. The Air Force is using these limits as one of the criteria for rotating aircraft into its overhaul facility.

Acknowledgments

The authors would like to thank McDonnell Douglas teammates T. L. Benton, C. L. Brooks, W. T. Fujimoto, R. T. Leist, J. C. Wenneker, P. Brake, and J. McFarland who contributed greatly to the analytical effort reported on in this paper. Appreciation is also extended to a special U.S.

Air Force advisory team of Dr. J. Lincoln, J. Resnicky, L. Smithers, and H. A. Wood who reviewed the work and offered valuable assistance, and to C. Wallace and C. Battle of the Warner Robins Air Logistics Center for consultation.

References

[1] Military Standard, "Airplane Structural Integrity Program, Airplane Requirements," MIL-STD-1530A (USAF), Dec. 1975.

[2] Military Specification, "Airplane Damage Tolerance Requirements," MIL-A-83444 (USAF), July 1974.

[3] Pinckert, R. E., "Damage Tolerance Assessment of F-4 Aircraft," AIAA Paper 76-904, American Institute of Aeronautics and Astronautics, Sept. 1976.

[4] Barrois, W., *Engineering Fracture Mechanics,* 1978, Vol. 10, pp. 115–176.

[5] Peterson, R. E., *Stress Concentration Factors,* Wiley, New York, 1974.

[6] Timoshenko, S. and Goodier, J. N., *Theory of Elasticity,* McGraw-Hill, New York, 1951, pp. 78–83.

[7] Crews, J. H., Jr., Hong, C. S., and Raju, I. S., "Stress Concentration Factors for Finite Orthotropic Laminates With a Pin-Loaded Hole," NASA TP 1862, National Aeronautics and Space Administration, May 1981.

[8] Impellizzeri, L. F. in *Cyclic Stress-Strain and Plastic Deformation Aspects of Fatigue Crack Growth, ASTM STP 637,* American Society for Testing and Materials, Philadelphia, 1977, pp. 153–175.

[9] Dill, H. D. and Saff, C. R. in *Fatigue Crack Growth Under Spectrum Loads, ASTM STP 595,* American Society for Testing and Materials, Philadelphia, 1976, pp. 306–319.

[10] Impellizzeri, L. F. and Rich, D. L. in *Fatigue Crack Growth Under Spectrum Loads, ASTM STP 595,* American Society for Testing and Materials, Philadelphia, 1976, pp. 320–336.

[11] Fujimoto, W. T., "Determination of Crack Growth and Fracture Toughness Parameters for Surface Flaws Emanating from Fastener Holes," presented at the AIAA/ASME/SAE 17th SDM Meeting, Valley Forge, PA, 5–7 May 1976.

Ian C. Howard[1]

Fracture of an Aircraft Horizontal Stabilizer

REFERENCE: Howard, I. C., **"Fracture of an Aircraft Horizontal Stabilizer,"** *Case Histories Involving Fatigue and Fracture Mechanics, ASTM STP 918,* C. M. Hudson and T. P. Rich, Eds., American Society for Testing and Materials, Philadelphia, 1986, pp. 259–276.

ABSTRACT: The paper shows how fracture mechanics can be used on data given in the accident report of the crash of a Boeing 707 at Lusaka to predict the fracture toughness of the material from which the cracked component was made, and to estimate the load transferred within the stabilizer by the arrest of this crack.

KEY WORDS: fracture, horizontal stabilizer, Boeing 707, crack modelling

Aircraft G-BEBP was a Boeing 707-321C airfreighter which, on 14 May 1977, was operated by Dan-Air Services Ltd. on a nonscheduled international cargo flight carrying palletized freight from London Heathrow to Lusaka International Airport. There were intermediate stops at Athens and Nairobi from which a fresh crew took off for Lusaka. The flight was normal all the way. On receiving final clearance for landing at Lusaka the pilot selected 50° flap which was followed six seconds later by a loud "break-up" noise on the cockpit voice recorder, recovered after the accident.

Eye-witnesses on the ground saw a piece of the aircraft fall away in flight and, in fact, after the crash the complete right hand stabilizer and elevator assembly (Fig. 1) was found some 200 m back along the flight path from the main wreckage. The six occupants of the plane were killed, probably immediately on impact of the plane with the ground. There were no other people injured.

The accident was notified by Dan-Air Services Ltd. to the Accidents Investigation Branch of the British Department of Trade some 2 h after its occurrence. After an initial investigation, the Zambian authorities delegated the whole of the inquiry to the United Kingdom, and the report [1] of the inquiry team is the major document upon which this case study is built.

The inquiry concluded that the origin of the accident was a fatigue failure

---

[1] Lecturer, Department of Mechanical Engineering, The University of Sheffield, Sheffield, U.K.

(*a*) General view with the location of the crack in G-BEBP arrowed.
(*b*) Structure.

FIG. 1—*Boeing 707-300 series/400 series horizontal stabilizer.*

of the top chord of the stabilizer. This had eventually propagated in overload through the structure of the stabilizer (Fig. 2), this fracture severing the stabilizer from the body of the aircraft as the stabilizer was bent downwards by the 50° flap maneuver on the approach to Lusaka. The propagation of the crack was, however, not simple since it consisted of a sequence of advances in fatigue interrupted by major static crack jumps.

This case study uses linear elastic fracture mechanics (LEFM) to investigate the first static jump of the crack in two ways. Firstly, the fracture toughness of the material of the crack was estimated from the estimated load [1] in the chord at the point of fracture. Good agreement is achieved with published [2] values of the aluminum alloy from which the chord was made. The major difficulty in this problem is due to the complex geometry of the fracture. Without access to three-dimensional finite element solutions, the assessment must involve a degree of approximation. This is accomplished by the use of published data [3] on the stress-intensity factor in a bar weakened by a quarter-circular corner crack. This solution is modified by a correction factor that takes approximate account of the elevation of the stress-intensity factor due to the geometry of the published solution being for a smaller crack than that in the failed chord.

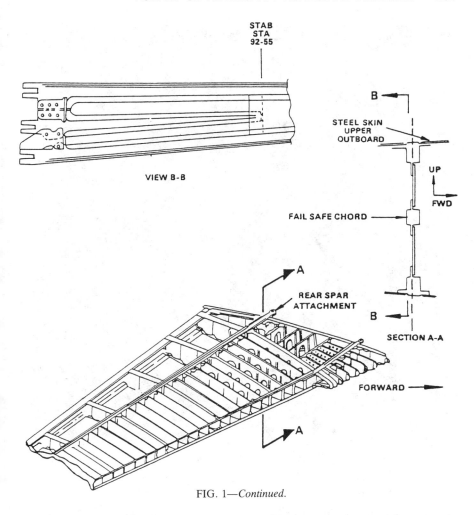

STAB
STA
92-55

VIEW B-B

B

STEEL SKIN
UPPER
OUTBOARD

UP

FWD

FAIL SAFE CHORD

A

REAR SPAR
ATTACHMENT

B

SECTION A-A

FORWARD

A

FIG. 1—*Continued.*

The first static jump stopped after some 20 mm of growth and the second part of this study shows that a reasonable cause of this would be a transference of load to the skin of the top surface of the stabilizer. This extra load is estimated by an assessment using fracture mechanics of the state of the chord at the end of the first static jump. This assessment estimates the value of a crack closure load operating across the top of the cracked chord, and then goes on to examine its nature in terms of the increase in stress in the skin in the vicinity of the crack.

The case study was evolved as part of an introductory short course [4] in fracture mechanics for engineers, metallurgists, and other technologists who require instruction in the subject. The paper ends with a review of the use of this study in our course and the experience of the course participants in coming to terms with it.

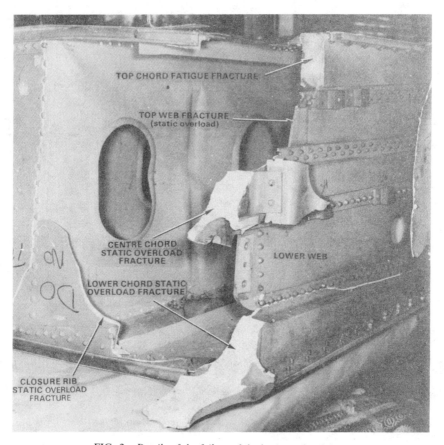

FIG. 2—*Details of the failure of the horizontal stabilizer.*

Cracked Top Chord

After preliminary investigation at the site of the accident, the complete right hand stabilizer and other components judged to be of importance at the time were transported to the United Kingdom for more detailed examination. The primary fracture surfaces were investigated by metallurgists at the Royal Aircraft Establishment, Farnborough, and at the Boeing Company. It appeared that the origin of the failure was at the upper edge of the eleventh fastener hole in the forward flange of the chord (Fig. 3). This fatigue crack propagated rearwards across the flange and into the main cross section of the chord. A secondary major fatigue crack originated in the forward side of the same fastener hole and propagated until it ran out of material at the forward edge of the flange (Fig. 4).

It was concluded that the rate of growth of the primary fatigue crack was very rapid for the first 2 mm of the resultant crack surface. At about 7 mm

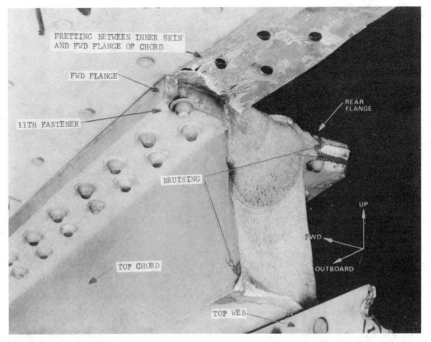

FIG. 3—*Details of the failure of the top chord.*

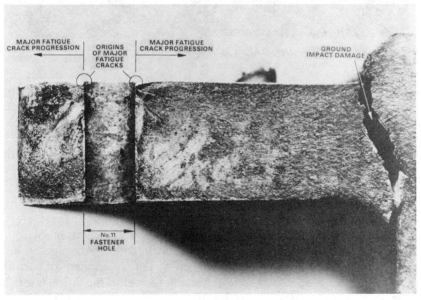

FIG. 4—*Origins of the fatigue crack in the forward flange.*

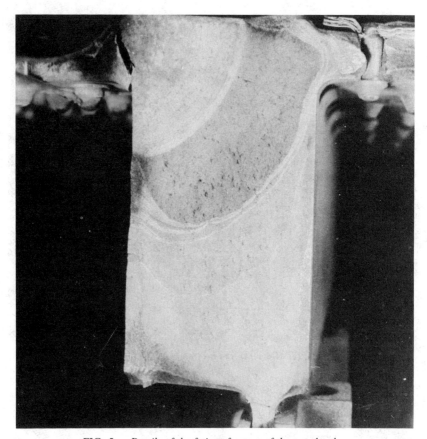

FIG. 5a—*Details of the fatigue fracture of the top chord.*

the growth rate settled down to about 125 flights per millimetre of growth. Eventually, Fig. 5, a static jump of about 21 mm took place after which the aforementioned rate of fatigue growth continued but was now punctuated by several smaller static jumps. A further large static jump ensued, followed by a short period of fatigue which led into the final static failure that caused the accident on the approach to Lusaka.

The investigation showed that the fatigue cracks initiated because higher fastener loads than those anticipated in the design were being carried by those fasteners in the vicinity of the crack in the top chord. These arose because of the diffusion of the stress in the skin into the chord of the spar which causes increased loads in those fasteners in the inboard parts of the chord (Fig. 6). This effect was exacerbated by the replacement of part of the alloy skin in the 707-100 series by stainless steel in the 707-300 series in order to increase the torsional stiffness of the stabilizer (Fig. 7). In principle, the fasteners at the end of the chord would carry the greatest

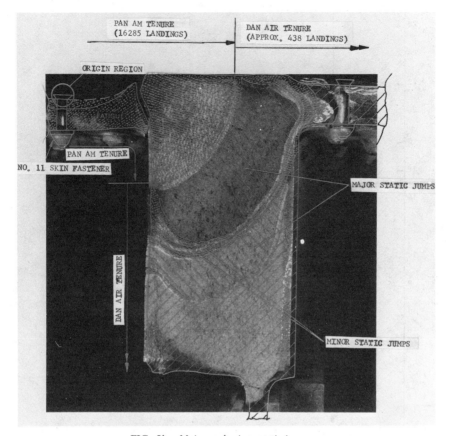

FIG. 5b—*Major and minor static jumps.*

concentration of load. However, investigation showed that the inboard fastener holes were distorted (Fig. 8), so relieving the high loads that would have existed if they had not yielded, and it was apparent that it was the 11th fastener, where there was no distortion of the hole, that maximum load was being carried.

After this investigation, a survey of the whole 707-300 fleet was undertaken. It revealed 38 planes with fatigue cracks in the top chord of the rear spar of the stabilizer. Four were sufficiently damaged as to require a replacement of the chord.

Analysis of the First Static Overload

Figure 9 (redrawn from Fig. 5) gives approximate dimensions of the crack before and after the major static jumps. Crack A-A propagated under the maximum stress exerted upon the chord during a flight of the aircraft, and it arrested at B-B. The shape of Crack A-A is roughly quarter-circular of

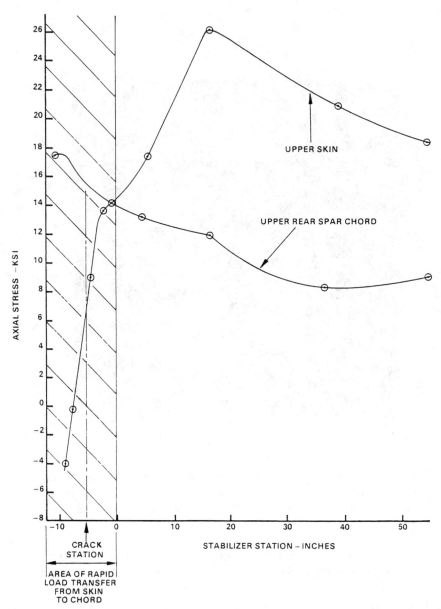

FIG. 6—*Structural test data obtained in a simulation of the loading on the stabilizer at its point of failure.*

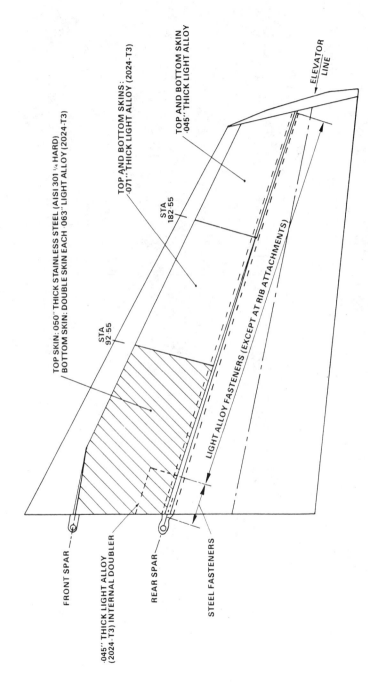

TOP SKIN: ·050″ THICK STAINLESS STEEL (AISI 301 ¼ HARD)
BOTTOM SKIN: DOUBLE SKIN EACH ·063″ LIGHT ALLOY (2024-T3)

TOP AND BOTTOM SKINS:
·071″ THICK LIGHT ALLOY (2024-T3)

TOP AND BOTTOM SKIN
·045″ THICK LIGHT ALLOY

ELEVATOR
LINE

STA
182·55

STA
92·55

LIGHT ALLOY FASTENERS (EXCEPT AT RIB ATTACHMENTS)

FRONT SPAR

·045″ THICK LIGHT ALLOY
(2024-T3) INTERNAL DOUBLER

REAR SPAR

STEEL FASTENERS

FIG. 7—Boeing 707-300/400 series. Details of the skins of the torsion box and the fasteners through the forward flange of the rear spar.

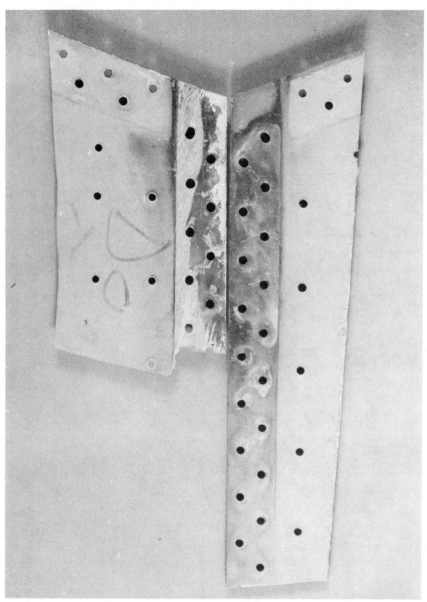

FIG. 8—*Doublers of the top skin of the torsion box of G-BEBP taken from the region near the inner end of the rear spar. Comparison should be made between the left* (undamaged) *and right* (separated) *stabilizers, the latter showing distortion of the fastener holes and fretting of the inner face of the skin.*

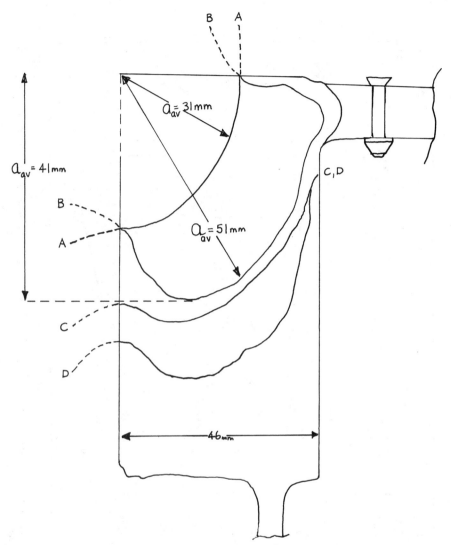

FIG. 9—*Details of the crack shapes to be analyzed.*

average radius 31 mm. The only calibration for this kind of geometry known to the writer is given in the compendium of Rooke and Cartwright [3] and the relevant diagram is reproduced as Fig. 10. This shows that the regions of the crack border near the intersection with the surface of the bar have a tendency to propagate first because the stress-intensity factor is greater there. This supposed tendency to straighten its border bears some resemblance to the shape of the central part of Crack B-B, although there is no

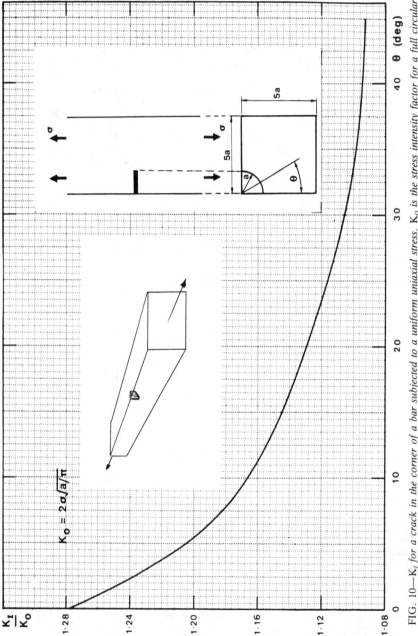

FIG. 10—K_I for a crack in the corner of a bar subjected to a uniform uniaxial stress. K_o is the stress intensity factor for a full circular crack perturbing a uniform stress.

causal connection between the starting of Crack A-A and its arrest at Crack B-B.

There are, unfortunately, significant differences between the geometry of the chord and the calibration of Fig. 10. We can, however, estimate the effects of these with the help of Fig. 11, which gives the effects of changing the ratio of crack length to bar width for the geometry of a flat, through-crack. The chord had a rectangular shape as distinct from the square of the calibration. The main effect of this is to stiffen the chord in its longer direction, an effect that would resist crack propagation in that direction. This assumption is borne out by the observation of Crack B-B being almost through in the short direction. Taking, now, crack propagation in the short direction as the most important, we see that the crack occupies about one half of the chord width instead of one fifth as assumed in the calibration of Fig. 10. We can estimate the effect of this by reading from Fig. 11 a multiplying factor of $(K)_{0.5}/(K)_{0.2} = 1.45/1.2 = 1.2$ from the curve of results where bending is restrained. This is because the presence of the skin on top of the chord will tend to have this effect.

It is most likely that the most severe loading on the top chord of the rear elevator occurred during the approach to landing with abrupt 50° flap. This loading of abrupt full up elevator was one of the conditions simulated in the test program undertaken by the Boeing Company after the accident, and data obtained in that simulation are given in Fig. 6. It can be seen that our best estimate of the stress carried by the chord at the instant of failure is around 15 ksi.

The operating stress is 15×6.9 MN m^{-2} = 103.5 MN m^{-2}. Taking the extremities as the point of initial fast fracture we read from Fig. 10 that

$$K = 1.28 \times 2 \times 103.5 \sqrt{0.031/\pi} = 26.3 \text{ MN m}^{-3/2}$$

and our estimate of the difference in geometry between the actual chord and the calibration increases this by 20% to give

$$K = 31.6 \text{ MN m}^{-3/2}$$

The chord was made from aluminum alloy, the specification being AISI 7079-T6. Quoted values [2] of the fracture toughness of this alloy lie in the range 30 to 35 MN m$^{-3/2}$, and the fact that our estimate above for the stress-intensity factor at the tip of Crack A-A at its point of instability lies within this range gives considerable confidence to the estimates of the load in the chord at the time of the failure.

Analysis of the First Crack Arrest

The arrested Crack B-B has a very complicated shape, a reflection both of the complicated geometry of the chord and the highly variable stresses

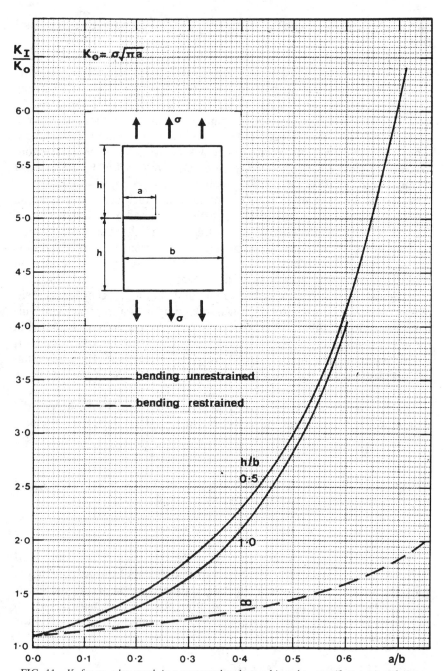

FIG. 11—K_I for an edge crack in a rectangular sheet subjected to a uniform uniaxial stress.

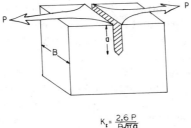

$$K_I = \frac{2.6\ P}{B\sqrt{\pi a}}$$

FIG. 12—K_I for an edge crack opened by forces P at its mouth.

in the chord and skin. The best that we can hope for in discussing this arrest is an *estimate* of the size of the forces involved. If our assumptions are correct, this estimate should be reconcilable with other information on the structure of the stabilizer.

If Crack B-B looks like anything simple, it is a flat crack of average length 41 mm. The stress-intensity factor at the crack tip at the operating stress is, from Fig. 11,

$$K_s = 1.5 \times 103.5\sqrt{0.041\ \pi} = 55.7\ \text{MN m}^{-3/2}$$

reading again from the curve for restricted bending with, now, $a/b = 0.5$. If the effect of the skin resembles a force P MN at the crack mouth, the stress-intensity factor due to this can be read from the calibration of Fig. 12 as

$$K_p = \frac{2.6\ P}{0.046\sqrt{0.041\ \pi}}\ \text{MN m}^{-3/2}.$$

We know that stress-intensity factors are additive, and, since the crack stopped at Crack B-B, the stress-intensity factor there must be just on the point of K_{Ic} (slightly less, in truth). This means that

$$K_s + K_p = K_{Ic}$$

or

$$55.7 + \frac{2.6\ P}{0.046\sqrt{0.041\ \pi}} = 31.6\ \text{MN m}^{-3/2}$$

that is

$$P = -\frac{24.1 \times 0.046\sqrt{0.041\ \pi}}{2.6} = -0.15\ \text{MN}$$

This force is negative, showing that it is compressive. It arises from extra stress in the skin diffused into it from the fasteners nearest the crack mouth, and also some effect of the unbroken flange on the right of Fig. 5a which also tends to close the crack.

In order to proceed further, we need to apportion this load between the members that carried it. They are the skin across the cracked flange on the left of Fig. 5a and the flange and skin on the right. The flange width was about 44 mm, its depth 12 mm, and the thickness of the skin was 1.25 mm (0.05 in.). At the station of the crack the skin was made from stainless steel. The chord was made from aluminum alloy. We denote values associated with the left and right sides of the chord of Fig. 5a by subscripts L and R. The analysis that follows assumes that each side of the chord at the cracked station receives the same extra strain e as the crack moves from Cracks A-A to B-B.

The modulus of steel is about three times that, E_A, for aluminium. The modulus of the right hand flange and skin assembly can be estimated by the rule of mixtures to be

$$E_R = \frac{1.25 \times 3\,E_A}{13.25} + \frac{12\,E_A}{13.25} = 1.19\,E_A$$

We take $E_L = 3\,E_A$.

The loads P_L and P_R are given in terms of the stresses σ_L and σ_R and the cross-sectional areas A_L and A_R by

$$P_L = \sigma_L A_L = \sigma_L l t_L = E_L e l t_L$$

$$P_R = \sigma_R A_R = \sigma_R l t_R = E_R e l t_R$$

where l is the length (44 mm) of the flange and t_L and t_R are the thickness of the skin and the skin and flange. We now may write

$$\frac{P_L}{P_R} = \frac{E_L t_L}{E_R t_R} = \frac{3 \times 1.25}{1.19 \times 13.25} = 0.24$$

Our estimate of the clamping load using fracture mechanics has already told us that

$$P = P_L + P_R = 150 \text{ kN}$$

and these two relations allow us to estimate that

$$P_L = 29 \text{ kN}$$

This means that the extra stress carried in the skin across the cracked flange is about

$$\frac{29 \times 10^3}{1.25 \times 44 \times 10^{-6}} \text{ N m}^{-2} = 527 \text{ MN m}^{-2}$$

Furthermore, if this extra load is to be transferred by the one fastener nearest to the cracked station, that fastener (of diameter about 6 mm) would have to sustain a shear stress of the order of

$$\frac{29 \times 10^3}{9\pi} = 1026 \text{ MN m}^{-2}$$

Although this analysis of the apportionment of load is very crude, it still contains the important conclusion that the crack was stopped at Crack B-B by loads in the skin and fasteners which were very near those that would cause them to suffer a simple tensile failure. Thus the continued operation of G-BEBP for some 100 flights after the static jump at Crack A-A was sustained only by increasing damage to the assembly of skins, flanges, and fasteners. It is likely that the static jump at Crack C-C and its arrest at Crack D-D occurred by similar mechanisms to those analyzed here, but now load was carried by fasteners further away from the cracked station due to failure of those fasteners nearer to it. The distortion of the skin away from the flange that is seen in Fig. 3 provides some evidence for the truth of this hypothesis. If this *is* true then it is highly probable that the small amount of fatigue between Cracks B-B and C-C was sufficient to stress the leading fasteners to their point of failure, whereupon the crack closing force developed in the skin would fall. Crack C-C would then propagate, but in doing so load would again be built back into the skin by transfer through fasteners further away from the cracked station. Eventually enough load diffused to hold the crack at Crack D-D, but a tiny amount of further fatigue was enough to make the whole structure thoroughly unstable, fracturing the chord entirely.

Use of the Case Study in Teaching Fracture Mechanics

This case study is used in a short course [4] which has now been running for six consecutive years. The course draws heavily on material that reinforces the principles of the subject by application to practical matters, and it is liberally scattered with tutorials, projects, and case studies. In fact, the participants attempt a project during the three weeks or so when they are away from Sheffield between part I and part II of the course.

The study of the accident at Lusaka forms the second of two miniprojects that take place in the second part of the course. It is the most difficult project attempted by the course participants. Their main difficulty is due to their inexperience in the art of modelling the geometry of the complicated cracks that they have to assess. In fact, most participants require assistance from the tutors at some stage of this modelling.

Most participants achieve an assessment of the static jump at Crack A-A with a value of the applied stress-intensity factor that is close enough to the fracture toughness of the material of the chord for them to feel comfortable. All of them require assistance in modelling the mechanics of the arrest of the crack at Crack B-B. Those whose background is on the materials side are usually reasonably happy to stop after the first part of the case study has been completed. Those who achieve some numerate view of the state at Crack B-B are mainly engineers who are familiar with mechanics modelling. A few are able to achieve a reasonably comprehensive assessment of the case study in the time (1½ to 2 h) that is available for it.

Many participants are happy to use the calibration of Fig. 10 on its own as their estimate of the effect of geometry on the instability of crack A-A. They arrive at a value for the applied stress intensity in the range 26 to 29 MN m$^{-3/2}$, depending on their reading of the applied stress from Fig. 6. A few continue their assessment by attempting to increase their initial value by a magnifying factor read from Fig. 11, but most of these, reading from the top two curves, estimate the multiplying factor to be about 2. Two points emerge from the discussion of this with them. Firstly, they misunderstood the way in which the stressing of the cracked chord would interact with the structure of the stabilizer and allowed the cracked chord to bend. Secondly, those who pursued this route realized that the massive change required by a multiplying factor of two makes this kind of approximation so unreliable as to be valueless. It is only when *ad hoc* approximations of this kind induce small changes that their inherent unreliability can be accepted.

Acknowledgments

C. N. Reid and K. Williams suggested to me that the official report [1] of the failure of G-BEBP might be a fruitful source for a case study in fracture mechanics. Of course, they bear no responsibility for any errors or omissions that may have invaded this paper.

Figures 1 to 8, 10, and 11 have been reproduced with the permission of the Controller of Her Britannic Majesty's Stationery Office.

References

[1] Department of Trade, Aircraft Accident Report 9/78, Boeing 707 321C G-BEBP, Report on the Accident Near Lusaka International Airport, Zambia, on 14 May 1977, HMSO, London, 1979.
[2] *Fracture Mechanics of Aircraft Structures*, H. Liebowitz, Ed. AGARDograph No. 176, NATO, 1974.
[3] Rooke, D. P. and Cartwright, D. J., *Compendium of Stress Intensity Factors*, HMSO, London, 1976.
[4] Brook, R., Demaid, A. P., Howard, I. C., Reid, C. N., and Williams, K., *Introduction to Fracture Mechanics*, Short Course give yearly from 1980. The University of Sheffield and the Open University, U.K.

Charles R. Saff[1] and M. A. Ferman[1]

Fatigue Life Analysis of Fuel Tank Skins Under Combined Loads

REFERENCE: Saff, C. R. and Ferman, M. A., **"Fatigue Life Analysis of Fuel Tank Skins Under Combined Loads,"** *Case Histories Involving Fatigue and Fracture Mechanics, ASTM STP 918,* C. M. Hudson and T. P. Rich, Eds., American Society for Testing and Materials, Philadelphia, 1986, pp. 277–290.

ABSTRACT: The superposition of high- and low-frequency loadings has caused numerous fatigue crack problems throughout the history of metal airframes. Fighter aircraft fuel tanks are subjected to the superposition of low-frequency (0.1 to 1 Hz) maneuver loads and high frequency (50 to 300 Hz) vibrations due to panel flutter and fuel slosh. This loading can lead to cracks, and leaks, in shorter times than can be predicted by using either load condition alone. This paper examines two methods for predicting fatigue lives under combined high- and low-frequency loadings. Tests of beam element specimens and a simulated fuel tank were used to examine prediction accuracy of both techniques. Results show that both techniques accurately predict the effects of combined load frequencies on life when adjusted to correlate predictions with the extreme (high- and low-) frequency data.

KEY WORDS: fatigue life analysis, load frequency effects, fuel tanks, vibrations

Fuel leaks in aircraft internal fuel tanks are potentially dangerous and certainly costly in aircraft downtime and maintenance manhours. A potential source for these leaks had been identified in McDonnell Aircraft Company (MCAIR) research on fluid-structure interaction dynamics from 1975 to 1978 [1]. This work revealed a dramatic influence of tank fluid oscillation on panel dynamics, vibration fatigue, and panel flutter. Air Force interest in tank fatigue and leakage culminated in a more detailed development [2]. The objective of that effort was to assess analytical capability for predicting fuel tank fatigue lives.

MCAIR's interest in this area was initiated by two problems encountered in F-4 aircraft in 1974 to 1975. These problems involved (*a*) fuel tank cracks that initiated during a slosh and vibration test (these cracks were caused by high stresses created by a match between a resonant response frequency and the vibration input frequency), and (*b*) cracks in a wing skin during

---

[1] Technical specialist, Structural Research, and senior staff engineer, Structural Dynamics, respectively, McDonnell Aircraft Company, St. Louis, MO 63166.

high speed flight, which were traced to panel flutter reinforced by fuel dynamics. In both cases, dry skin tests showed no problems. However, the addition of fuel in contact with the skins produced a significantly different response. This study produced a fundamental explanation for both cases, which couples dynamic fluid and structure motions. In one case, the fluid-structure interaction produced a factor of five reduction in frequency and a factor of four increase in dynamic strain, for a tank with 0.635 cm (¼-inch)-thick skins.

An analytical method was developed and verified by tests. The technique employs a linear small amplitude vibration analysis and panel strain response solution, based on a Rayleigh method for fluid-structure interaction dynamics. It includes a fatigue life estimation procedure. Nonlinear effects due to membrane forces and fluid response at high "g" levels are included via a semi-empirical modification. The approach was fully corroborated in laboratory tests, using fluid depths up to 27.9 cm and 25.4 × 40.6 cm (11 in. and 10 × 16 in.) fixed edge panels of 0.08, 0.10, and 0.16 cm (0.032, 0.040, and 0.063 in.) thicknesses.

The test results show that panels under as little as 10.16 cm (4 in.) of water exhibit considerably shorter fatigue lives than do dry panels at the same input excitation. The combined effects of preload, low frequency loads, and dynamic excitation cause a significant reduction in fatigue life over cases with either loading applied independently.

Our results suggest that fuel tank design criteria should include a coupling of: (1) dynamic vibration, (2) lower frequency, maneuver loads, and (3) static loads, such as those due to fuel pressure.

The fluid-structure interaction is described in detail in Ref 2. This paper examines the procedures used to predict fatigue lives under spectra which couple markedly different frequency regimes.

Analytical Methods

The analytical method for predicting fuel tank durability is based on the theory that the skin oscillation causes a fluid oscillation, which in turn produces added mass loading in each vibration mode. The effect of the added mass loading is to lower frequencies to regions where larger sources of excitation are present. In addition, the dynamic fluid motions produce larger panel strains in moving base excitation tests simulating environmental vibration. Thus, wetted skins, particularly bottom panels, have larger strain amplitudes than dry skins. Though lower frequencies exist in the wet case and result in fewer fatigue cycles for any test time, the sharp increase in strain more than compensates to reduce fatigue life over the dry case.

The overall analytical method is summarized in Fig. 1. Environmental input is combined with the fluid-structure vibration, using the moving base concept. This produces the strain response. From the strain response, fatigue life is predicted.

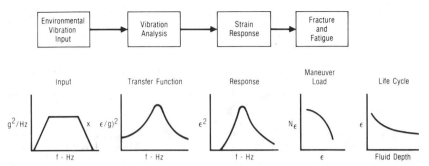

FIG. 1—*Overall aspects of fuel tank durability with fluid structure interaction dynamics.*

Fatigue Life Analysis Methods

The fatigue life prediction techniques used in this program are based on cumulative damage methods developed by Impellizzeri [3]. These methods are similar to many cumulative damage analyses which use Neuber's rule for computation of local stress-strain behavior, and use damage parameters sensitive to local stress ratio as well as strain range. The primary developments presented here are a variation of the damage parameter to account for the strain gradient under bending, and a simple method for accurate life estimation under random load histories.

Development of Bending Fatigue Damage Parameter

One damage parameter, which correlates axial fatigue data very well, is loosely based on the energy input to the material per cycle during loading. Expressed as an effective strain, it can be formulated as

$$\epsilon_{\text{eff}} = \sqrt{\frac{f_{\max}\Delta\epsilon}{2E}} \tag{1}$$

where

$f_{\max}$ = maximum stress at the notch,
$\Delta\epsilon/2$ = strain amplitude at the notch, and
E = Young's modulus.

This parameter can correlate life data for a wide range of stress ratios with and without high initial prestrain cycles sometimes used to shakedown cyclically unstable materials (Fig. 2). In addition, this parameter appears loglinearly related to fatigue life, simplifying subsequent life predictions. This parameter correlates life significantly better than strain range alone (see Fig. 2).

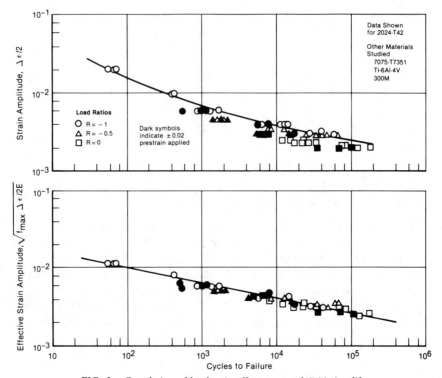

FIG. 2—*Correlation of load ratio effects on crack initiation life.*

Comparison of beam bending fatigue data with previously developed axial fatigue data shows bending fatigue lives are almost twice as long as axial fatigue lives. This trend is expected because in bending, strains, and stresses at the crack tip decrease as the damage crack grows through thickness. This does not happen under axial fatigue.

Another result of the difference between axial and bending stress distributions is a change in the influence of constant static preload (or prestrain) on life. In this analysis, the bending fatigue data for reversed loading was used as the basis for life prediction. Increasing preload or prestress shows more effect in the bending tests than that predicted by axial fatigue test methodology.

This difference in behavior is caused by a change in crack-tip plasticity behavior affecting small flaws initiating at the specimen surface. In bending, due to the reduced stress and strain at the specimen interior, plasticity does not extend into the specimen interior as far as it would under axial fatigue loading. The reduced plasticity means more of the strain range is effective in producing damage at positive stress ratios. To account for this change in behavior, we revised our damage parameter to encompass the entire positive portion of the strain range. The revised parameter for bending is

$$\epsilon_{\text{eff}} = \sqrt{\frac{f_{\text{max}}\Delta\epsilon}{E}} \qquad (2)$$

where $\Delta\epsilon$ is the difference between the maximum strain and the minimum strain, or zero, if the minimum strain is less than zero.

Another difference occurs in damage growth for bending versus axial fatigue. This can be seen by comparing the fracture surfaces for beam specimens failed with and without preload (Fig. 3). When testing includes preload, the flaw initiates on the tension side, as expected, and grows past the midplane of the beam prior to failure. Without preload, flaws initiate and grow equally to failure from both sides. Preload does not affect the failure mode in axially loaded fatigue specimens. The damage parameter used in bending is more sensitive to changes in preload than that used for axial fatigue because it uses the entire positive portion of the strain range.

Life analyses using this damage parameter correlate the lives of the beam tests very well (Fig. 4). The life is computed from an effective strain versus life curve, similar to that shown in Fig. 4 for 7075-T6 material. This curve was developed from the beam bending life data without preload.

Random Loads Spectrum Fatigue Life Analysis

This life analysis technique is based on the generation of a simulated strain history developed from the power spectral density (PSD) of the strain response, including both vibration and maneuver loads. The load history is cycle-counted to determine the most damaging strain cycles. Then these cycles are analyzed using the effective strain parameter to determine damage per cycle. Palmgren-Miner's rule is then used to sum damage and predict life.

Life predictions are compared with narrowband random fatigue test results for beam specimens in Fig. 5. Correlation is very good.

The routine developed to compute fatigue life based on a simulated strain history requires a significant amount of computer memory for spectrum generation. Therefore, a simplifed technique was developed to predict life without using the strain-time history.

Simplified Random Spectrum Fatigue Life Analysis

Development of a simplified spectrum fatigue life analysis was based on an estimation of an equivalent constant amplitude loading which can be used to predict fatigue lives. These fatigue lives are comparable to those predicted by the larger routine for narrowband random strain histories. At

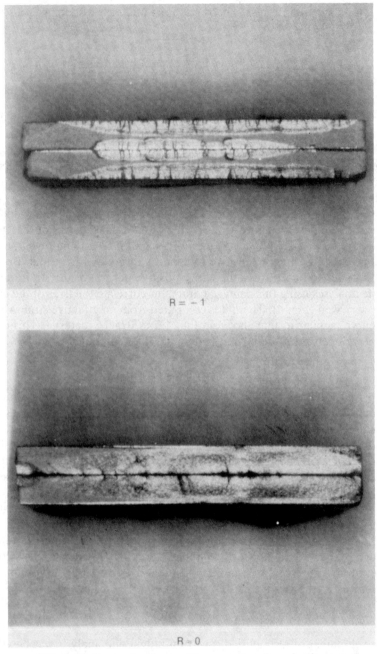

FIG. 3—*Failure modes change with stress ratio in beam bending tests.*

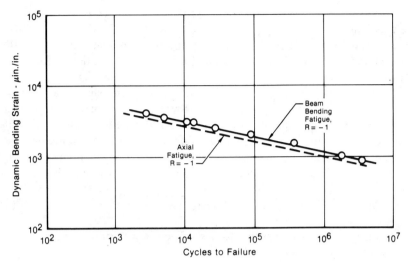

FIG. 4—*Difference in life between axial and bending fatigue tests.*

present, the use of this routine is limited to narrowband random spectra, since it is only for these spectra that any test verification exists.

The technique is described by Fig. 6. The root mean square dynamic strain (ϵ_{rms}) is computed from integration of the PSD over the entire range of frequencies. Comparisons of life predictions from the larger routine for the same PSDs and for constant amplitude loadings showed that a constant amplitude cycle having a maximum strain twice ϵ_{rms} gave best correlation with those results. Once this equivalent constant amplitude cycle has been

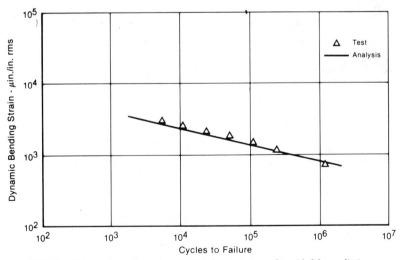

FIG. 5—*Comparison of random fatigue beam test results with life predictions.*

determined, life analysis is the same as that used for constant amplitude life prediction.

Life predictions using this technique are compared in Fig. 7 with the narrowband random fatigue test results for beams. Correlation is good, but not quite as good as that found using the larger routine, Fig. 5.

While the random fatigue test results shown previously did not couple high and low frequencies, preloads- and low-frequency loads were applied in the constant amplitude dynamic tests. As discussed previously, preloads were accommodated by linear superposition in the damage parameter. Correlation of the simplified analysis for superimposed low- and high-frequency sine waves was based on defining the dynamic strain response as

$$\epsilon_{RMS} = \sqrt{\epsilon_{RMS_1}^2 + \epsilon_{RMS_2}^2} \tag{3}$$

and treating the ϵ_{rms}-value as that for a single cycle with or without prestrain, as defined by the loading conditions.

Experiments

Beam bending coupon fatigue tests were run to establish strain-to-failure data for 7075-T6 material for various levels of prestrain- and low-frequency sine input. Panel tests were run to obtain fatigue life data for 25.4 by 40.6 cm (10 by 16 in.) panels of three thicknesses. Fluid depths up to 27.9 cm (11 in.) were used with a wide range of vibration levels.

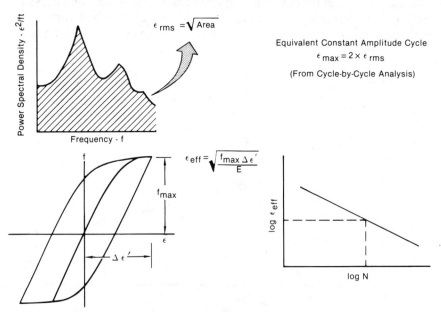

FIG. 6—*Simplified strain-life analysis.*

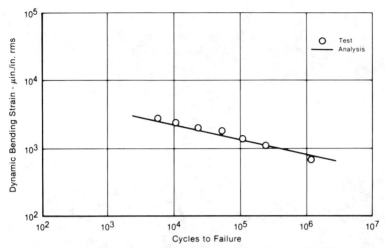

FIG. 7—*Comparison of random fatigue beam test results with simplified life predictions.*

Beam Data

Six sets of fatigue tests were conducted on aluminum beam coupons including:

(*a*) Narrow band random noise excitation applied at a 10 hz bandwidth about the resonant frequency of the beam without preload.

(*b*) Sine excitation without preload.

(*c*) Sine excitation with a static preload of 500 μ cm/cm at the root of the beam.

(*d*) Sine excitation with a static preload of 1000 μ cm/cm at the root of the beam.

(*e*) Sine excitation with a superimposed low-frequency load varying from 500 to 1000 μ cm/cm (750 ± 250) at a rate of 2.5 Hz.

(*f*) Sine excitation with a superimposed low-frequency load varying from +250 to −250 μ cm/cm at a rate of 2.5 Hz.

The test setup, as it was configured for the coupled sine and preload, is shown in Fig. 8. Failure of a coupon was defined as either a 20% reduction in resonant frequency or total failure at the root. Power spectral density curves of strain, tip acceleration, and input acceleration for several of the test coupons excited with random noise were summarized in Ref *3*.

Panel Tests

Test Fixture and Panel Description—The test tank is shown in Fig. 9. It is constructed of 1.27 cm (½-in)-thick aluminum plate welded together, with the top and bottom open. The top was covered by a 2.54 cm (1-in.)-thick

FIG. 8—*Typical coupled oscillatory preload test setup* (*Condition 6*).

plexiglass panel (for viewing purposes) bolted securely to the tank. The test panel was bolted to the bottom using an aluminum "picture frame" retainer. The tank and retainer were sufficiently rigid to insure a fixed edge boundary condition.

One test panel was stiffened by mechanically fastening two "C" channel beam stiffeners. The life of this panel was longer than expected from the test data for an unstiffened panel tested at the same strain levels.

Two dry panels were tested to demonstrate that dry panels would not fail at long test times (1 000 000 cycles), while similar panels having fluid levels of 10.16 and 20.32 cm (4 and 8 in.) failed at lives as short as 80 000 cycles under the same excitation level. This dramatizes the loss of durability from fluid-structure interaction.

Correlation Between Theory and Experiment

Both beam and panel test results were used to assess the analytical prediction methods. Initially, the beam bending tests were used to develop the analytical method.

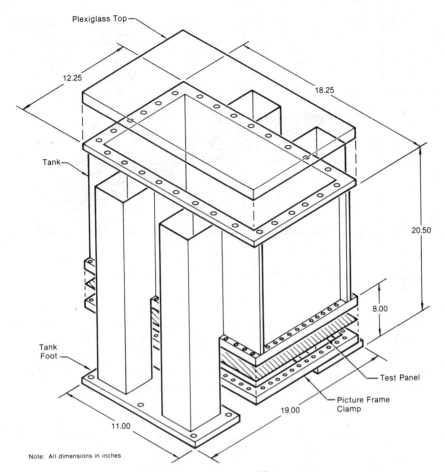

FIG. 9—*Test tank for panel fatigue tests.*

Beam Bending Fatigue

A comparison of measured and predicted beam bending curves are summarized in Fig. 10 for dynamic excitation low-frequency input and preload. In the tests, static preloads were varied through 1000 μ cm/cm while theory is shown through 2500 μ cm/cm. This indicates the trends because these higher preloads were used in later tests. Good correlation is shown with the fatigue life predictions.

As noted earlier, we conducted beam bending tests employing a basic excitation of a higher frequency sine wave, while modulating the static preload with a low-frequency sine wave. This was done to explore life predictions for the combination of different frequency loadings and preload expected to occur in panel tests including fluids. Two tests series were

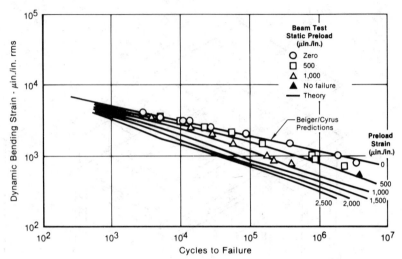

FIG. 10—*Comparison of measured and predicted beam fatigue-sine excitation.*

performed with a low-frequency (2.5 Hz) constant amplitude loading superimposed on the dynamic excitation (50 Hz typically). A static preload was also applied as shown in Fig. 11. In the first case, no preload was applied. In the second case, a static preload of 750 μ cm/cm was applied. Good correlation is seen between the theory and experiments.

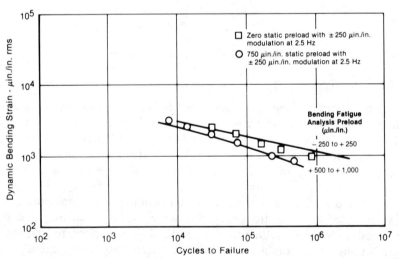

FIG. 11—*Comparison of measured and predicted beam bending fatigue results with both high- and low-frequency sine input.*

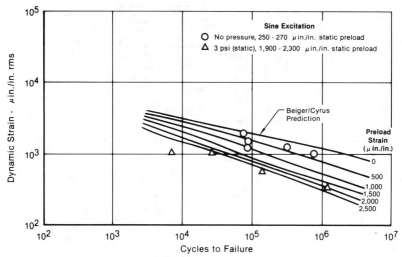

FIG. 12—*Panel fatigue results, effect of internal pressure (panel thickness = 0.16 cm (0.063 in.)).*

Panel Fatigue

Test results for the effects of static fluid pressure on panel fatigue life are compared with fatigue life predictions in Fig. 12. These results clearly indicate the sharp reduction in fatigue life due to static preload. The measured prestrain with zero pressure was between 240 to 268 μ cm/cm. With 20.68 kPa (3 psi) gage pressure, the prestrains were between 1880 and 2340 μ cm/cm as shown by the correlation. Our fatigue life predictions are reasonably accurate.

Results of all of the unpressurized panel fatigue tests are shown in Fig. 13. Predictions based on measured strains for sine inputs of with and without preloads, and for random input at 1000 μ cm/cm preload, are also shown in the figure. The predicted lives bracket the sine test data for the 0.063 panels. However, they are too long for the thinner panels, which have a more complex wave form response. The random load prediction at 1000 μ cm/cm preload shows good correlation. The panel test results fall in between the two predictions as they should because complex sine response is somewhere between unimodal sine and unimodal random. The complex sine is a little closer to random load condition.

Conclusions

Fluid-structure interaction dynamics significantly lower skin frequencies and increase the strain response due to vibration loads as compared to the dry state. These altered conditions reduce fatigue life compared to the dry state.

Symbol	Panel Thickness (in.)	Static Preload Range (μin./in.)
○	0.063	250 - 270
□	0.040	300 - 770
△	0.032	300 - 380
◇	0.032 Stiffened panel	430
◻△	No failure	
▽	0.032 Random excitation	720

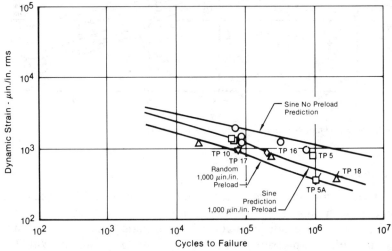

FIG. 13—*Panel fatigue results, without internal pressure.*

Combinations of vibration loads and static preloads, caused by internal presure, further reduce fatigue life. The combination of low-frequency loads and vibration loads also significantly reduces fatigue life from that predicted under vibration loads alone.

These findings suggest that fuel tank design techniques based only on low-frequency spectrum loads, such as from maneuvers, are inadequate. Techniques such as those described herein, which combine maneuver spectrum loads with the fluid-structure induced vibration loads, can be used to develop these spectra. Laboratory tests are required to validate predictions.

References

[1] Ferman, M. A. and Unger, W. H., "Fluid-Structure Interaction Dynamics in Aircraft Fuel Tanks, *Journal of Aircraft*, Vol. 16, No. 12, Dec. 1979.
[2] Ferman, M. A. et al, "Fuel Tank Durability with Fluid-Structure Interaction Dynamics," AFWAL-TR-82-3066, Sept. 1982.
[3] Impellizzeri, L. F., *Effects of Environment and Complex Load History on Fatigue Life, ASTM STP 462*, American Society for Testing and Materials, Philadelphia, 1970, pp. 40–68.

Anthony G. Denyer[1]

Aircraft Structural Maintenance Recommendations Based on Fracture Mechanics Analysis

REFERENCE: Denyer, A. G., **"Aircraft Structural Maintenance Recommendations Based on Fracture Mechanics Analysis,"** *Case Histories Involving Fatigue and Fracture Mechanics, ASTM STP 918,* C. M. Hudson and T. P. Rich, Eds., American Society for Testing and Materials, Philadelphia, 1986, pp. 291–310.

ABSTRACT: The paper discusses the implementation of the Force Management (FM) phase of the Aircraft Structural Integrity Program (ASIP) for a USAF utility transport/trainer aircraft. The analyses, performed in support of the Force Management tasks, use the principles of linear elastic fracture mechanics with emphasis on subcritical flaw growth. The durability and damage tolerance lives, generated with a crack growth computer program, are the basis for structural maintenance recommendations and inspection intervals presented to the USAF in the Force Structural Maintenance Plan (FSMP).

The case study considered herein shows how the T-39 wing, designed to the fatigue requirements of the late 1950s, performs to the structural criteria of the 1980s as defined by the military specifications MIL-STD-1530 and MIL-A-83444.

The initial FM task was the durability and damage tolerance assessment (DADTA) of the structure based on the current service usage as defined by the loads/environment spectrum survey (L/ESS). An outline of the necessary spectrum development is provided, as is the crack growth analyses leading to estimates of the economic life and structural life enhancement recommendations.

The DADTA was followed by the institution of the individual aircraft tracking (IAT) program, the purpose of which is to compute the rate at which the available structural life of each aircraft is being used and to establish inspection intervals to ensure safety. Tracking is accomplished by compiling flight records, collected by means of pilot logs containing mission information for each flight. The paper will show the economic crack growth procedure used, both to ascertain the accumulated damage based on the flight records and to estimate the remaining structural life.

The conclusion of the paper will discuss the maintenance recommendations and the advantages of using fracture mechanics based aircraft tracking in support of force management.

KEY WORDS: fracture mechanics, durability, damage tolerance, spectrum, service loads, structural life, force management, structural integrity

During the early 1970s the United States Air Force (USAF) established MIL-STD-1530 [1] to provide the overall requirements necessary to achieve

---

[1] Senior engineering specialist, Rockwell International, Los Angeles, CA 90009.

structural integrity in USAF aircraft. The specification is designed to ensure that both the durability and damage tolerance of the structure will be an integral part of the design and will be maintained during the service life of the airplane. The tasks to implement the aircraft structural integrity program (ASIP) on the T-39 utility trainer included the preparation of a preliminary durability and damage tolerance assessment, development and verification testing, and the development of the force management program.

The various tasks in the force management program, shown in Fig. 1, provide for the collection and generation of data required to manage force operations in terms of inspections, modifications, and damage assessments. The loads/environment spectra survey (L/ESS) provides the cyclic fatigue load spectrum for the durability and damage tolerance assessment (DADTA) which defines the critical structural locations. The individual aircraft tracking (IAT) program monitors the usage of each aircraft and provides the maintenance recommendations that are transmitted to USAF in the force structural maintenance plan.

History and Background of T-39

The T-39 is an 8200 kg (18 000 lb) gross weight twin jet utility transport/ trainer with passenger capacity of seven. It was initially designed as a four passenger aircraft to a load factor of 4 g for use as a radar profile trainer, transition trainer, and as a VIP transport. Subsequent modifications increased the capacity to seven passengers with a load factor of 3.0. The airplane, designed in the late 1950s went into service in 1961 with the final aircraft of the 149 airplane fleet delivered in 1963. The T-39 preceded the application of MIL-STD-1530 and the advent of fracture mechanics as a

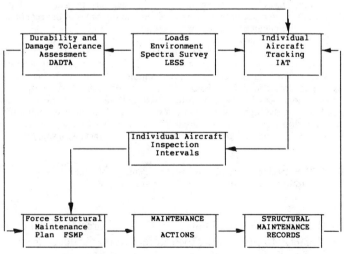

FIG. 1—*Force management program.*

structural analysis tool. Based on fatigue analysis using the design spectrum and supported by a full scale fatigue test, the life of the T-39 airframe was established as 22 500 flight hours.

In 1977, when the lead aircraft had accumulated 18 000 flight hours, it was obvious that a substantial proportion of the force would reach the established fatigue life capability by 1982/1983. The USAF plans at that time were to use the aircraft as a pilot support training vehicle into the 1990s. It was decided to apply the principles of MIL-STD-1530, using fracture mechanics methodology, for the purpose of identifying the capability of the airframe, with modifications if necessary, to meet a life extension to a 45 000-h service life.

Loads/Environment Spectra Survey (L/ESS)

In 1976, 10% of the T-39 force were equipped with a flight loads data recording system (FLDRS). The load magnitudes and frequencies recorded with this system provided the fatigue load spectrum for the durability and damage tolerance analysis of the structure. The FLDRS consisted of an airborne multichannel flight data MXU 553 tape recorder, an associated signal converter/multiplexer unit, and various flight parameter sensors and accelerometers. For each flight, two distinct types of data were recorded. The first type included identifying descriptors such as aircraft serial number, date of flight, mission and base codes, and takeoff fuel and gross weights. The second type of data consisted of the time histories of key flight profile parameters, weight, indicated airspeed, altitude, and dynamic pressure which were used to define the mean flight load levels. Cyclic loads about the mean levels, resulting from pilot induced maneuvers and atmospheric turbulence were provided by aircraft load factor time histories. In addition a ground/air transfer event parameter was used to flag and separate ground taxi events from flight data.

A mission analysis computer program was used to validate the data, calculate and accumulate the key flight profile parameters, and count the load factor exceedance data. The final output of the mission analysis program was a data base containing composite flight profiles, each representing average flight conditions, mission distribution description, and the accumulated load factor exceedance curves for maneuver and gust conditions.

The study of over 15 000 h of flight data resulted in the identification of five composite profiles, shown in Table 1, with a usage distribution as shown in Table 2. Load factor exceedance curves, examples of which are shown in Figs. 2 and 3, were accumulated for the conditions in Table 3.

Rockwell's automated stress spectrum program described in Refs 2 and 3 has the capability of generating load factor or stress spectra representing any location in the primary airframe structure. When used to generate T-39 spectra the input to the computer code consisted of the flight profiles,

TABLE 1—*Composite flight profiles representing T-39 usage.*

Flight Segment	Time, min	Gross Weight kg	(lb)	Mach No.	Altitude m	(ft)	No. of T & G Landings
MISSION 1: HIGH ALTITUDE CROSS COUNTRY FLIGHTS (2.0 H)							
Taxi		7700	(17000)	. . .		. . .	
Climb	25.0	7500	(16500)	0.6	6700	(22000)	
Cruise	60.0	6800	(15000)	0.8	10600	(35000)	
Descent	35.0	6100	(13500)	0.5	5500	(18000)	0
Taxi		5900	(13000)	. . .		. . .	
MISSION 2: MEDIUM ALTITUDE CROSS COUNTRY FLIGHTS (1.2 H)							
Taxi		7700	(17000)	. . .		. . .	
Climb	15.0	7500	(16500)	0.45	4600	(15000)	
Cruise	32.0	6800	(15000)	0.7	7600	(25000)	
Descent	25.0	6100	(13500)	0.45	3600	(12000)	0
Taxi		5900	(13000)	. . .		. . .	
MISSION 3: HIGH ALTITUDE TRAINING FLIGHTS (2.0 H)							
Taxi		7700	(17000)	. . .		. . .	
Climb	25.0	7500	(16500)	0.5	6700	(22000)	
Cruise	62.0	6800	(15000)	0.75	10600	(35000)	
Descent	35.0	6100	(13500)	0.5	5500	(18000)	0
Taxi		5900	(13000)	. . .		. . .	
MISSION 4: LOW ALTITUDE TRAINING FLIGHTS (1.6 H)							
Taxi		7700	(17000)	. . .		. . .	
Climb	15.0	7700	(17500)	0.5	4300	(14000)	
Cruise	16.0	7500	(16500)	0.5	5200	(17000)	
Descent	30.0	7300	(16000)	0.4	2700	(9000)	
Go around	35.0	6800	(15000)	0.3	900	(3000)	2
Taxi		6400	(14000)	. . .		. . .	
MISSION 5: PILOT PROFICIENCY FLIGHTS (1.2 H)							
Taxi		7700	(17000)	. . .		. . .	
Go around	72.0	7300	(16000)	0.3	900	(3000)	5
Taxi		6800	(15000)	. . .		. . .	

the distribution of mission types and the load factor exceedance data. The output was a flight segment-by-flight segment stress spectrum for 100 flights in random sequence correlated with the distribution of mission types shown in Table 2. The relationship between airplane load factor and local stress for each mission segment was derived using industry accepted methodology for static external loads development with appropriate aerodynamic and

TABLE 2—*T-39 Mission mix.*

Mission Title	Missions, %	Hours, %	Flight Length, h
High altitude cross country	43.0	53.0	2.0
Med. altitude cross country	18.0	12.0	1.2
High altitude training	10.0	11.0	2.0
Low altitude training	14.0	13.0	1.6
Pilot proficiency	15.0	11.0	1.2

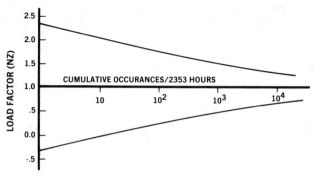

FIG. 2—*Gust load factor data (altitude 0 to 1500 m) based on 2353 h of recorded data.*

stiffness data and the generation of internal loads by finite element modeling (FEM) [4]. A sample wing location stress trace for a 2.0 h cross country flight is given in Fig. 4 and for a 1.4 h training mission in Fig. 5.

Durability and Damage Tolerance Analysis

Candidate critical locations in the primary airframe structure were selected after a survey of the design and fatigue analysis as well as fatigue test data. For the purposes of this paper the wing only will be discussed. Figure 6 presents the stress contour map of the wing lower cover from the NASTRAN internal loads solution for the design limit load condition (4.0 g symmetric maneuver). Analysis of the higher stressed areas of the wing, supported by the fatigue test evidence, showed that the most critical detail is the wing skin at the attachment to the front spar between butt plane 0 and butt plane 15. The stress spectrum to which this part is subjected is shown for two sample flights out of 100 in Figs. 4 and 5.

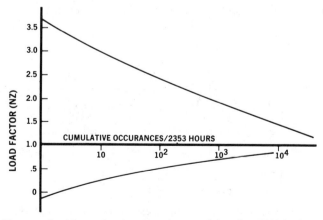

FIG. 3—*Maneuver load factor data (go around segment) based on 2353 h of recorded data.*

TABLE 3—*Conditions of measured load factor data (Nz).*

Title	Flight Hours Represented By Data
Taxi	19440 taxi periods
Gust, 0 to 1500 m (0 to 5000 ft)	2353.0
Gust, 1500 to 3000 m (5000 to 10000 ft)	533.0
Gust, 3000 to 4500 m (10000 to 15000 ft)	1457.0
Gust, 4500 to 7600 m (15000 to 25000 ft)	6196.0
Gust above 7600 m (25000 ft)	4395.0
Climb maneuver (cross country)	2450.0
Climb maneuver (training)	4353.0
Cruise maneuver (cross country)	2977.0
Cruise maneuver (training)	759.0
Descent maneuver (cross country)	1009.0
Descent maneuver (training)	1037.0
Go around maneuver	2353.0

The durability of the structure is the capability to withstand the repeated load spectrum for the service life without becoming functionally impaired or causing uneconomic maintenance problems. Damage tolerance is the ability of the structure to continue safe operation with damage (fatigue cracks) in the primary structure. Durability and damage tolerance lives are defined in Fig. 7 which presents an analytical crack growth curve from an initial flaw, representing the "as manufactured" condition, to the critical crack length.

The as manufactured condition is defined by an equivalent initial flaw size (EIFS) which, when used as the initial flaw in the crack growth analysis, permits correlation with the fatigue life of nominally unflawed structure. Evaluation of the EIFS is based upon the generation of crack growth analysis to match the crack growth records generated from the striation counts of the fracture surfaces from fatigue tests. The equivalent initial flaw is computed by regressing the crack growth analysis to the beginning of the test. This procedure, performed on nine cracks which propagated during T-39 fatigue test program, correlated well with data derived from other USAF

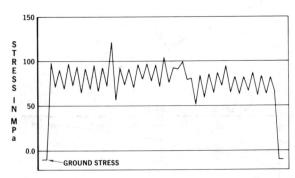

FIG. 4—*Wing stress: typical high altitude cross country flight (2.0 h).*

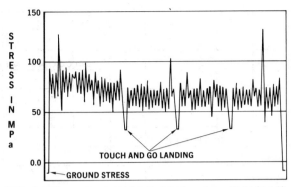

FIG. 5—*Wing stress: typical low altitude training flight (1.4 h).*

sponsored DADTA programs and resulted in the evaluation of EIFS as 0.0635 mm (0.0025 in.).

The T-39 wing skin is an integrally stiffened panel highly sculptured for weight-strength optimization, a design precluding the use of bolted-on reinforcement patches and confining economic repairs to reaming existing bolt holes in padded areas. This restraint results in an economic repair limit of 0.672 mm (0.03 in.) crack size for durability purposes.

The initial flaw size for damage tolerance analysis was established as a 1.27 mm (0.05 in.) radius corner crack, the minimum flaw size which can be reliably detected with current nondestructive inspection (NDI) techniques. The maximum allowable crack size is the crack length at which rapid unstable crack growth occurs when the wing skin is subjected to design limit load.

The crack growth analysis for both durability and damage tolerance was generated using Rockwell's crack growth computer code (CRKGRO) [5]

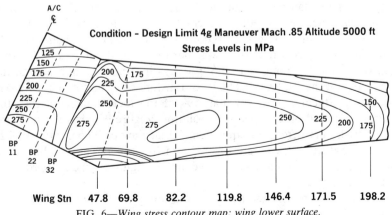

FIG. 6—*Wing stress contour map: wing lower surface.*

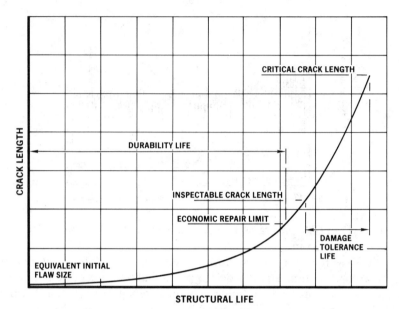

FIG. 7—*Structural life definitions.*

CRKGRO is based on the principles of linear elastic fracture mechanics wherein the stress state at the crack tip can be characterized by a single parameter, the stress-intensity factor (k). Furthermore the subcritical flaw growth can be characterized as a function of the cyclic range of the stress-intensity factor. The CRKGRO computer code integrates the growth rate per cycle (da/dN) which is characterized by the Walker growth rate equation [6] to obtain the predicted crack growth curve (a versus flights) from EIFS to the critical crack length.

The structural detail of the critical wing location is shown in Fig. 8 as are the appropriate materials properties. The wing skin being a monolithic integrally stiffened panel requires a crack growth life of two service lives to achieve the damage tolerance required by MIL-STD-1530 and MIL-A-83444. The crack growth analysis was performed with an assumed corner crack, with an $a/2c$ geometry of 0.5, propagating from the bolt hole to the leading edge of the wing skin. The load transfer at any bolt in the wing skin to spar cap attachment is less than 5% of the local skin load, thus the analysis was based on an open hole solution. Figure 9 shows the resulting crack growth curve from an equivalent initial flaw of 0.0635 mm (0.0025 in.) to the critical crack length of 4.57 mm (0.18 in.). The durability and damage tolerance lives are designated in the figure at their appropriate crack size intervals.

The analysis demonstrated adequate durability for the service life extension of 45 000 h based on the force average usage defined by the L/ESS

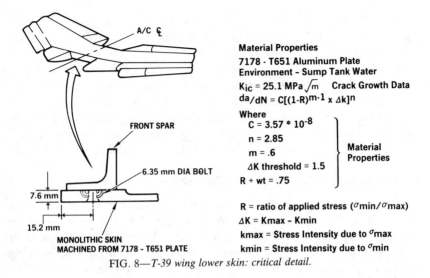

Material Properties
7178 - T651 Aluminum Plate
Environment - Sump Tank Water
K_{ic} = 25.1 MPa $\sqrt{m}$ **Crack Growth Data**
$da/dN = C[(1-R)^{m-1} \times \Delta k]^n$
Where
$C = 3.57 * 10^{-8}$
$n = 2.85$
$m = .6$ **Material Properties**
ΔK threshold = 1.5
R + wt = .75

R = ratio of applied stress ($\sigma min/\sigma max$)
$\Delta K = Kmax - Kmin$
kmax = Stress Intensity due to σmax
kmin = Stress Intensity due to σmin

FIG. 8—*T-39 wing lower skin: critical detail.*

program. The longer durability service life is a result of a benign fatigue spectrum, dominated by the VIP transport mission profile rather than the more severe radar training mission profile anticipated during the T-39 design-fatigue test phase. However the wing is deficient in damage tolerance when measured against the criterion of MIL-A-83444 [7] which requires two life-times of crack growth in order to assure adequate safety with minimum structural inspections. The primary cause of damage tolerance defi-

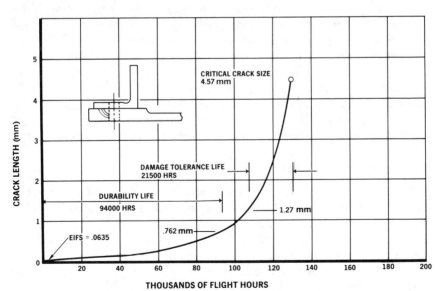

FIG. 9—*T-39 wing DADTA crack growth analysis.*

ciency is the single load-path monolithic wing box design coupled with the choice of material (7178-T6) which has K_{Ic} equal to 25.1 MPa $\sqrt{m}$ (23 ksi $\sqrt{in}$.) resulting in a very small critical crack length.

In order to meet the intents and purposes of MIL-STD-1530 two options were available for the T-39 wing: wing skin replacement with improved fracture mechanics material properties, or continual periodic inspections to maintain safety with the existing damage tolerance capability. The latter supported by an individual aircraft tracking (IAT) program was selected by USAF as the approach to allow continued operation of the T-39 Force.

Individual Aircraft Tracking (IAT) Program

The purpose of the IAT is to compute the rate at which the available structural life of each aircraft is being consumed and to estimate the remaining structural life based on future usage being extrapolations of the recent past usage. The crack growth analyses, upon which the life estimates are based, are performed for both crack size regions as defined by the durability and damage tolerance criteria.

The durability life can be used for projecting the useful service life of the airframe, assuming that the necessary actions are taken to assure adequate damage tolerance safety throughout the projected durability lifetime. The remaining durability lives can be used by USAF for economic comparisons, trading off maintenance of the T-39 force in operation against the procurement of new aircraft. Finally, when it becomes necessary for USAF to phase-out the T-39 force, the IAT remaining durability lives will be a valuable statistic for establishing a retirement phase-out schedule by aircraft serial number. The same data will provide a convenient selection tool in the event that stored aircraft are returned to active flight status.

Damage tolerance tracking is necessary to assure the structural safety of each component on each aircraft. The IAT damage tolerance lives will establish inspection intervals for each critical component on each aircraft. In the event that cracks are detected during the resulting inspections, using prescribed nondestructive inspection (NDI) techniques, preplanned repairs can be implemented, thus maintaining a high level of force readiness.

Tracking Computer Program

The IAT computer code provides an analytical estimate of accumulated damage at critical structural locations on four major components including the wing lower skin at the front spar. The structural damage is developed in terms of ever increasing crack lengths calculated within the computer code and based on the usage of each individual aircraft.

The aircraft usage is available to the tracking program in terms of mission parameters recorded on AFTO FORM 166 flight log reporting forms (Fig. 10). The forms are completed by the aircrew at the conclusion of each flight,

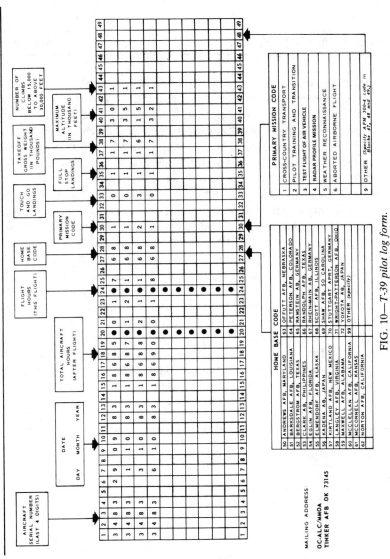

FIG. 10—T-39 pilot log form.

TABLE 4—*Pilot log mission/analysis mission correlation.*

Pilot Log Mission	Analysis Mission Type
Cross country transport	cross country profiles
Pilot training and transition	training profiles
Test flights	training profiles
Radar profile mission	training profiles
Weather reconnaissance	cross country profiles
Aborted airborne flight	cross country profiles

sent to Tinker Air Force Base at Oklahoma City where they are compiled, copied to computer tapes, and mailed to the contractor. An algorithm relates the flight log data (mission description, flight length, take off gross weight, maximum flight altitude, number of full stops, and touch and go landings) to an assumed flight profile. The first step is to assign a primary mission type to each pilot log mission description per Table 4. The necessary flight profiles were selected after a study of the L/ESS data. Table 5 provides the available flight profiles for each primary mission type. The IAT program contains a library of precalculated crack growth curves each representing a specific flight profile in Table 5 and selected by mission description, maximum flight altitude, and take off gross weight. The curves, generated with Rockwell's crack growth program CRKGRO [5] for all tracked structural locations, are stored as crack length (a) versus crack growth rate ($da/dFlt$).

The crack growth calculations for the detail tracked locations on an individual aircraft are obtained by performing a graphical integration procedure [8] using the stored crack growth curves as outlined next.

The current crack lengths (a_c) for both durability and damage tolerance are extracted from the aircraft records library for the selected aircraft and component. Based on the pilot log data (the mission type and maximum flight altitude) a series of crack growth rate curves are selected covering

TABLE 5—*Available flight profiles for which crack growth curves are generated.*

Mission Type	Take Off Gross Weight kg	(lb)	Max Flight Altitude m	(ft)	Flight Length, H 3.0	2.0	1.5	1.0	0.5	0.1
Cross country	7700	(17000)	10600	(35000)	x	x		x		x
			7600	(25000)	x	x		x		x
			4500	(15000)	x	x		x		x
			2400	(8000)	x	x		x		x
Training			10600	(35000)	x		x		x	x
			7600	(25000)	x		x		x	x
			4500	(15000)	x		x		x	x
	7700	(17000)	2400	(8000)	x		x		x	x
Touch and go	6100	(13500)	600	(2000)						x

the range of flight lengths. Knowing the current crack length, values of da/dF are extracted. The value of da/dF and thus da for the particular flight is obtained by interpolation using the actual flight length supplied on the pilot log. The crack length at the end of the flight ($a_c + da$) is returned to the library. The described graphical integration procedure replaces, without significant loss of accuracy, the expensive cycle-cycle integration computer codes which would be prohibitive for the number of tracked locations and aircraft in the force (that is, 6IAT locations and 149 aircraft at 360 flights per year).

In addition to updating the analytical crack lengths, the program estimates the remaining structural life by applying the same crack growth procedure to an assumed aircraft usage model and growing the cracks to the allowable lengths defined as the end of the life. The usage model is assumed to be identical to that flown for that individual aircraft over the previous two years. As a result, the remaining lives reflect the most recent aircraft usage. When an aircraft has not flown a sufficient number of flights (100) during the previous two years to form a stable statistical usage, the remaining life is based on the aircraft's total usage history records.

Backup Data for Accumulated Flight Hours

By its nature the IAT program must account for all flight hours of each aircraft. Like all data collection systems, the pilot logs do not return 100% of the flights flown nor are the returned logs 100% correct. Logs are obviously missing when a change in accumulated flight hours are unaccounted for by individual flight records. After sorting the cards by aircraft serial number, date, and accumulated flight hours, it is frequently possible to correct entries for aircraft serial number, date, base code, and accumulated flight hours for logs which failed the program validity checks.

A backup system to the pilot logs is a force-wide USAF reporting system known as AFR 65-110/G033B [9]. This sytem provides a monthly summary of accumulated flight hours as well as a distribution of flights by mission type. The number of sorties, number of touch and go landings, and average flight length is also given for each mission type.

The accumulated flight hours for each airplane is established after considering both the G033B and the pilot card evidence; thus, it is known how many flight hours are unaccounted for by pilot cards. If for a complete calendar month the pilot cards are unavailable, the G033B data is substituted using Table 6 to convert the G033B mission descriptions to the T-39 IAT missions. Dummy pilot cards are created using historic statistics to complete the pilot card information. Lost flight hours for less than a month are accounted for by creating dummy pilot logs based on historic statistics to fill exactly the flight hour gaps.

TABLE 6—*AFR65-110/GO33B/T39 IAT mission type correlation.*

AFR65-110/G033B System		T-39 IAT Analysis Mission
Code	Description	
S1	Administrative personnel transportation	cross country mission
S2	Personnel transportation	
S3	Material transportation	
S4	Logistics transportation	
S5	Special transportation missions	
O1	Special transportation missions	
A1	Special transportation missions	
S6	Radar check flights	training mission
S7	Aircrew qualification	
S8	Support training	
T1	Student training	
T2	Combat crew training	
T3	Operation training	
O4	Test engineering test of aircraft	
O5	Direct test support	
O6	Indirect test support	
O7	Special mission	
O8	Maintenance tests	

Baseline Tracking Report

The T-39 force had been in service for 20 years before the implementation of the IAT program. Consequently, it was necessary to establish the usage and analytical damage condition of each aircraft at that time. The only available G033B records, accumulated to provide that baseline, were for two and one half years. It was thus assumed that the usage distribution and flight lengths for the last two and one half years would be representative of the entire lifetime and that the most recent G033B data provided the accumulated flight hours for each aircraft as of that date.

A baseline tracking report was issued showing the status of each tracked structural location on each aircraft at the implementation of the IAT program. The damage was assessed by creating dummy pilot cards to represent all past defined usage and using the tracking program with the preproduced crack growth curves. The initial flaw sizes prior to first flights of each aircraft were those defined in the DADTA analysis.

Periodic Tracking Report

Tracking reports, issued at six month intervals, display the usage statistics for each aircraft, the damage accumulation for each structural location and significant maintenance actions affecting the durability and damage tolerance capability of the component (Table 7). Two types of summary are used: the force wide variation of crack growth is shown in Fig. 11, and the most severely damaged aircraft are highlighted as shown in Table 8.

TABLE 7—*Sample individual aircraft tracking log sheet period June 1983 to December 1983.*

Mission/design series	CT-T39A
Aircraft serial number	603483
Command	MAC
Home base	Scott AFB
Component	Wing

Key maintenance action—wing skin inspection at 18450 h (Apr 1983)

Usage Statistics			Mission Type Distribution, %		
Data	Period	Accum.	Mission Type	Period	Accum.
Total mission	198	12624	cross country	92.9	94.6
Total flight hours	235	18770	training	7.1	5.1
Total landings	278	17033	test flights	...	.3
Average take off gross weight	18.1	17.2	radar profile	...	...
Average flight length	1.2	1.5	weather reconnaissance	...	...

DAMAGE DATA

Initial durability crack length	0.0635 mm (0.025 in.)
Current durability crack length	0.0668 (0.00263 in.)
Economic repair limit	0.762 (0.03 in.)
Remaining useful life	>100000 Hrs

Initial damage tolerance crack length	1.27 mm (0.05 in.)
Current damage tolerance crack length	1.292 mm (.05089 in.)[a]
Critical crack length	4.57 mm (0.18 in.)
Damage tolerance life	20906 h

[a] Damage tolerance crack size was reset to 1.27 mm (0.05 in.) after the inspection at 18450 h declared the part crack free.

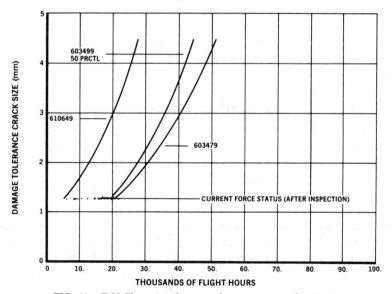

FIG. 11—*T-39 Fleet-wing damage tolerance status and projection.*

TABLE 8—*Ten most damaged aircraft by durability crack size.*

Aircraft Serial Number	Accumulated Flight Hours	Durability Crack Size		Remaining Useful Life, h
		mm	(in.)	
603508	21721	0.0965	(0.00380)	>100000
610650	22390	0.0947	(0.00373)	>100000
610634	22466	0.0922	(0.00363)	>100000
610681	21515	0.0909	(0.00358)	>100000
624455	20386	0.0901	(0.00355)	>100000
610645	21076	0.0891	(0.00351)	>100000
610637	21514	0.0886	(0.00349)	>100000
624487	20413	0.0886	(0.00349)	>100000
624464	20824	0.0840	(0.00331)	>100000
610668	21112	0.0833	(0.00328)	>100000

Implementation of Maintenance Recommendations

Due to the lack of criteria defined damage tolerance, coupled with the aircraft having been subjected to its design service life, a one time force wide inspection of the wing lower cover was recommended. This plan was implemented in the form of a time compliance technical order (TCTO) requiring fastener removal and eddy current inspection of all bolt holes in the wing lower cover in the zones shown in Fig. 12. All of the 137 aircraft in service were inspected between June 1982 and June 1983.

One aircraft was reported as having a crack in the critical area, that being in the skin at the front spar attachment 1 in. from the centerline of the aircraft. All evidence of the crack was removed with a reaming operation, enlarging the hole diameter from 9.52 mm (0.375 in.) to 11.09 mm (0.437 in.), demonstrating the capability of in-service NDI methods of detecting cracks in the order of 1.27 mm (0.05 in.), at which size economic repair is possible.

Reset of Crack Sizes and Inspection Intervals

As the inspection results were received for updating the IAT program, the damage library was modified to reflect the status of the fleet. Specifically the damage tolerance crack sizes were reset to the initial flaw size of 1.27 mm (0.05 in.). Continued accumulation of crack growth from the reset initial size then provides current estimates of the damage tolerance lives which are used as the basis for recalculation of future inspection intervals. Future inspection intervals are transmitted to the USAF in the force structural maintenance plan (FSMP) as recommended "inspect before" dates. Since a bolt hole crack in the wing skin could become critical without becoming a through crack and consequently without causing a "tell tale" fuel leak, the inspection interval was set as the damage tolerance life divided by four. A sample of the applicable page from the FSMP is shown in Table 9.

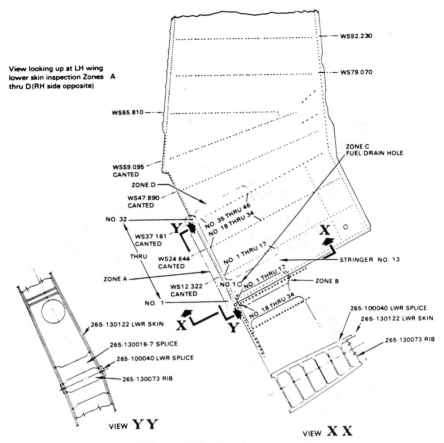

FIG. 12—*T-39 wing inspection zones.*

Maintenance Reporting System

In order for the IAT program to correctly reflect the status of the force, a maintenance reporting system has been instituted. A sample maintenance report is shown in Fig. 13 with a partial "affected item" list. The prime reporting requirements are maintenance actions on the affected parts or components and the accumulated flight hours at the time of the maintenance. Typical of the maintenance actions affecting the structural life of the critical parts is the interchange of components such as wings, from one aircraft to another. The IAT program, using the maintenance data, assigns the analytically accumulated damage in the component to the tail numbered aircraft receiving the interchanged part. Replacement of critical parts, such as the wing lower skin, with an unused part will require the tracking program damage library to be updated to reset both the durability and damage tolerance crack sizes to appropriate initial values. If the replacement parts

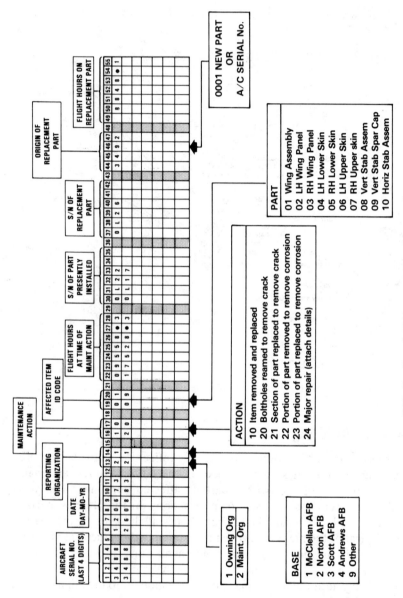

FIG. 13—Sample T-39 maintenance reporting form.

TABLE 9—*Inspection calculation method for FSMP.*

Wing Skin Inspection Schedule

Aircraft Serial Number 603483

Damage tolerance life (IAT Report Jan. 1982 to June 1982)   20111 h
Damage tolerance life (IAT Report July 1982 to Dec. 1982)   19603 h
Damage tolerance life (IAT Report Jan. 1983 to June 1983)   19884 h
Damage tolerance life (IAT Report July 1983 to Dec. 1983)   20906 h
Average damage tolerance life   20126 h
Inspection interval is 20126/4 = 5023 h

Aircraft usage (Jan. 1982 to June 1982)   393 h
Aircraft usage (July 1982 to Dec. 1982)   380 h
Aircraft usage (Jan. 1983 to June 1983)   460 h
Aircraft usage (July 1983 to Dec. 1983)   235 h

Assume the yearly usage is twice the maximum usage for any six months in the last two years
Yearly usage is 920 h
Inspection period is 5023/920 = 5 years
Based on last inspection being accomplished in April 1983, the next inspection must be
implemented before April 1987

are of a different geometry or different material the crack growth curve library must be modified with appropriate precalculated crack growth curves.

Conclusion

The introduction of fracture mechanics based IAT to the T-39 maintenance procedures results in two improvements. A more systematic and coherent quantitative procedure has been established to compute the inspection intervals necessary to assure that potential cracks in critical locations do not propagate to critical length constituting a safety problem. A guide is provided to the sequence of removal of aircraft from service in the event of the T-39 being phased-out of the USAF inventory.

References

[1] Military Standard, "Aircraft Structural Integrity Program, Airplane Requirements," Report MIL-STD-1530A, Department of the Air Force, Sept. 1972.
[2] Denyer, A. G. and Maynard, D. P., "Generation of Fatigue Load Spectra from Recorded Flight Data, Materials, Experimentation, and Design in Fatigue," *Proceedings Fatigue 81,* Society of Environmental Engineers, Ed., U.K., March 1981.
[3] Denyer, A. G. in *Service Loads Monitoring, Simulation, and Analysis, ASTM STP 671,* P. R. Abelkis and J. M. Potter, Eds., American Society for Testing and Materials, Philadelphia, 1979, pp. 158–175.
[4] "NASA Structural Analysis Theory Manual," NASA Structural Analysis (NASTRAN) Program, NASA Report SP-221/03, National Aeronautics and Space Administration, Washington, DC, March 1976.
[5] Chang, J. B., Szamossi, M., and Liu, K.-W., "A Users Manual for a Detailed Level Fatigue Crack Growth Analysis Computer Code—The CRKGRO Program," AFWAL-TR-81-3093, Air Force Wright Aeronautical Laboratory, WPAFB, OH, Nov. 1981.
[6] Chang, J. B., "Improved Methods for Predicting Spectrum Loading Effects," NA-78-491-5, Rockwell International, North American Aircraft Division, Los Angeles, CA.

[7] Military Standard, "Airplane Damage Tolerance Requirements," Report MIL-A-83444 (USAF), Department of the Air Force, July 1974.

[8] Denyer, A. G. in *Fracture Mechanics Thirteenth Conference, ASTM STP 743,* Richard Roberts, Ed., American Society for Testing and Materials, Philadelphia, 1981, pp. 288–302.

[9] Air Force Regulation, "HQ USAF Requirements for AFR 65-110 Utilization Data Reporting System," AFR 65-110/G033B.

Thomas P. Rich[1] and James G. Orbison[1]

Analysis of Two Metal-Forming Die Failures

REFERENCE: Rich, T. P. and Orbison, J. G., **"Analysis of Two Metal-Forming Die Failures,"** *Case Histories Involving Fatigue and Fracture Mechanics, ASTM STP 918*, C. M. Hudson and T. P. Rich, Eds., American Society for Testing and Materials, Philadelphia, 1986, pp. 311–335.

ABSTRACT: This case study presents the application of fracture mechanics to analyze the failures of two dies. One is the failure of a die used to swage fittings onto wire ropes. This failure is traced to very high stresses in the die cavity and relatively low fracture toughness; fatigue plays no role. In the second die, the failure is caused by the growth of a crack from the die cavity and is solely a fatigue problem. This die is used to mold gears in a powder metallurgy process.

KEY WORDS: dies, failure analysis, fatigue, finite element analysis, fracture, metal-forming, powder metallurgy, swaging, tool steels, wire rope

Two industries in northern Pennsylvania make extensive use of dies as part of their manufacturing process. They are the wire rope and the powder metallurgy industries. The production of these regional industries is of national significance with the Pennsylvania wire rope manufacturing facilities producing about 40% of the wire rope made in North America. Powder metallurgy manufacturing is also a major industry in northern Pennsylvania.

Within the wire rope industry, dies are used in various stages of production from the drawing of steel wire (from which the ropes are wound) to the attachment of fittings which are used to connect the ropes to other structural components. Examples of the application of wire rope include uses in elevators, mining operations, power shovels, naval drag and tow lines, bridges, and ski lifts. In addition, a growing number of wire rope applications is developing in the off-shore oil industry.

The particular die of interest in this first case study is used in the wire rope industry to swage tubular aluminum alloy fittings onto the ends of rope. As shown in Fig. 1, the swaging process consists of inserting the end of a wire rope into the open end of a fitting. In this illustration the fitting

[1] Associate professor, Mechanical Engineering Department, and assistant professor, Civil Engineering Department, respectively, Bucknell University, Lewisburg, PA 17837.

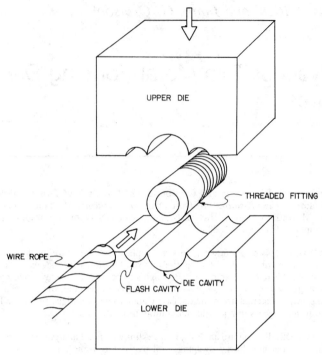

FIG. 1—*Wire rope swaging die.*

is a threaded connector. Note that the outer diameter of the fitting is greater than the diameter of the die cavity. After the rope has been inserted, the upper and lower dies are pressed together under a large force. This results in plastic flow of the fitting within the die cavity to tightly surround the rope and form the attachment. Excess material from the fitting is squeezed into the flash cavities and sheared-off by the dies. In many swaging operations the total deformation of the fitting is carried out in several successive presses. The piece is rotated within the dies between presses and the load increased until the attachment is made. Upon completion of the swaging, the dies are fully separated and the wire rope/fitting assembly is removed.

Dies are also critical components in the powder metallurgy industry. They form the mold for producing powder metallurgy parts ranging from standard machine parts, like gears, to custom parts for special machine applications. Through the use of precision-cut dies, parts can be quickly, accurately, and inexpensively produced with powder metallurgy methods that would be costly if fabricated with conventional machining techniques.

This second case example will focus upon a die insert used in the production of bevel gears. Figure 2 shows an illustration of this die. The lower die insert contains a cavity that has been cut by an electric discharge machining process to the exact shape and dimensions of the desired powder-molded gear. The lower die is fixed within the lower platen of a press. At the start of the process, a metal powder is injected through a nozzle into the cavity. Upon filling of the cavity, an upper platen is lowered into place and a large force is applied to press the powdered metal and compact it into the mold cavity. The nozzle is designed to avoid interference with the pressing stage of the process. When compaction is complete, the upper platen rises, and the molded gear is ejected from the cavity by a piston which rises through the center of the lower die cavity. The entire cycle takes place in a matter of seconds, and results in a rapid, efficient method to produce complex machine parts. After molding, the gear is then put through a sintering process where other metals are diffused into the gear, and heat treating is used to produce the desired strength levels.

In either the wire rope swaging application or the powder metallurgy molding process, breakage of a die results in production shutdown and its associated costs. The remainder of this paper presents the failure of both types of dies, and fracture mechanics techniques are used to explain the causes and suggest general guidelines for prevention of future failures.

A Fracture Mechanics Approach to Failure Analysis

Several aspects of the operation and nature of these dies suggests a potential for failure resulting from the growth of cracks in the components.

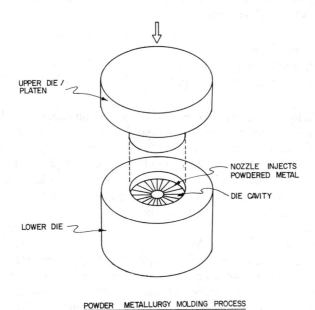

POWDER METALLURGY MOLDING PROCESS

FIG. 2—*Powder metallurgy die.*

First, during normal operations, the dies are subjected to large, repeated loads. Second, the cavities in both dies create regions of high stress concentration where cracks can initiate and grow resulting in catastrophic failure. Third, in order to produce parts of well-controlled shape and dimensions, the dies are made of high-strength, hardened materials. As such they may be prone to brittle fracture.

It was found that several of these factors contributed to the failure of both dies. A complete description is given in the following sections. In order to provide a general format for the fracture analysis of these dies, the following information can be identified as essential to this as well as any structural failure analysis:

1. Evidence
    (a) Geometry of component
    (b) Magnitude and spectrum of loading
    (c) Environmental factors
    (d) Examination of failure surface
2. Material properties for fracture and fatigue
3. Failure hypothesis
4. Stress analysis
5. Initiation, shape, and size of flaws

In addition, it is helpful to think of the ultimate failure in terms of three main stages of crack growth:

1. Crack initiation
2. Slow, stable crack propagation
3. Rapid, catastrophic crack propagation

In the following analyses, attention is focused on seeking evidence and analytical information to identify and quantify the roles of each crack growth stage in the ultimate die failures. In the case of the swaging die it became apparent that the failure was dominated by stage (3) rapid fracture with no evidence of slow stable fatigue crack growth. On the other hand, the powder metallurgy die showed evidence of considerable slow stable crack growth through fatigue for the majority of the service life. Each of these cases represent an extreme situation where one crack growth stage totally dominates the component's life. As seen in the following sections, the macroscopic evidence on the failure surfaces plus the die performance as reported by the operators were sufficient to identify the principal crack growth stage for a fracture mechanics analysis. It was concluded that in both these cases a detailed microscopic metallographic analysis would provide little additional information of practical importance.

Failure Analysis of a Wire Rope Swaging Die

Evidence

The wire rope swaging die investigated in this failure analysis was used to swage tubular 6061-T1 aluminum alloy fittings onto steel wire rope. The die consists of identical upper and lower halves, each with a length along the longitudinal axis of the die cavity of 168 mm (6.63 in.), a maximum height of 110 mm (4.34 in.) and a width of 155 mm (6.12 in.). The die cavity in each half is semi-cylindrical in shape with a diameter of 44 mm (1.74 in.). The photograph in Fig. 3 shows the remains of the lower half of the die, which fractured into two pieces along the longitudinal axis of the cavity. The cavity is seen in the upper right quadrant with the smaller flash cavity next to it.

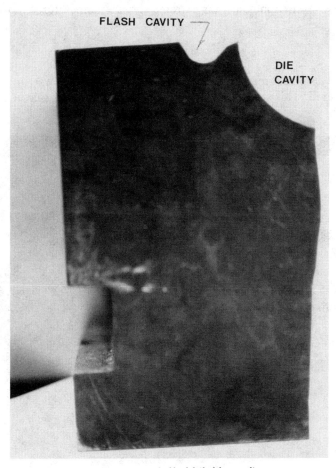

FIG. 3—*One half of failed lower die.*

The die was installed in a press with a 4.45 MN (500 ton) capacity. The die failed upon first application of the load following initial installation in the press. Discussions with the die manufacturer familiar with this particular installation revealed that the die was probably loaded to near-capacity of the press, and was being used to swage a tubular aluminum alloy fitting onto a wire rope at the time of failure. Figure 4 shows a side elevation of the failure surface, with the die cavity appearing at the top of the figure. The failure surface is grainy with no evidence of significant plastic deformation. The brittle fracture started at the root of the die cavity approximately 51 mm (2 in.) from the left edge of the die as seen in Fig. 4. From there, the fracture progressed quickly through the die body. Figure 5 is a closeup of the failure origin; there is no evidence that a crack visible to the human eye existed prior to loading. Further, examination of the die and discussions with the die manufacturer revealed that the die was not exposed to a corrosive environment, nor extreme service temperatures, during its life.

Material

The die was made from a standard AISI Type S2 tool steel, water-quenched and tempered. This steel has a carbon content of 0.50%, a silicon content

FIG. 4—*Lower die failure surface.*

FIG. 5—*Failure initiation site in lower die.*

of 1.00%, a molybdenum content of 0.50%, and a manganese content of 0.40%; all percentages are by weight. Hardening due to quenching will typically not occur at depths greater than 32 mm (1¼ in.) from the surface with Type S2 steel. This results in components the size of the swaging die having a hard, very high strength outer case with a relatively soft interior.

This die was quenched and then tempered back to a minimum surface hardness of Rockwell 55C. As determined from the technical data supplied by the steel producer, this results in a 0.2% offset yield strength of approximately 2000 MN/m² (290 ksi) and a 7% elongation [in a 51 mm (2 in.) gage length] for the material within 6 mm (¼ in.) of the surface. The information from the producer also indicated that material from the interior of the die should exhibit a Rockwell hardness of approximately 25 to 28C, with a yield strength of 725 to 830 MN/m² (105 to 120 ksi). Two tension specimens were machined from the interior of the failed die and tested. Both specimens exhibited an elastic modulus of 203.4 GN/m² (29 500 ksi), with yield strengths of 693 and 786 MN/m² (100.5 and 114 ksi), thus supporting the production specifications.

The plane strain fracture toughness (K_{Ic}) of tool steels is known to be a function of the hardness and yield strength of the material and is thus sensitive to quenching depth characteristics. For example, AISI Type M7 steel will exhibit a typical K_{Ic} value of 25.3 MN/m³ᐟ² (23 ksi $\sqrt{\text{in.}}$) at a Rockwell hardness of 52C, and an average K_{Ic} value of 11 MN/m³ᐟ² (10 ksi $\sqrt{\text{in.}}$) at a Rockwell hardness of 69C [1]. The plane strain fracture toughness of Type S2 steel has not been found in the literature. Therefore, a test program to determine the toughness of this material at various hardness levels was initiated, and is continuing at this time. However, based upon K_{Ic} values for other types of tool steel, it is estimated that the surface material of the die has a fracture toughness in the range of 11 to 33 MN/m$^{3/2}$ (10 ksi $\sqrt{in.}$ to 30 ksi $\sqrt{in.}$).

Failure Hypothesis

The load history and physical examination of the failed die indicated the occurrence of a brittle failure with no significant plastic deformation during fracture. Both the load history and examination eliminated the possibility of a fatigue-type failure characterized by slow, controlled crack growth. The failure surface indicated that fracture started from a very small defect at the root of the die cavity. This defect, when the die was fully loaded, underwent catastrophic growth leading to immediate die fracture. Several finite element analyses were performed to quantify the stress state at the root of the die cavity. Of particular interest was the tensile stress (tangent to the cavity surface) at the root. With the analytical results, an estimate of the required critical flaw size was determined.

Stress Analysis

Several plane strain linear elastic finite element analyses were performed on various models of the lower die. The ANSYS finite element program was used, and all element meshes were generated using four-node quadrilateral 2D elements. In each case, as the load and model are symmetric about a vertical plane through the center of the die cavity, only one half of the lower die was modeled.

Five different models were developed for analysis. In each, the horizontal displacements along the plane of symmetry were restrained, and only the vertical displacements at the base of the model were restrained. Additional analyses performed with both the vertical and horizontal nodal displacements restrained at the model base produced essentially identical results.

The five analysis models are shown in Fig. 6. In the first model, an (elastic) aluminum alloy fitting was assumed to have filled the cavity, and a uniform pressure applied to the top of the fitting. The flash cavity was not modeled in this analysis. In the second model, a uniform pressure was applied to the inside of the empty die cavity; the flash cavity was again not modeled. In the third model, a uniform pressure was applied to the cavity, and the flash cavity was incorporated into the model. In the fourth model, a single inclined concentrated load was applied to the top edge of the die cavity to model the early contact and load transfer between the aluminum fitting and the

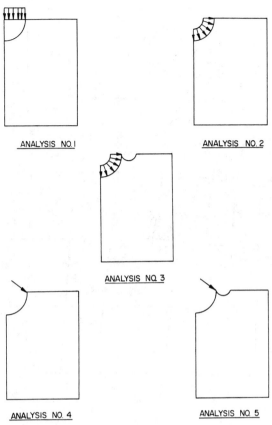

ANALYSIS NO. 1

ANALYSIS NO. 2

ANALYSIS NO. 3

ANALYSIS NO. 4

ANALYSIS NO. 5

FIG. 6—*Finite element analysis models.*

die. The flash cavity was not modeled in this analysis. The fifth model employed the same loading as the fourth model, but the flash cavity was included in the model. The element mesh, geometry, and boundary conditions used in the third analysis are shown in Fig. 7.

The results of the five analyses are shown in Fig. 8. The total load applied to the die by the press is plotted versus the tangential tensile stress in the die at the root of the die cavity. As seen in the figure, the five analyses produced differing estimates of the tangential tensile stress at the cavity root under a given load level. The first analysis employed a model with the cavity filled with an elastic aluminum material. As this material undergoes large plastic deformations in the swaging operation, it is considered to act more nearly as an incompressible fluid rather than an elastic material. Thus, of the first two models, the second (that with a uniform pressure along the die cavity surface) is considered to be the more accurate representation of the actual behavior. The third model employed the same pressure loading

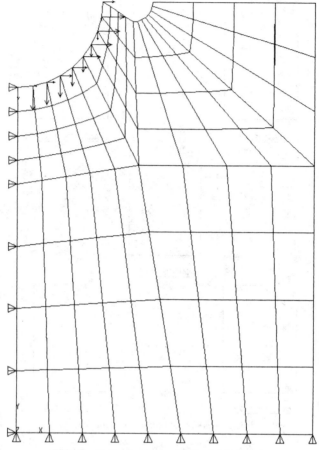

FIG. 7—*Finite element model for third analysis.*

used in the second model, but is a more accurate representation of the die geometry as the smaller flash cavity is modeled as well. The fourth and fifth analyses are considered to accurately model the loading, but only at low levels of press load. This model essentially reflects only the initial and low-load contact pressure between the fitting and the die. As the load is increased to near-working levels, the aluminum fitting is deformed fully into the die cavity. Once this occurs, the point loading used in this low-load model is clearly inaccurate. Thus, of the five analyses and models employed, the third model is considered to be the most representative of the die behavior in service. This model incorporates an accurate geometry with a reasonable model of the load transferred from the fitting to the die at working load levels. The results of this third model were used for the remainder of the

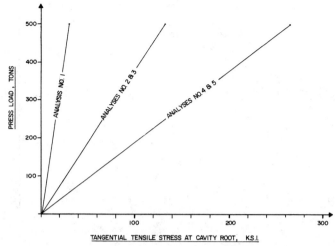

FIG. 8—*Finite element analysis results (1 ton = 8.90 kN, 1 ksi = 6.895 kN/m²).*

failure analysis. Note that the cavity root tensile stress predicted by Analysis No. 3 is essentially identical to that predicted by Analysis No. 2. Further, the tensile stress from the fourth analysis is nearly identical to that from the fifth analysis. Clearly, the presence or absence of the flash cavity in the models had little effect on the stresses at the root of the die cavity. The tangential tensile stresses at the die cavity root predicted by each analysis under a 4.45 MN (500 ton) press load are tabulated below.

A plot of the x-direction normal stress contours is shown in Fig. 9. The maximum stress (MX) is shown in Table 1; the minimum stress (MN) is -1020 MN/m² (-147.8 ksi). Note that the MX stress is tangent to the root of the die cavity. A plot of the y-direction stress contours is shown in Fig. 10. The maximum compressive value for this stress is -495 MN/m² (-71.8 ksi) which occurs at the die cavity root. The maximum tensile value for this stress is 443 MN/m² (64.3 ksi). In addition, another analysis of the third model using four times the number of elements in the same pattern produced a tensile stress at the die cavity root of 915 MN/m² (132.7 ksi), which is within 0.4% of that shown in the previous table.

The third finite element analysis predicted a von Mises yield equivalent stress of 1229 MN/m² (178.2 ksi) at the root of the die cavity. This is less than the surface material's yield strength as indicated in the producer's technical data. A plot of the equivalent stress contours is shown in Fig. 11. The minimum equivalent stress is 3.5 MN/m² (0.51 ksi). Note that in Figs. 9 through 11, the ANSYS program indicates the maximum stress by the label "MX," and the minimum stress by the label "MN."

FIG. 9—*Normal stress distribution in the X-direction.*

Determination of Critical Flaw Size

The critical flaw size for die failure was determined using a tangential tensile stress of 912 MN/m² (132.2 ksi) at the root of the die cavity predicted from the third finite element model under a press load of 4.45 MN (500 tons). The critical flaw size, or depth, a_c, was determined from the following

TABLE 1—*Tangential tensile stress at cavity root, under a 4.45 MN (500 ton) press load.*

Analysis No. 1	203 MN/m²	(29.5 ksi)
Analysis No. 2	912 MN/m²	(132.2 ksi)
Analysis No. 3	912 MN/m²	(132.2 ksi)
Analysis No. 4	1840 MN/m²	(266.9 ksi)
Analysis No. 5	1822 MN/m²	(264.3 ksi)

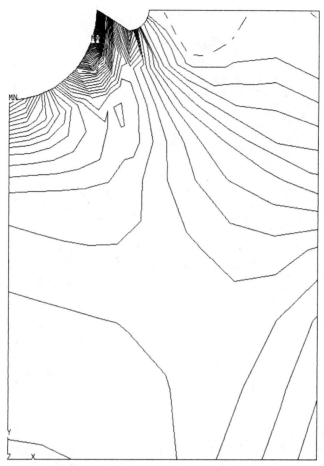

FIG. 10—*Normal stress distribution in the Y-direction.*

expression relating the material fracture toughness, K_{Ic}, to the tangential tensile stress, S, and the critical flaw size a_c [2]

$$K_{Ic} = 1.12S \sqrt{\pi a_c/Q} \qquad (1)$$

In this equation, Q is a shape factor based on the assumption of a semi-elliptical crack with a depth a_c and a width (perpendicular to the tensile stress) of $2c$. Values of Q for various ratios of a_c to $2c$ are tabulated in Table 2.

Neglecting the free-surface-correction factor of 1.2, Eq 1 can be rearranged to yield

$$a_c = \frac{Q}{\pi} [K_{Ic}/S]^2 \qquad (2)$$

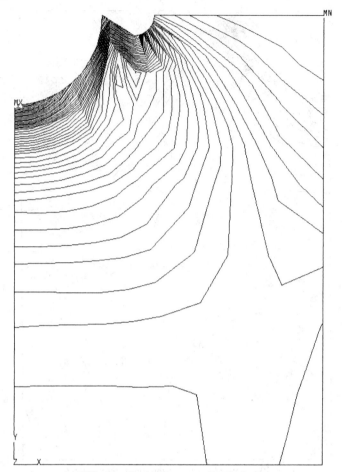

FIG. 11—*Equivalent stress distribution.*

TABLE 2—*Shape factors for elliptical flaws.*

	$a_c/2c$ Ratio					
	0.0	0.1	0.2	0.3	0.4	0.5
Shape Factor Q	1.00	1.09	1.30	1.59	1.96	2.40

NOTE—the $a_c/2c$ ratio of 0.0 represents a through crack of constant depth a_c, while an $a_c/2c$ ratio of 0.5 represents a semi-circular crack of radius a_c.

TABLE 3—*Critical flaw depth, a_c, in.*

K_{Ic}, ksi√in.	$a_c/2c$ Ratio					
	0.0	0.1	0.2	0.3	0.4	0.5
10	0.002	0.002	0.002	0.003	0.004	0.004
20	0.007	0.008	0.009	0.012	0.014	0.017
30	0.016	0.018	0.021	0.026	0.032	0.039
50	0.045	0.050	0.059	0.072	0.089	0.109

NOTE—1 ksi √in. = 1.099 MN/m$^{3/2}$, 1 in. = 25.4 mm.

In Eq 2, a value of S of 912 MN/m^2 (132.2 ksi) was used along with the values of Q tabulated in Table 2. The critical flaw depth, a_c, determined for various $a_c/2c$ ratios is tabulated in Table 3 for different fracture toughness (K_{Ic}) values. The critical flaw depth versus the $a_c/2c$ ratio for the fracture toughness values ranging from 11 to 33 MN/m$^{3/2}$ (10 to 30 ksi √in.) is also shown in Fig. 12.

The critical flaw sizes tabulated are quite small for K_{Ic} estimates of 33 MN/m$^{3/2}$ (30 ksi √in.) and less, and cracks of this depth (and typically smaller width) would not be detected during the manufacturer's visual inspection. As shown in Table 3, even for a material with a fracture toughness of 55 MN/m$^{3/2}$ (50 ksi √in.) which is well beyond the expected range, the critical flaw depth is at best only about 0.1 in. Therefore it is reasonable to conclude that failure was due to the presence of a small flaw in the root of the die cavity, which caused catastrophic failure upon initial loading. It is noted that a tougher material in the interior of the die may not arrest unstable crack propagation that began at a small surface flaw. Rapid crack growth requires a greater material toughness to arrest than that required

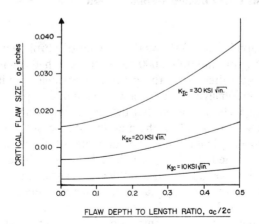

FIG. 12—*Critical flaw size versus flaw depth to length ratio for differing material toughness values (1 ksi √in. = 1.099 MN/m$^{3/2}$, 1 in. = 25.4 mm).*

to prevent initial crack propagation as unstable growth is a dynamic phenomenon driven by stored elastic energy. Further, the stress intensity factor increases with increasing flaw size (see Eq 2), and the tensile stress, S, can be also expected to increase, in this case, as the flaw depth increases. Thus, it is the fracture toughness of the surface material, and the tangential tensile stress, S, that are of central importance in the analysis of this die failure.

Recommendations to Increase Service Life

Clearly some modification is required to extend the service life of these dies. Two changes can be recommended based upon the results of the analysis.

First, a reduction in surface hardness should be considered by tempering at a higher temperature. The required dimensional accuracy of the die cavity is not high in this application, and the present yield strength of 2000 MN/m^2 (290 ksi) of the surface material is excessive, in light of the peak equivalent stress of 1229 MN/m^2 (178.2 ksi) determined from the finite element analysis. Tempering back to a Rockwell hardness of 48C will produce a surface material with a yield strength of approximately 1520 MN/m^2 (220 ksi) and a corresponding increase in fracture toughness. The results of the current fracture toughness test program will better quantify the improvements expected from this change in heat treatment.

Second, an attempt to reduce the high stress magnitudes that occur at the root of the cavity is indicated. It is recommended that changes in the die geometry or componentry be investigated in an attempt to reduce the cavity root stresses.

Failure Analysis of a Powder Metallurgy Die

Evidence

The photograph in Fig. 13 shows the remains of a broken die that reached this state after approximately 5000 to 6000 gears had been molded. The ultimate failure resulted from the growth of a single dominant flaw that traveled circumferentially around the lower die until the upper section sheared away. Closer inspection showed that the single flaw which caused the final failure actually was the result of the merging of at least three smaller flaws. These flaws originated along the inside diameter of the die where the upper end of the ribs joined the upper section. Note that these ribs in the die cavity form the bottom lands of the molded bevel gear. In order to maintain a close tolerance in the molded gears, this structural detail produces a sharp reentrant corner in the die cavity geometry and hence is a potential spot for significant tensile stress concentration. It was reported that some limited cracking was observed at this location almost immediately after the die was put into service.

FIG. 13—*Failure surface of powder metallurgy die.*

The die is essentially cylindrical in shape with the ribs distributed uniformly around the cavity. Figure 14 shows the dimensions of an axial cross section with both a typical rib and valley shown. The cross-hatched section represents the ejection piston in the center of the die cavity.

It was given that the maximum load on the die was 2.455 MN (552 kips) from the upper platen. This corresponds to a hydrostatic pressure of 396 MN/m^2 (57.4 ksi) for the 89 mm (3.50 in.) diameter die cavity. This pressure was applied cyclically with each cycle corresponding to the molding of one gear. An important implication of this loading is the potential for powder to be forced into any crack which initiates from the cavity wall. This pressurizes the crack and can accelerate growth. Finally the die was set tightly into the lower platen of the press, which provided support and some restraint to deformation of the die. As the die was operated in air at room temperature, thermal or chemical aspects of the environment had no significant effect upon the failure of this die.

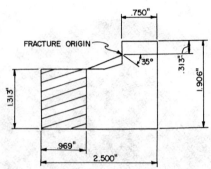

FIG. 14—*Radial section of powder metallurgy die (1 in. = 25.4 mm).*

Material

The die was made from a standard AISI A2 tool steel hardened to a measured value of Rockwell 63C. This was higher than the 54 to 56C value that was specified for this application, and is close to the as-quenched hardness. In fact the material was too hard to machine standard tensile and fracture toughness specimens from the die with conventional machine shop equipment.

For purposes of analysis it was assumed that the modulus, $E = 206.85$ GN/m^2 (30 * 10^6 psi), and the fatigue properties for the Paris equation could be taken from Rolfe and Barsom for martensitic steel [2]

$$\frac{da}{dN} = C_p \, \Delta K^n \tag{3}$$

where

a = crack length in inches,
N = number of load cycles,
ΔK = change in the stress intensity factor from zero to peak load in ksi $\sqrt{\text{in.}}$,
C_p and n = empirical material properties for martensitic steels determined experimentally [2] for the English units,
$C_p = 0.66 * 10^{-8}$, and
$n = 2.25$.

Flaws

As mentioned earlier, at least three separate fracture origins were observed when looking at the fracture surface (shown in Fig. 13). The presence of a small thumbnail flaw could be seen at each origin site. Coarse fatigue striations are clearly observable on the fracture surface coming from the thumbnail flaw and progressing to the final fracture. The visible coarse striations number approximately 30, but the majority of the fatigue life of the die would have occurred while the crack was growing within the thumbnail region.

Also, note that while the final stages of fatigue crack growth took place on a plane roughly parallel to the top surface of the die, the initial development of the thumbnail flaw occurred on a plane oriented at a 35° angle to the top surface as shown in Fig. 14. Finally the thumbnail flaw is approximately 5 mm (0.2 in.) deep along the 35° plane.

Stress Analysis

To estimate the cycles to failure for the die, knowledge of the stress field was required. The finite element method was employed and the mesh used is shown in Fig. 15. A Bucknell finite element program was used, and the element mesh was generated using three-node axisymmetric finite elements with a closed form stiffness representation [3]. Each of the quadrilateral elements shown in Fig. 15 is comprised of four of these three-node elements. Care was taken to develop a well-formed mesh with the greatest density of elements in the region of the crack plane (see Fig. 14). As the problem was modeled as an axisymmetric solid, this had the effect of treating the ribs and valleys within the die cavity as a solid, tapered volume. Since the actual cavity would provide less stiffness to the wall of the die than the solid, tapered model, the mesh was constructed with a separate zone of three quadrilateral elements in the rib region. This is seen in Fig. 15. In this

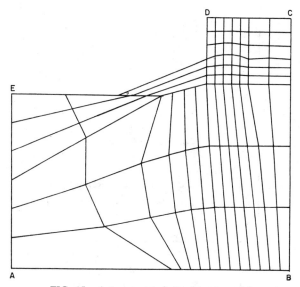

FIG. 15—*Axisymmetric finite element model.*

manner the elements in the rib region could be given a lower Young's modulus to model the reduction in stiffness.

The baseline solution for this analysis was performed with the following boundary conditions (refer to Fig. 15):

Side	Condition
AB	no vertical displacements
	free to move horizontally, no friction
BC	no horizontal displacements
	shear free surface
CD	load free surface
DE	hydrostatic pressure of 6.895 kN/m² (1.0 psi)
EA	symmetry conditions
	no horizontal displacements
	no vertical loads

In addition, the baseline solution was obtained assuming that all material had the modulus of steel including that in the rib region. The normal stress distribution along the 35° plane is given in Fig. 16. The original data for this solution are given in Ref 4 by McKernan including variations from easing constraints on boundary BC and reducing the elastic modulus in the rib zone of elements.

The graph shows that with radial restraint the maximum tensile stress normal to the 35° plane occurs at the crack initiating/fracture origin and

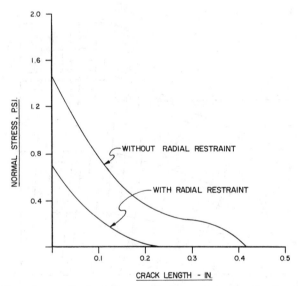

FIG. 16—*Normal stress distribution perpendicular to crack plane (1 psi = 6.895 kN/m², 1 in. = 25.4 mm).*

has a value of about 4.83 kN/m^2 (0.7 psi) for the 6.895 kN/m^2 (1.0 psi) internal die pressure. As long as the die remains elastic, the results can be linearly scaled with the internal pressure. Thus at the maximum operating pressure the peak normal stress would be 0.7 × 396 MN/m^2 ≈ 276 MN/ m^2 (40 ksi). From Ref 4, Fig. 16 shows that if side BC is permitted to move freely in all directions, bending is enhanced and the peak stress could reach 579 MN/m^2 (84 ksi) with an equivalent stress of about 752 MN/m^2 (109 ksi) at the fracture origin. However, with the lateral constraint that the lower platen exerts on the die, it is not probable that the stresses can rise to the 579 MN/m^2 (84 ksi) level except right at the sharp corner.

From the theory of elasticity the equivalent stress σ_{eq} is given by the following formula for an axisymmetric state of stress and has been defined such that when it reaches the yield strength of a given material in tension at a particular point in a component, yielding is predicted to occur under a general stress state

$$\sigma_{eq} = \frac{1}{\sqrt{2}} \{(\sigma_r - \sigma_z)^2 + (\sigma_z - \sigma_\theta)^2 + (\sigma_\theta - \sigma_r)^2 + 6\tau_{rz}^2\}^{1/2} \qquad (4)$$

where

σ_r = normal stress in radial direction in the die,
σ_θ = normal stress in hoop direction in the die,
σ_z = normal stress in axial direction in the die, and
τ_{rz} = shear stress in the die.

A typical value of the yield strength of Type A2 tool steel with the hardness of the die is approximately 1590 MN/m^2 (230 ksi). The exact yield strength for the die in question is unknown. Note that no appreciable evidence of plastic deformation was present in the broken die.

McKernan also showed that a reduction in Young's modulus within the rib zone elements to a value 1% of that within the rest of the die resulted in an increase of the equivalent stress at the fracture origin of about 10%. This effect is of the same order as the overall accuracy of the finite element method for this solution and thus will be ignored in this case study analysis.

Conclusions from the stress analysis that can be applied to the failure analysis are:

1. The finite element analysis indicated a tensile stress perpendicular to the crack plane at the reentrant corner where the rib joins the inner wall of the die cavity. The theoretically infinite stresses associated with a reentrant corner would have been sufficient to initiate and grow a small crack in this region.

2. The radial constraint supplied by the lower platen of the press was probably sufficient to limit the normal stresses along the 35° plane to the baseline data of Fig. 16 (with radial restraint).

The number of cycles that it took for the thumbnail flaw to develop can be estimated from the baseline data of Fig. 16. The following section presents the fatigue calculation and includes the effect of pressure within the crack.

Fatigue Analysis

Equation 4 presents the fatigue crack growth equation of Paris. To estimate the cycles to develop the thumbnail flaw, an estimation of ΔK is required for a crack growing down the 35° plane. Even though the actual shape of the thumbnail flaw is semi-elliptical, a first approximation for the stress intensity factor can be taken from the solution of an edge crack in a semi-infinite plate under uniaxial tension. This leads to a simple integration of the Paris equation and provides an order of magnitude estimation of the cycles required to grow the observed thumbnail flaws. For an edge crack growing into a semi-infinite solid under a normal, tensile stress, S, aligned perpendicular to the crack surface, the stress intensity factor is known to be [2]

$$K = 1.12 \, S \, \sqrt{\pi a} \tag{5}$$

Hence, for a load cycling between zero and some maximum value, which causes the stress in the vicinity and normal to the crack to vary between zero and some maximum value, S^*

$$\Delta K = 1.12 \, S^* \, \sqrt{\pi a} \tag{6}$$

Therefore, Eq 3 becomes

$$\frac{da}{dN} = C_p \, [1.12 \, S^* \, \sqrt{\pi a}]^n \tag{7}$$

Integration of this expression leads to an estimation of the cycles to produce a crack of a given size, a_c. To perform the integration, some value for S^* is needed. One approximation would be to assume that the normal stress distribution shown in Fig. 16 reflects the dominant stress component at any subsequent crack length during fatigue. McKernan employed this approximation and using a numerical integration of Eq 8 calculated that it would take at least 50 000 cycles to propagate a crack from around 0.25 to 10 mm (0.01 to 0.40 in.) along the 35° plane. This was for the case where enhanced bending could take place. Radial constraint would increase the number of cycles even more.

A more realistic approximation includes the effect of pressure getting into the crack tip region as it propagates. This will accelerate the crack growth and lead to fewer cycles to failure. In this approximation the driving stress would be the maximum tensile stress in the die at the crack tip area plus the opening pressure. Consider writing the driving stress in terms of the internal die pressure

$$S^* = C_0 p \tag{8}$$

where

p = pressure inside the die cavity and
C_0 = proportionality constant based upon the die geometry.

Substitution into Eq 7 results in

$$\frac{da}{dN} = C_p [1.12 \, C_0 p \, \sqrt{\pi a}]^n$$
$$= C_p \, [1.12 \, C_0 p \, \sqrt{\pi}]^n \, a^{n/2}$$
$$\int_0^N dN = \left[\frac{1}{C_p \pi^{n/2}} \right] \left[\frac{1}{1.12 \, C_0 p} \right]^n \int_{a_i}^{a_c} \frac{da}{a^{n/2}} \tag{9}$$

where N = the number of cycles needed to grow a crack from an initial size a_i to a final size a_c.

$$N = \left[\frac{1}{C_p \pi^{n/2}} \right] \left[\frac{1}{1.12 \, C_0 p} \right]^n \left[\frac{2}{2-n} \right] [a_c^{(2-n)/2} - a_i^{(2-n)/2}] \tag{10}$$

Substitution of the fatigue properties given earlier for martensitic steels results in

$$N = 2.59 \times 10^8 \left[\frac{1}{C_0 p} \right]^{2.25} [a_i^{-0.125} - a_c^{-0.125}] \tag{11}$$

If it is assumed that the driving stress equals the maximum stress at the fracture origin from Fig. 16 ($S = 0.7$ psi) plus the opening pressure ($p = 1.0$ psi), C_0 then equals = 1.7. For a total operating pressure of 396 MN/m² (57.4 ksi), the number of cycles needed to grow a crack to the 5 mm (0.2 in.) thumbnail flaw depth becomes

$$N = 2.59 \times 10^8 \left[\frac{1}{1.7 \times 57.4} \right]^{2.25} [a_i^{-0.125} - 0.2^{-0.125}] \tag{12}$$

Thus the fatigue life becomes a function of the assumed size of the initiated crack at the fracture origin. The following Table 4 shows the cycles for a few possible values of a_i.

Even with the approximations made in estimating the stress intensity factor and the Paris fatigue law properties, this example demonstrates the power of fracture mechanics in predicting the cycles to develop the observed thumbnail flaw. The answer is sensitive to the initiated crack size, but the order of magnitude is reasonable for the likely range of values chosen.

Concluding Remarks

These two die failures illustrate the extremes in the effect of the crack growth stages upon the life of the components. In the wire rope swaging die, the combination of high stresses at the root of the cavity and the relatively low toughness of the surface material resulted in a catastrophic fracture at a small critical flaw size. Small pre-existing cracks are sufficient to cause rapid fracture without any cyclic, slow stable fatigue crack growth.

On the other hand, in the case of the powder metallurgy die, failure was not a result of rapid fracture from a small crack coming from high stresses and relatively low fracture toughness. Instead, most of the life was spent in growing a fatigue crack until the die virtually separated into two pieces.

Based upon these observations the following work is being undertaken to improve the life of the dies. In the case of the wire rope swaging die, an improved fracture toughness through changes in heat treatment would help the situation. A second approach is to explore changes in geometry or loading to reduce the high stresses at the root of the cavity.

Since the fracture toughness does not play a critical role in the fatigue life of the powder metallurgy die, a change in materials or heat treatment probably would have a minor effect on the service life. Therefore, in this case changes in geometry and loadings are being sought in order to reduce the maximum tensile stress at the fracture origin. As reflected by S^* in Eq 8, and by Eq 11, this would increase the number of cycles to failure.

TABLE 4—*Cycles versus initial flaw size.*

a_i	N, cycles
0.0025 mm (0.0001 in.)	16 785
0.0127 mm (0.0005 in.)	11 797
0.0254 mm (0.0010 in.)	9 940
0.127 mm (0.0050 in.)	6 200
0.254 mm (0.0100 in.)	4 807
1.27 mm (0.0500 in.)	2 002

Acknowledgments

This work was carried out under the joint sponsorship of the United States Small Business Administration/Pennsylvania Small Business Development Center Program and the State of Pennsylvania's Ben Franklin Partnership Program/North East Tier Advanced Technology Center. The authors wish to thank the Muncy Machine and Tool Company of Muncy, Pennsylvania, and the Pennsylvania Pressed Metals Company of Emporium, Pennsylvania, for permission to publish these case study examples.

References

[1] *Tool and Die Failures Source Book,* S. Kalpakjian, Ed., American Society for Metals, Metals Park, OH, 1982.
[2] Rolfe, S. T. and Barsom, J. M., *Fracture and Fatigue Control in Structures,* Prentice-Hall, Inc., Englewood Cliffs, NJ, 1977.
[3] Rich, T. P., *International Journal for Numerical Methods in Engineering,* Vol. 12, No. 1, 1978, pp. 59–65.
[4] McKernan, J. B., "Failure of a Powder Metallurgy Die Insert," Report of the Consortium on Computer-Aided Mechanics and Materials Design and Testing, Bucknell University, Lewisburg, PA, 1984.

C. Kendall Clarke[1]

Analysis of a Failed Saw Arbor

REFERENCE: Clarke, C. K., "**Analysis of a Failed Saw Arbor,**" *Case Histories Involving Fatigue and Fracture Mechanics, ASTM STP 918*, C. M. Hudson and T. P. Rich, Eds., American Society for Testing and Materials, Philadelphia, 1986, pp. 336–343.

ABSTRACT: Stress analyses incorporating stress concentration, fatigue, and fracture calculations should be an important aspect of failure analyses. The following description of a saw arbor failure provides a simple example of this necessity. The failure was a classical stepped shaft in bending fatigue failure. Stress concentration at the change in diameters was calculated to be a factor of five. However, stress calculations revealed normally expected stresses to be well below the endurance limit even with the stress concentration effects. Additional investigation revealed evidence for severe imbalance loads from one saw blade. It was hypothesized that some of the teeth from the carbide tipped saw had been knocked off, and the resulting imbalance loads were more than sufficient to initiate fatigue cracking.

KEY WORDS: fatigue, stress concentration, stress analysis, mutual hardness calculations

A saw arbor holding a 0.56 m (22 in.) diameter saw blade in a lumber mill trimming saw machine failed while in operation. The spinning carbide tipped blade then cut through guards and seriously injured a worker. Visual observation of the fracture revealed that fatigue cracking had initiated at and propagated from a change in diameters on the arbor. No significant radius was observed at the change in diameters.

This type of shaft failure is often written up in failure analysis literature as a stress concentration induced failure caused by a change in shaft diameters. The purpose of this paper is to show that routine calculations reveal that this particular failure, and probably many other similar failures, are not always simple. Stress calculations involving stress concentration and fatigue calculations provide a powerful tool with which to arrive at a more complete analysis of failure. The underlying principle is that results of stress, fatigue, and fracture calculations should be consistent with the proposed mode of failure.

The trimming saw consisted of eleven identical, independent saws and was used to trim lumber to finished lengths. Figure 1 shows a side view of

[1] President, Metallurgical Consulting, Inc., Mobile, AL 36609.

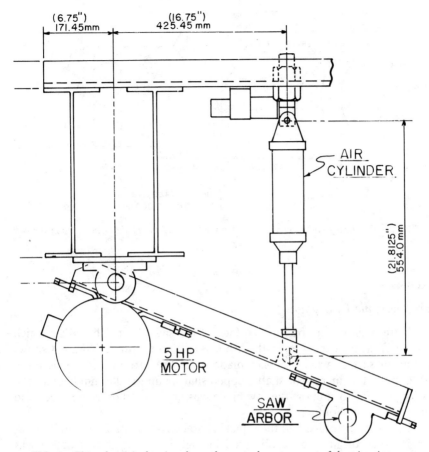

FIG. 1—*This schematic drawing shows the general arrangement of the trimming saw.*

one of the identical trimming saw systems. The arbor design and crack origin location are shown in Figure 2. This design had been in service for several years without any reported problems or failures. The specific equipment involved in the accident had been in service nearly two years prior to the accident.

Trimming saws were arranged every 0.6 m (2 ft)[2] so that lengths from 0.6 m (2 ft) to 6 m (20 ft) could be trimmed. Each saw was equipped with a 3730 J (5 HP) electric motor turning the saw blade at 2200 rpm. All of the saws were normally turned on each day of operation. However, the saw which failed was at the 5.5 m (18 ft) location and only cut wood about 2% of the time.

---

[2] All measurements and calculations were originally in English units.

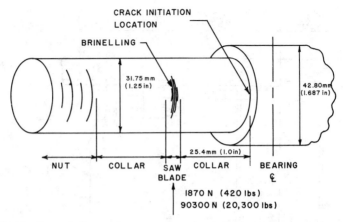

FIG. 2—*Two heavy collars and a nut were used to hold the blade in place upon the saw arbor.*

Results

Fractographic Evaluation

Figure 3 shows the fracture surface of the saw arbor. The collar which butted up against the step in the arbor is still on the shaft. Figure 4 shows the arbor surface where the saw blade rode. Fracture occurred where the 31.75 mm (1 ¼ in.) arbor shaft stepped sharply up to 42.8 mm (1 ¹¹⁄₁₆ in.). A large collar to support the saw blade was press fitted onto the arbor and up against the shoulder.

The fracture surface had the typical characteristics of a fatigue failure. A single crack origin can be observed in Fig. 3. No significant defect was observed at the origin. No evidence for fretting was observed at the crack origin. The very initial cracking propagated straight across the arbor diameter for approximately 2.5 mm, as if a bending moment were applied only at the crack origin location. Cracking then changed abruptly to a manner more typical of rotating bending. (The point of maximum bending stress changed with shaft rotation.) The final fracture region was small, thus indicating low service stresses at failure.

Stress Calculations

Stress calculations were made assuming normal operating conditions to determine if stresses capable of initiating fatigue cracking were possible at the shoulder (change in diameters where cracking initiated) in the arbor. These calculations included stress concentration effects. When these initial calculations revealed low stresses, additional calculations became necessary to justify fatigue crack initiation.

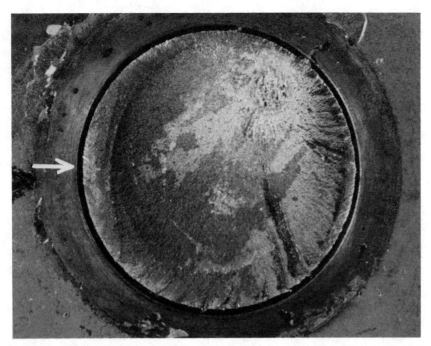

FIG. 3—*The complete arbor fracture surface is shown here. An arrow points to the origin. No obvious material defect was observed.*

A determination of the stress concentration factor was the first calculation made. The radius at the base of the shoulder on the cold-rolled steel shaft was measured to be 0.13 mm (0.005 in.) by using wire feeler gages and a low power microscope. This combination of radius and ratio of diameters produced stress concentrations of 4.7 and 5.1 using charts from Peterson [1]. A shear stress concentration factor was also found from the same charts to be 4.0 to 4.5. The collar on the arbor was reportedly press fitted up against the shoulder where the fracture occurred. This contribution to stress concentration was ignored for lack of data.

Torsion stresses were calculated to be very low on the shaft. The torque on the arbor with no load was calculated to be

$$T = \frac{(396000)\ (HP)}{2\pi\ rpm} = \frac{(396000)\ (5)}{2\pi\ (2200)} = 143 \text{ in.} \cdot \text{lb or } 16.1 \text{ N} \cdot \text{m} \quad (1)$$

Or under load:

$$T = \frac{(396000)\ (5)}{2\pi\ (1100)} = 286 \text{ in.} \cdot \text{lb or } 32.3 \text{ N} \cdot \text{m}$$

FIG. 4—*The large washer or collar was press fitted onto the shaft and up against the shoulder. A brinelled area can be seen* (arrow) *where the unbalanced saw blade deformed the shaft.*

The torsional stress on the shaft is therefore

$$\tau = \frac{Tr}{J} = \frac{(143 \text{ to } 286) \ (0.625)}{(0.782 \text{ in.}^4)} \tag{2}$$

$$= 114 \text{ to } 228 \text{ psi } (7.86 \times 10^5 - 1.57 \times 10^6 \text{ Pa})$$

Including stress concentration effects, these stresses are too low to have initiated a failure.

The basic saw dimensions are shown in Figs. 1 and 2. The air cylinder could provide up to 3560 N (800 lb) in downward force. A simple statics analysis was made assuming the primary force on the arbor was the air cylinder. The nut clamping force was ignored because torquing requirements were low. (The wood feed force was much less.) The maximum force on the arbor was 1870 N (420 lb), and this force decreased as the saw arm dropped. This force translated into a bending moment on the 31.75 mm (1 ¼ in.) arbor of 47 500 N · mm (420 in. · lb). The maximum or outer fiber bending stress was, therefore

$$\sigma = \frac{Mc}{I} = 15.0 \text{ MPa } (2180 \text{ psi}) \tag{3}$$

A stress concentration factor of 5 would not elevate this stress to anywhere near the endurance limit for the steel.

No material properties other than hardness were available for the arbor because destructive testing had not been permitted. Therefore, a lower limit on the fatigue endurance limit was estimated from the hardness. The cold rolled steel shaft had a hardness of 143 BHN which produces a tensile strength of approximately 482 MPa (70 000 psi). The endurance limit could be conservatively estimated to be 35% of the tensile strength or 169 MPa (24 500 psi). This number is well above the calculated bending stresses in the shoulder region including effects of stress concentration. Therefore, some other factor was required to have raised stresses high enough to have initiated a fatigue failure.

Careful examination of the arbor revealed brinelling at one point on the steel shaft where the saw blade fit (Figs. 2 and 4). This was caused by a force on the saw blade high enough to deform the shaft in one location. The circumferential positions of the brinelled area and of the crack initiation site coincide.

The principle behind hardness testing was used to estimate the force required to deform the saw arbor. Hardness tests are run by pressing with a known force an indenter of known geometry and hardness into an unknown material. The hardness or plastic flow stress of the unknown material is calculated on the basis of the area of the indention and the known force. However, in this case, the hardness of the arbor and the indention area were known. Reference 2 provides a method for determining the force required to make the indention. This method provides factors which take into account the indentor geometry and hardness.

Hardness measurements on the shaft yielded a hardness of 143 BHN. The hardness of the saw blade was assumed to be comparable. A local yield stress can be calculated using a factor of 2.8 to account for the Brinell ball shape and hardness

$$\frac{143 \text{ BHN}}{2.8} = 51.1 \text{ kg/mm}^2 = 500 \text{ MN/m}^2 \text{ (72.6 ksi)} \tag{4}$$

Similar hardnesses and a cylinder inside a cylinder geometry yield a hardness factor of 1.4^2. This factor times the calculated yield stress from Eq 4 yields a constrained flow stress of approximately 700 NM/m² (Eq 5). The indentation area

$$(500 \text{ MN/m}^2) \ 1.4 = 700 \text{ MN/m}^2 \text{ (102 ksi)} \tag{5}$$

$$(700 \text{ MN/m}^2) \ (1.29 \times 10^{-4} \text{ m}^2) = 0.0903 \text{ MN (20.4 kips)} \tag{6}$$

was measured to be 1.29×10^{-4} m², and this area times the constrained flow stress yields a force required to produce the indentation of approximately 90 300 N.

This calculated force is relatively high and yields a maximum bending stress at the site of crack initiation of 730 MPa (106 ksi). This is at or near the failure stress of the steel in bending. Assumptions regarding the hardness of the saw blade and the probability that the total measured indented area developed as a result of increasing blade movement as the blade hole deformed certainly would affect the magnitude of this force. (The actual instantaneous contact area could have been half or less of the total measured area of damage.) The point is that a source for higher stresses has been found, and that refinements to the calculations could be made if the data were available.

The possibility of an overloaded saw blade was confirmed when a check with local saw repair shops revealed that the particular saw mill in question had experienced problems with pieces of occasional steel in the lumber knocking carbide teeth from the saw blades. The resulting imbalance would have overloaded the saw arbor. In this particular case, half of the teeth missing would produce just about the level of force calculated in Eq 6.

A closer approximation of the actual stress on the shaft might be made by considering the fracture surface. Cracking initially propagated about 2.5 mm straight along the shaft diameter. Then the crack front changed to one more typical of rotating bending. It was known that the saw on the arbor at failure was in good condition and was about 1.5 mm wider than the one which initiated the failure. Thus the change in crack propagation can be explained by a saw blade change. When the blade was changed is not known. However, the damaged saw ran for at least several minutes before being shut down and the blade changed. Therefore, the initial crack growth on the arbor would have occurred over the period of a few thousand cycles to over a million cycles. Arbor stresses can then be considered to be bounded by the endurance limit of 170 MPa and the 1000 cycle limit. One thousand cycle life limits can be estimated by the following equation [3]

$$\sigma = \sigma_f'(2\,N)^b \qquad (7)$$

where

σ = predicted stress for N cycle life,
σ_f' = fatigue strength coefficient (estimated = 900 MPa),
b = exponent (estimated = -0.13), and
σ = 900 (2000)$^{-0.13}$ = 335 MPa (48.6 ksi).

The estimated brinelling force is certainly in this order of magnitude and would probably fall within this range if more data were available.

Conclusion

The failure of the saw arbor is hypothesized to have started when a piece of steel in lumber being trimmed knocked off several carbide teeth on the saw blade. The resulting imbalance load was adequate to initiate fatigue cracking, but it was not adequate to cause blade failure before the blade was changed. (A new blade was put on the arbor and cracking continued at a much slower rate.) Calculations or the force required to cause the brinelling were on the high side. Actual forces were most probably lower. Therefore, the imbalance hypothesis was consistent with the stresses required for fatigue fracture while stress concentration alone was not.

This case provides an excellent illustration of the necessity for performing stress calculations as part of a failure analysis. Conclusions from the fractographic analysis should be consistent with the stress calculation results. Sophisticated fatigue and fracture calculations were not required for this particular case. However, these calculations are often very helpful in more complex cases such as pressure vessel failures.

References

[1] Peterson, R. E., *Stress Concentration Factors*, Wiley, New York, 1974, Figs. 78, 78a, 79 and 79a.
[2] Atkins, A. G. and Felbeck, D. K., *Metals Engineering Quarterly*, Vol. 14, No. 2, May 1974, pp. 55–61.
[3] Fuchs, H. O., and Stephens, R. I., *Metal Fatigue in Engineering*, Wiley, New York, 1980, p. 155.

Peter J. Tubby[1] *and J. Graham Wylde*[2]

Role of Fracture Mechanics in Assessing the Effect on Fatigue Life of Design Changes in Welded Fabrications

REFERENCE: Tubby, P. J. and Wylde, J. G., "**Role of Fracture Mechanics in Assessing the Effect on Fatigue Life of Design Changes in Welded Fabrications,**" *Case Histories Involving Fatigue and Fracture Mechanics, ASTM STP 918*, C. M. Hudson and T. P. Rich, Eds., American Society for Testing and Materials, Philadelphia, pp. 344–360.

ABSTRACT: Experience has shown that in many instances designers have reacted to service fatigue failures in welded structures in the wrong way, with the result that design changes made have not had the desired effect. It is the intention of the present paper to demonstrate that fracture mechanics techniques, which are now well established, can be used to provide an appreciation of those aspects of a design which are important with respect to the fatigue performance of the structure. This understanding provides an excellent basis for evaluating the benefit of proposed design changes when service problems are encountered.

Analysis is based on stress-intensity solutions derived by two-dimensional linear elastic finite element modelling. Fatigue lives are predicted by integration of the appropriate fatigue crack growth law on the basis that the life of a welded joint is dominated by the propagation of small pre-existing defects.

An important consequence of this feature of their behavior is that the fatigue lives, and hence fatigue strengths, of joints are strongly influenced by factors which affect the rate of crack propagation. Results are given which illustrate that factors such as weld profile, the tensile strength of the parent material, and stress relief by post weld heat treatment have little effect on fatigue strength, although historically designers have often looked to modify one of these in the face of service problems. Other factors such as the detailed design of joints and the reduction of stress concentration factors are shown to be more significant.

It is also demonstrated that even the most fundamental step in the design process, that is, the selection of scantlings, may not have the expected effect on fatigue life since the fatigue strength of a joint may be reduced by increasing the plate thickness.

KEY WORDS: fatigue, welded joints, crack propagation, finite elements, fracture mechanics, weld defects

When a structure or component fails in service the pressure on design, materials, or welding engineers to find a quick and reliable solution may

[1] Senior research engineer, The Welding Institute, Abington Hall, Abington, Cambridge, U. K.

[2] Manager of engineering, Edison Welding Institute, Columbus, OH 43212.

be considerable. Under these circumstances it is tempting to grasp at al-
ternatives which at first sight may appear to solve the problem, only to find
in due course that a permanent solution has not been effected. In welded
construction, fatigue cracking at welded joints is by far the most common
cause of failure. Experience has shown that when faced with this problem
engineers may be tempted by crude repair procedures. If repair is not
possible, parameters such as the parent material, heat treatment, welding
procedure, or welding consumable may be varied in manufacturing a replacement item, often with little or no appreciation of the true mechanism
of failure, or the effect such changes are likely to bring about.

Fracture mechanics is an invaluable aid in such circumstances. Perhaps
the simplest illustration of this fact is the example of a cylindrical pressure
vessel which developed a through-thickness crack in service running parallel
to a longitudinal butt welded seam. A single weld run made from the outside
(Fig. 1) sealed the leak and allowed operation of the vessel to continue,
but left the original fatigue crack penetrating the section to a depth of about
one third of the wall thickness. With only the slightest appreciation of the
mechanics of fatigue crack growth, and without recourse to detailed calculations, it is obvious that, due to the exponential characteristic of crack
growth, through-thickness cracking would occur again within a very short
proportion of the time taken to produce the original failure.

In addition to its more obvious application in the assessment of fitness
for purposes of cracked or defective structures, fracture mechanics provides
a useful model of the fatigue behavior of nominally sound joints. This results
from the fact that in steels the weld toe, which is the site of failure in the
majority of joints, invariably contains sharp planar slag intrusions entrapped
in material which either remained pasty or was molten only for an instant
during welding. Typically these defects have been found to be between 0.15
and 0.4 mm deep [1]. They can be regarded as small pre-existing cracks
which will propagate virtually from the start of cyclic loading, thus eliminating the need to consider a fatigue crack initiation period. As a result,
factors which influence crack propagation have a significant influence on
the fatigue life. In the present paper we describe briefly the basis of the

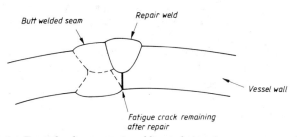

FIG. 1—*Example of poor repair welding technique in a pressure vessel.*

fracture mechanics model and use it to illustrate the extent to which various factors affect the fatigue strengths of joints.

Fracture Mechanics Model

The rate of fatigue crack growth in materials has been found to follow an expression of the form of [2]

$$\frac{da}{dN} = C(\Delta K)^m \tag{1}$$

where

da/dN = crack growth rate,
ΔK = range of stress intensity factor, and
C and m = constants dependent upon the material and environment.

The stress intensity solution for a crack in an infinite plate takes the form

$$\Delta K = \Delta \sigma \sqrt{\pi a} \tag{2}$$

where $\Delta \sigma$ is the remotely applied stress range.

For more complex geometries a correction factor $Y(a)$ is required to allow for the crack shape and its location, thus Eq 2 becomes

$$\Delta K = Y(a)\Delta \sigma \sqrt{\pi a} \tag{3}$$

Combining Eqs 1 and 2 we obtain

$$\frac{da}{dN} = C\{Y(a)\Delta \sigma \sqrt{\pi a}\}^m \tag{4}$$

Separating the variables and integrating between the limits of the initial and final crack sizes, a_i and a_f, we obtain the predicted life, N, as follows

$$\int_{a_i}^{a_f} \frac{da}{(Y(a)\sqrt{a})^m} = C\Delta \sigma^m \pi^{m/2} N \tag{5}$$

To apply the model to a particular joint geometry it is necessary to derive a suitable stress intensity solution, that is, to evaluate $Y(a)$, and to perform the integration as described further below.

Stress-Intensity Solution

Stress-intensity solutions may be derived by cracked body finite elements analysis. For each joint geometry to be investigated, a finite element model is developed using either two or three dimensional elements as appropriate, and is solved for a range of crack sizes. The program developed at The Welding Institute uses third order isoparametric quadrilateral elements which have been shown to yield results of reasonable accuracy when compared with a range of published solutions. Details of the element formulation, crack tip modelling, and the derivation of stress intensities are described elsewhere [3]. The solutions for a large number of joint geometries are given in by Smith and Hurworth [4].

Figure 2 shows a typical two-dimensional mesh used to model a plate with full penetration welded non-load-carrying attachments, and the resulting stress-intensity solution when the joint is subjected to a uniform remote tensile stress transverse to the weld.

Crack Growth Integration

A versatile numerical integration program has been developed at The Welding Institute to allow crack growth predictions to be made for a wide

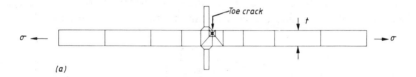

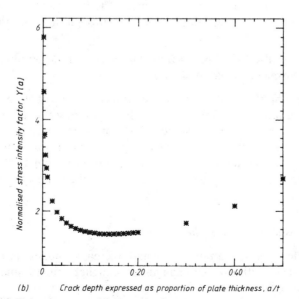

FIG. 2—(a) *Finite element model of non-load-carrying transverse joint subjected to remote tensile stress; and* (b) *stress intensity solution (after Ref 4).*

range of input data. The program carries out a cycle by cycle integration of the crack growth law using the appropriate stress intensity solution obtained as described previously. Both constant and variable amplitude loading may be analyzed, provided the variable amplitude spectrum is expressed as a histogram of stress range and number of cycles. Other parameters which may be varied are the initial and final crack sizes, crack shape, the crack growth law, and the threshold stress intensity.

In the present study we compare the influence of various parameters on the predicted fatigue strength, defined here as the stress range, $\Delta\sigma_0$, for 2 × 10^6 cycles, which is derived as follows.

Considering Eq 5, it is apparent that for a particular joint geometry, that is apparent for a particular $Y(a)$ solution, and for fixed values of m, a_i, and a_f, the left hand side reduces to a constant, yielding an expression of the form

$$\Delta\sigma^m N = \text{constant} \tag{6}$$

A single integration is carried out with an arbitrary stress range, $\Delta\sigma$, to obtain the corresponding life N. $\Delta\sigma_0$ may be then evaluated from

$$\Delta\sigma_0 = \left\{ \frac{\Delta\sigma^m N}{2 \times 10^6} \right\}^{1/m} \tag{7}$$

In the cases described below, either a continuous or semielliptical defect at the weld toe, 0.15 mm deep, is assumed. The limit to crack propagation is assumed to be attainment of a crack 50% through the plate thickness.

Examination of the Effect of Various Parameters on the Predicted Fatigue Performance of Welded Joints

Using the simple model described the predicted effect of various material and design parameters on the fatigue performance of welded joints may be examined rapidly. In many instances the results of experimental investigations are available to confirm the validity of the predictions, but the most useful feature of the model is clearly its ability to examine the extent to which parameters not yet investigated in practical studies are likely to be significant.

Effect of Tensile Strength of the Parent Material

In unwelded steels, the fatigue strength increases roughly in proportion to the tensile strength, so that changing to a higher strength material is an option often considered when service failures arise. For welded joints, the fracture mechanics model predicts that only factors which affect fatigue crack propagation have a significant influence on fatigue strength. Examination of crack growth data suggests that steels with widely differing tensile strengths have similar crack growth characteristics, so that one would not expect joints in high-strength steels to have significantly higher fatigue strengths than those in mild steel. In fact, a review conducted by Gurney [5] indicated a tendency for high strength steels to give higher growth rates in the more important low ΔK regime.

The fracture mechanics model may be used to examine the effect of such variations in crack growth characteristics. Taking the stress-intensity solution for a single joint type, and assuming constant values of a_i and a_f, the integration may be performed for a range of pairs of C and m. This has been carried out for a simple transverse joint of the type shown in Fig. 2 in 12.5-mm-thick plate. The C and m values used were those for structural and high-strength steels with tensile strengths from 400 to 1450 N/mm^2 cited by Gurney [5]. The results are shown in Fig. 3 in terms of the predicted fatigue strength versus tensile strength of the parent material. The range of experimentally derived fatigue strengths for similar joints in structural steels with tensile strengths varying from 400 to 600 N/mm^2 [6] are plotted for comparison and show good agreement with the predictions.

While there is considerable scatter in the data in Fig. 3 they indicate a tendency for fatigue strength to decrease with increasing tensile strength of

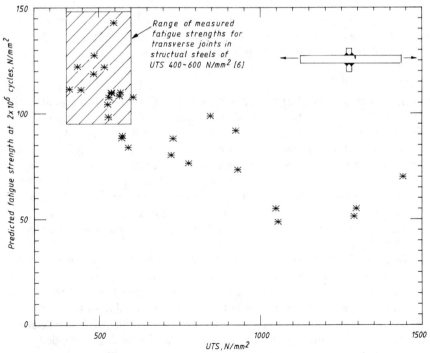

FIG. 3—*Predicted variation of fatigue strength with tensile strength (resulting from different crack growth parameters C and* m) *for a range of structural and high strength steels.*

the parent material. Practical investigations of the fatigue strengths of welded high-strength steels have been relatively few but, in the main, have found little difference from those of lower strength structural steels [7].

Of course, it should be borne in mind that in the as-welded condition a single joint contains a range of microstructures with widely differing strengths. If there were significant differences between the crack growth characteristics of these regions, one would expect nominally identical joints in the same parent material to have very different fatigue strengths according to whether crack propagation was initially in parent material, heat affected zone, or weld metal. In reality the scatter is not so broad, suggesting that the different microstructures have similar crack growth characteristics. This was confirmed by Maddox [8] who measured similar crack growth rates in a range of weld metals, heat affected zones, and parent steels (Fig. 4). The mean of the data in Fig. 4 may be represented by the constants $C = 1.83 \times 10^{-13}$, $m = 3$ (N,mm units). These values have been used in the subsequent analyses.

Therefore it is apparent that, in the event of failure of a welded joint, substituting a higher strength parent steel will achieve little or nothing in terms of fatigue strength. If for some other reason an alternative material has to be chosen, selection should be made on the basis of fatigue crack growth characteristics rather than tensile strength, materials giving low crack growth rates at low ΔK being the best option.

It should be noted however that in cases where joints are subjected to a very high mean applied stress or a stress spectrum including high-stress cycles, high-strength steels may be used to advantage.

Fracture Toughness

In instances where fatigue crack growth is followed by fast fracture, the fracture toughness of the material at the crack tip governs the onset of final failure. Fracture occurs when a critical stress intensity or crack-tip opening is exceeded. It is a common assumption that the fatigue life may be extended by the use of materials of high toughness since this has the effect of delaying the transition from stable fatigue crack growth to rapid fracture. In many instances, however, this is not a valid assumption since, in the final stages of fatigue crack growth, the rate of crack extension is such that a large increase in the maximum tolerable defect size has only a small influence on the fatigue life.

This may be demonstrated using the crack growth model by comparing a prediction of crack development, that is, crack depth versus cycles, against the toughness required to resist fracture at each crack length (Fig. 5). The particular case considered here is that of a joint similar to that shown in Fig. 2, in 12.5-mm-thick plate, subjected to an axial pulsating tensile stress range of 200 N/mm^2. A semi-elliptical initial defect 0.15 mm deep was assumed. The toughness parameter considered is the crack-tip opening dis-

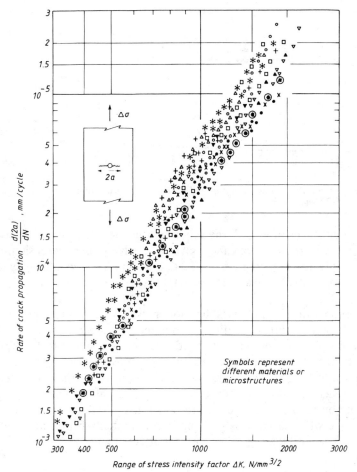

FIG. 4—*Fatigue crack propagation data for structural carbon-manganese steel weld metals, heat affected zones and parent plate (after Ref 8).*

placement (CTOD), estimated as follows [9]

$$\text{CTOD} = \frac{1}{E} \left\{ [Y(a)]^2 \, 2\pi a \sigma_y \left[\frac{\sigma_{\text{total}}}{\sigma_y} - 0.25 \right] \right\} \tag{8}$$

where

σ_y = yield strength of parent material,

$\sigma_{\text{total}} = \sigma_{\text{max}} + \sigma_{\text{res}}$,

σ_{max} = maximum applied stress,

σ_{res} = residual stress (which is here assumed equal to σ_y), and

E = Young's modulus.

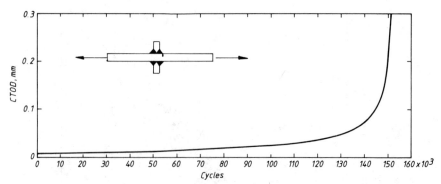

FIG. 5—*Prediction of toughness required to resist brittle fracture at a crack growing at the weld toe in 12.5-mm-thick plate.*

In the final stages of growth the required toughness rises very rapidly such that a large increase in toughness gives a relatively small extension in life. For example a 50% increase from 0.2 to 0.3 mm CTOD would delay failure by only about 1000 cycles in 150 000, that is, an increase in life of less than 1%.

A corollary of this analysis is that, for cracks growing in reasonably tough material, there is little benefit, in terms of the accuracy of the predicted fatigue life, in carrying out a detailed fracture assessment to establish the upper limit to fatigue crack extension. In a uniform direct stress field, the difference in fatigue life (expressed as a proportion of the total) between a crack reaching a depth of say 50 or 100% of the wall thickness is marginal. It is, of course, necessary to ensure that fracture will not intervene at an early stage where a crack is growing in a potentially brittle material such as the coarse grained region of the heat affected zone.

Stress Relieving Treatment

Joints in the as-welded condition contain residual stresses due to expansions and contractions resulting from the thermal cycle imposed by welding. Measurements made near weld toes indicate that residual stresses here are invariably tensile and may be as high as the yield strength of the parent material. When a cyclic load is superimposed on the residual stress distribution, cyclic plastic deformation occurs locally, until, after the first few cycles, regardless of the nominal applied stress ratio, the stress at the point of fatigue cracking pulsates downwards from yield stress tension. This behavior is incorporated in the fracture mechanics model by assuming that the whole of the applied ΔK range contributes to fatigue crack propagation irrespective of whether the applied stress range is wholly tensile or partially compressive.

As a consequence of these assumptions, the model predicts identical lives for a given stress range, irrespective of the applied stress ratio. This behavior

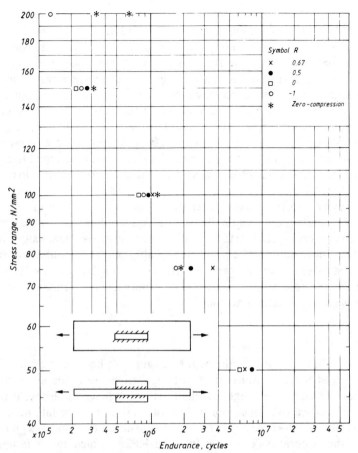

FIG. 6—*Fatigue test results for longitudinal non-load-carrying joints tested at different stress ratios in the as-welded condition (10). The material is BS 4360 Grade 50B steel.*

is largely borne out by fatigue test results for steel joints in the as-welded condition. For example, Maddox [*10*] has obtained test results for joints subjected to stress ratios ranging from $R = -1$ to $+0.67$, including some tests under zero-compression loading (Fig. 6). There is no apparent influence of stress ratio on fatigue strength, even for joints subjected to fully compressive cycles, although it should be pointed out that in these specimens the cracks did not propagate to final failure but arrested once they had grown away from the residual stress field.

For stress relieved joints, the fracture mechanics analysis must be modified to allow for the fact that local stresses at the failure site are not pulsating downwards from yield. If the stress relieving treatment were entirely effective, the resulting static stress would be zero. With this assumption, and assuming only the tensile part of the ΔK range contributes to crack extension, the model would predict identical fatigue strengths for all positive applied stress ratios (pulsating tension cycles), and increasing fatigue strengths for increasingly negative stress ratios, as a result of the increasing compressive portion of the ΔK range. It is rarely possible, however, to achieve complete relief of residual stresses. Some level of tensile stress remains at the potential failure site, especially in large structures which are not readily heat treated. Even in simple laboratory specimens, stress levels greater than 100 N/mm^2 have been measured after thermal stress relieving treatment. Incorporating this feature in the analysis, fatigue strengths predicted by the model are now dependent on the level of residual stress assumed to remain after treatment, which is a true reflection of the behavior obtained in practice [10].

We may therefore conclude that stress relief will only be of benefit in terms of extending the fatigue life of a welded structure in instances where the applied stress cycle is partly or wholly compressive. The extent of the benefit which results can be predicted reliably only if measurement of the residual stresses remaining can be carried out, for example by X-ray diffraction or the hole drilling method [11]; otherwise, it is prudent to assume that no significant benefit accrues.

Weld Profile

There is an increasing tendency for designers to insist that critical joints shall meet stringent requirements with regard to weld shape on the assumption that a higher fatigue strength will be achieved thereby. It is also common practice to blame poor weld profiles if premature fatigue failures are encountered in service. This attitude is fostered by some design codes, notably the American Petroleum Institute RP2A, which allows fatigue assessment for some joint types to be made according to a higher S/N curve for welds which are designated as "profiled," which in essence means those with a concave weld face merging smoothly into the adjoining base metal. Assessment of the extent to which this has been adequately achieved is a difficult job for the inspector, and aids such as the "dime test" have been employed to assist in this task. A dime is held on-edge against the weld face. Any gaps between the coin and the weld metal should not be large enough to allow a 1-mm-diameter wire to be passed through.

It is obvious that such idealized weld shapes cannot be achieved without a great deal of additional effort on the part of the welder, so that in large complex structures considerable additional expense may be involved. Therefore, it is of interest to examine the return on this investment by estimating the extent to which additional fatigue strength may be expected.

Smith and Hurworth [4] have predicted the fatigue strengths of transverse nonload-carrying joints with weld shapes varying from concave, with a low toe angle, to sharply convex with a toe angle approaching 90°. Figure 7

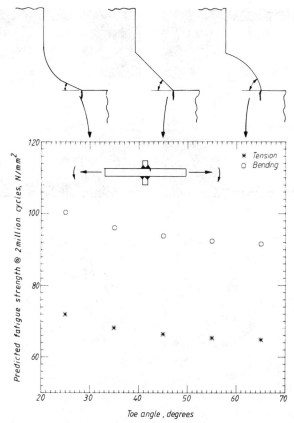

FIG. 7—*Effect of toe angle on predicted fatigue strength of transverse non-load-carrying joints (after Ref 4).*

shows predictions for toe angles in the range 25 to 65° for joints subjected to tension or bending. Intuitively one would expect that a more favorable stress distribution, and hence a lower stress concentration, would occur in the joint with the most concave weld face and the smoothest transition between weld metal and parent material, that is, the lowest toe angle. The results do show this trend to some extent, but the predicted variation in fatigue strength is relatively small, with only a 10% increase as the toe angle reduces from 65 to 25°.

These predictions may be compared with the results of fatigue tests reported by Gurney [12] in which the fatigue strengths of K butt welded specimens with welds ground to a smooth concave profile, which satisfied the dime test, were compared with those of untreated specimens (Fig. 8). It should be noted that no attempt was made to remove material at the weld toe where the inherent crack-like flaws referred to earlier are situated, which is known to give significant benefit (see for example the results pub-

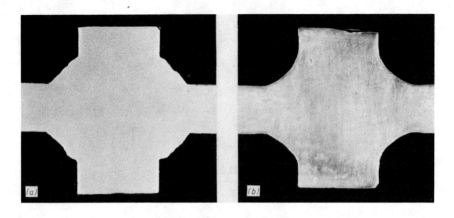

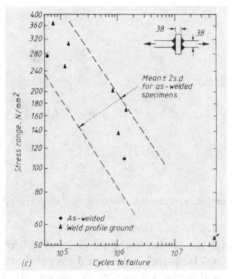

FIG. 8—*Test results showing effect of weld profiling* (12): (a) *as-welded profile*, (b) *after grinding, and,* (c) *comparison of fatigue test results for as-welded and profiled specimens.*

lished by Knight [*13*]). While the mean of the test data suggests a greater dependence on profile than predicted by the model, in two cases there was virtually no improvement in fatigue strength, and all the results lay within the scatter band for joints of that type in the untreated condition.

It therefore appears that the costly process of profiling joints is of limited benefit. At best the improvement in fatigue strength achieved is relatively small; at worst, it may be insignificant. Even where profiles are achieved by grinding, benefit cannot be relied upon unless the toe is specifically treated. A "cosmetic" treatment of the weld face may be totally ineffective.

Stress Concentration

From Eq 6 it will be seen that

$$N \propto \frac{1}{\Delta\sigma^m} \tag{9}$$

Since for structural steels m is approximately three, it will be appreciated that the fatigue life is highly sensitive to the applied stress range.

The stress parameter $\Delta\sigma$ is the nominal applied stress range in the vicinity of the joint. In evaluating the appropriate value of $\Delta\sigma$ it is not necessary to include the stress raising effect of the weld itself since this is modelled in the finite element solution; however, the effect of stress concentrations resulting from the design must be taken into account. Thus, for a joint located at the edge of an open hole, for example, the predicted fatigue strength will be approximately one third, and the fatigue life one third cubed, that is, $\frac{1}{27}$, of that for a joint in a member without the stress raiser.

Therefore it is apparent that fatigue strength is strongly dependent on detail design. In fact, this parameter is far more significant than those discussed earlier. Very significant improvements in fatigue strength may be achieved by avoiding or reducing stress concentrations where possible, locating welded joints away from structural discontinuities, and avoiding features which introduce secondary stresses.

While it is acknowledged that such principles were recognized long before fracture mechanics models of joint behavior were developed, an understanding of the principles of the model and the discipline of visualising failure in terms of crack propagation serves to improve appreciation of such effects.

Plate Thickness

In the event of a welded joint failing prematurely it is frequently necessary to increase the section size despite the additional weight and cost penalties incurred. Recently it has become apparent that even this simple remedy is not entirely straightforward since fatigue strength is related to section size, joints in thicker materials having lower fatigue strengths. This "size effect" had been appreciated for unwelded components for many years, but it was not until fracture mechanics models of fatigue behavior were developed that it was realized that the same may apply to welded joints [14]. It was found that fatigue lives predicted for a series of joints of identical geometry but of different scale decrease with increasing plate thickness as shown in Fig. 9. The key to this rather surprising prediction is the fact that the characteristic initial defect size is the same for all plate thicknesses, since it is influenced only by metallurgical parameters such as the composition

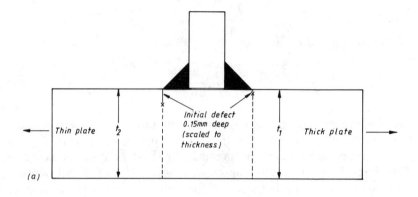

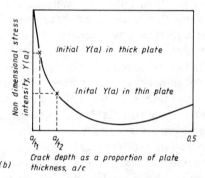

FIG. 9—*Variation of predicted fatigue strength with plate thickness for transverse non-load-carrying joints (after Ref 4).*

of the parent material, weld metal, slag etc. As shown in Fig. 10, the normalized stress-intensity factor, $Y(a)$, (which may be considered for simplicity as a stress concentration factor) varies with a/t. Thus, as indicated, the initial toe defect in the thicker joint is subjected to a much higher stress and, hence, grows faster than that in the thinner joint. Even though in the thicker plate the crack must grow further, it does so faster and hence causes failure earlier than that in the thinner plate.

Fatigue tests subsequently confirmed these predictions and allowed correction factors to be drafted [15]. For design purposes the fatigue strength is taken to be inversely proportional to the fourth root of plate thickness, that is

$$\Delta\sigma_0 \propto \frac{1}{t^{0.25}} \tag{10}$$

Referring again to the problem of increasing the section to avoid premature

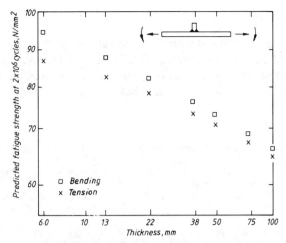

FIG. 10—*Schematic diagram showing the influence of plate thickness on the stress intensity field at the initial crack tip in a transverse joint loaded in tension:* (a) *relative size of initial toe defect and* (b) *influence of plate thickness on initial stress intensity.*

failure, it can be shown that, for joints loaded in tension, to increase the fatigue life from N_1 to N_2 requires an increase in thickness by a factor of $(N_2/N_1)^{1/m}$ when fatigue strength is assumed to be independent of thickness, or $(N_2/N_1)^{4/3m}$ when the thickness correction (Eq 10) is considered. Thus, for example, assuming $m = 3.0$, in order to double the life it is necessary to increase the section by 36% allowing the thickness correction, rather than 26% estimated when the correction is not considered.

Concluding Remarks

Fracture mechanics analysis is being used increasingly by engineers and designers in industry. This step is certainly to be encouraged since the discipline of considering the fatigue failure of welded joints in terms of crack propagation provides a valuable insight into the influence of materials and design parameters for both defective and nominally sound joints.

Research into the accuracy of the available models continues. In particular improvements are needed in the modelling of initial defects and the variation in crack shape during the life. In many instances cracking starts at a number of points along the weld as a series of separate semi-elliptical defects which subsequently coalesce, rather than a single defect as assumed at present. Improvement in the finite element modelling of cracks in complex geometries is also required to study joints such as tubular connections, for which two-dimensional analysis is not appropriate. With regard to materials data there is a lack of information on crack growth in the near threshold regime which limits the accuracy of predictions for joints subjected to spectrum loading with many cycles of small amplitude.

Acknowledgments

This work was carried out with the joint support of the United Kingdom Department of Industry and Research Members of The Welding Institute.

The authors are indebted to their colleagues at The Welding Institute, in particular Dr. S. J. Maddox for many helpful suggestions.

References

[1] Signes, E. G. et al, *British Welding Journal,* Vol. 14, No. 3, 1967, pp. 108–116.

[2] Paris, P. C., *Proceedings,* 10th Sagamore Conference, Syracuse University Press, 1965.

[3] Smith, I. J., *Proceedings,* 2nd International Conference on Numerical Methods in Fracture Mechanics, Swansea, 1980.

[4] Smith, I. J. and Hurworth, S. J., "The Effect of Geometry Changes Upon the Predicted Fatigue Strength of Welded Joints," Welding Institute Research Report 244/1984, July 1984.

[5] Gurney, T. R., "The Application of Fracture Mechanics to Fatigue of Welded Joints," Fracture Mechanics Seminar, ITBTP, St-Rémy-lès-Chevreuse, June 1982.

[6] Gurney, T. R. and Maddox, S. J., *Welding Research International,* Vol. 3, No. 4, 1973, pp. 1–54.

[7] Gurney, T. R., *Fatigue of Welded Structures,* 2nd Ed., Cambridge University Press, Cambridge, U.K., 1979.

[8] Maddox, S. J., *Welding Research International,* Vol. 4, No. 1, 1974, pp. 36–60.

[9] Burdekin, F. M. and Dawes, M. G., "Practical Uses of Yielding and Linear Elastic Fracture Mechanics with Particular Reference to Pressure Vessels," International Mechanical Engineering Conference on the Practical Application of Fracture Mechanics to Pressure Vessel Technology, London, May 1971.

[10] Maddox, S. J., "Some Aspects of the Influence of Residual Stresses on the Fatigue Behaviour of Fillet Welded Joints in Steel," Welding Institute Research Report 123/1980, Sept. 1980.

[11] Parlane, A. J. A., "The Determination of Residual Stresses: A Review of Contemporary Measurement Techniques," Welding Institute Conference, London, Nov. 1977.

[12] Gurney, T. R., *Metal Construction,* Vol. 15, No. 1, pp. 37–44.

[13] Knight, J., *Welding Research International,* Vol. 8, No. 6, 1978.

[14] Gurney, T. R. and Johnston, G. O., *Welding Research International,* Vol. 9, No. 3, 1979.

[15] Gurney, T. R., 2nd International Conference on Behaviour of Offshore Structures, Imperial College, London, Aug. 1979.

Claus Mattheck,[1] Bernhard Kneifel,[1] and Peter Morawietz[1]

Fatigue Crack Growth and Crack Arrest in the Nails Used for Intramedullary Fixation of Femur Fractures

REFERENCE: Mattheck, C., Kneifel, B., and Morawietz, P., **"Fatigue Crack Growth and Crack Arrest in the Nails Used for Intramedullary Fixation of Femur Fractures,"** *Case Histories Involving Fatigue and Fracture Mechanics, ASTM STP 918,* C. M. Hudson and T. P. Rich, Eds., American Society for Testing and Materials, Philadelphia, 1986, pp. 361–376.

ABSTRACT: The fractures of large bones are often fixed by use of intramedullary nails. Sometimes these nails show cracks originating at the end of their slot and also at perforations which are necessary for the screws of interlocking nails. In this paper it is shown that notch stresses near these perforations are responsible for crack initiation. A fracture mechanical analysis is carried out in order to calculate the stress-intensity factor for such cracks under physiological loading. Fatigue considerations are presented, and conditions for crack arrest are shown. A new nail profile is proposed which reduces the notch stresses significantly.

KEY WORDS: fracture of bones, intramedullary nail, notch stresses, stress intensity factor

Intramedullary fixation is a well known and accepted method for the treatment of certain fractures of the long bones. Implants with a partially slotted, partially closed design have a high incidence rate (20 to 30%) of circumferential cracks initiating in the transition zone (Fig. 1 and Fig. 4), with very few cases, however, proceeding to complete failure [1].

In this paper two categories of intramedullary nails are considered: (*a*) the simple intramedullary nail (Fig. 2) (fully slotted or partially slotted tube, whereby different types of cross section are realized) and (*b*) the interlocking nail (Fig. 3) (fully slotted or partially slotted tube with perforations in the upper and lower part for the interlocking screws).

Sometimes the interlocking nails fail due to cracks originating from holes (Fig. 5) in the nail shaft. In order to get a first impression of the physiological

---

[1] Kernforschungszentrum Karlsruhe, Institut für Reaktorbauelemente, Arbeitsgruppe Zuverlässigkeit und Schadenskunde im Maschinenbau, Karlsruhe, West Germany. Mr. Morawietz is now at RW-TÜV Essen, West Germany.

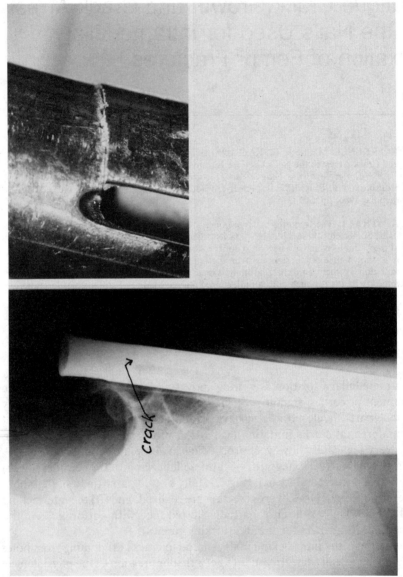

FIG. 1—*Crack at the end of the slot in an intramedullary nail (X-ray graph and detail).*

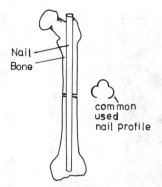

FIG. 2—*Intramedullary nail in human femur.*

loading situation the model of the "one leg stand" [2] (Fig. 6) is used for the human femur treated as a bone-nail composite in a simplified FEM-model (Fig. 7). From these rough considerations the load acting on the nail is known and can be then used for more refined separate modelling of the nail with a slot and holes. This will give information about the notch stresses which are responsible for the crack initiation at the slot end as well as at the perforations in the interlocking nail. A fracture mechanical considera-tion will show under which conditions crack arrest may happen.

Finally some further developments in the nail profile are proposed to avoid crack initiation.

FEM Analysis of the Bone-Nail Composite

Figure 7 shows the finite-element method (FEM) model of the upper (proximale) part of the femur. The part considered is shown in detail in the circle of Fig. 6, which is loaded as sketched. The nail is assumed to be a circular tube without slots and holes. This relatively simple model has

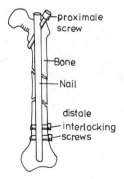

FIG. 3—*Interlocking nail in human femur.*

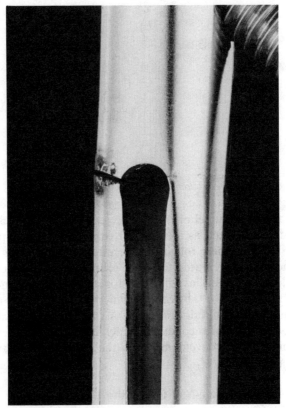

FIG. 4—*Crack originating from the end of the slot in intramedullary or interlocking nail with drilling hole at the slot end.*

been chosen for the bone because of the large range of individual variations of the femur shaft.

Therefore, only a simple model is used which is accurate enough for the conclusions drawn from it, not more.

The assumed Young's moduli for the three sections sketched in Fig. 7 are $E_I = 2 \times 10^5$ N mm^{-2} (steel), $E_{II} = 14\,200$ N mm^{-2} (cortical bone), and $E_{III} = 380$ N mm^{-2} (cancellous bone).

The nail is assumed to be surrounded by a very soft film of $E_{IV} = 20$ N mm^{-2} and a thickness of $0.03\,d_N$, whereby $d_N = 15$ mm, is the nail diameter.

Figure 8 shows the axial stresses along the cuts B-B' and C-C' of Fig. 7. It can be seen that the nail is loaded mainly by bending. This result is not changed qualitatively if the cancellous bone is assumed to behave as a cortical one ($E_{II} = E_{III} = 14\,200$ N mm^{-2}).

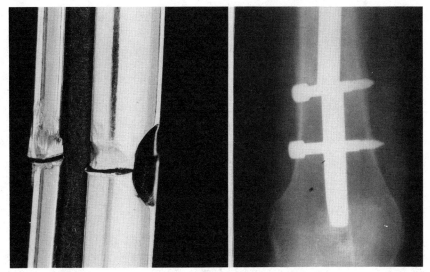

FIG. 5—*Crack originating from a hole, without screw, in an interlocking nail and X-ray graph of a cracked nail with screws.*

With this assumption the whole range of mechanical stiffness variations of the cancellous bone should be covered. The nail is loaded mainly by bending stresses when exposed to physiological conditions. Therefore, only this case of loading has to be considered when the nail is studied in more detail.

Notch Stresses at the End of the Slot in Intramedullary Nails

There is no doubt that the end of the slot will induce notch stresses so that the axial stresses will be amplified near the slot end. Because of the large number of differently shaped slot ends, a three-dimensional analysis would be too expensive. Therefore, the problem is reduced to a plane one (Fig. 9). Simple axial tension loading was used in order to have symmetrical loads with respect to the slot axis. This also reduces the FEM-structures by 50%. Of course, these simplifications will result in stress distributions other than what the physiological conditions would give. But the principal notch effect of the various slot shapes can be demonstrated sufficiently with these simplifications. Figure 10 shows the plots of isolines of axial stresses, whereby the location of the maximum axial stress is drawn as a black dot. The factor "f" is defined as follows

$$f = \frac{\text{maximum stress for the actual slot end geometry}}{\text{maximum stress in the fully slotted nail}}$$

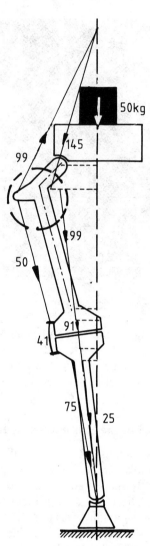

FIG. 6—*Physiological load situation of "one leg stand."*

Figure 10 indicates that by drilling a hole (Fig. 10*c*) at the slot end or some manufacturing notches (Fig. 10*d*) the notch stresses may amplify significantly.

From all these considerations it immediately follows that the best design solution, in order to avoid cracking at the slot end, is to use a fully slotted nail.

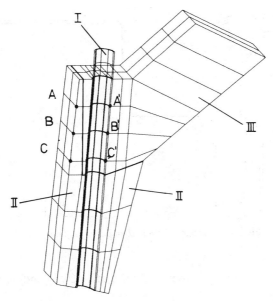

FIG. 7—*FEM-model of the upper (proximale) part of the human femur.*

Notch Stresses at the Holes in an Interlocking Nail

Figure 11 shows the simplified flow of force in an interlocking nail, which is used for fractures in the very upper and lower part of the bone or in very complex fractures. Again the nail is loaded mainly by bending, and will cause notch stresses in the neighborhood of the holes. The cracks originating from those holes have been also observed where no interlocking screw is brought into the holes at one end of the nail (dynamic nailing).

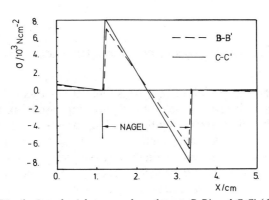

FIG. 8—*Distribution of axial stresses along the cuts B-B' and C-C' (from Fig. 6).*

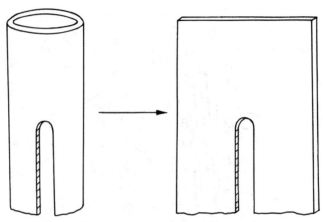

FIG. 9—*Transition from a 3D-model to a plane model.*

Therefore, in the present paper, only the effects of bending loads will be studied.

The contact problem between screw and nail is not the subject of this paper, because it is not important to know how the axial force flow is induced in the nail. The conclusions drawn from the present consideration would not be altered because of the screw contact, for it also initiates an axially directed force flow.

The geometry in Fig. 12a was used in order to calculate the stress distribution for external bending loads. The related FEM-model is plotted in Fig. 12b. Figure 13 shows the stress distribution normalized on the maximum bending stress applied at the cross section without a hole. The axis of bending is located under physiological loading conditions. The presence of the hole amplifies the stresses by a factor of about 3.5. In some cases this may cause crack initiation.

Fracture Mechanical Analysis of the Cracks Originating from the Holes in the Nail

In Ref. *3* the calculation of the stress-intensity factors for the crack geometry in Fig. 12 is given in detail for the loading case of axial tension and two types of bending. Here only the bending case of Fig. 13 is of interest, which represents a good approximation of the situation of physiological loading. The method for calculating the related stress-intensity factors [3] will be only described here in principle. In order to get the weight functions for this crack problem, the reference crack opening displacement field u_r has been calculated for four different crack depths. The reference loading case was constant pressure σ_0 acting internally on the crack faces. The

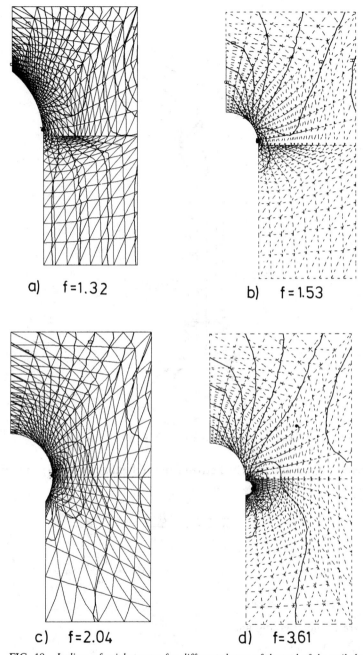

a)   f=1.32

b)   f=1.53

c)   f=2.04

d)   f=3.61

FIG. 10—*Isolines of axial stresses for different shapes of the end of the nail slot.*

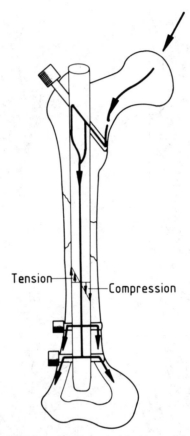

FIG. 11—*Simplified force flow in an interlocking nail-bone composite.*

displacement approximation by Petroski and Achenbach [4] was generalized as follows

$$u_r(x,a) = \frac{\sigma_0}{E\sqrt{2}} \left[4F(a)\sqrt{a}\sqrt{a-x} + \frac{G(a)}{\sqrt{a}} \right.$$
$$\left. \times (a-x)^{3/2} + \frac{W(a)}{a^{3/2}}(a-x)^{5/2} \right] \quad (1)$$

where

σ_r = reference loading and
E = Young's modulus.

The unknown functions $F(a)$, $G(a)$, and $W(a)$ have been expressed as a

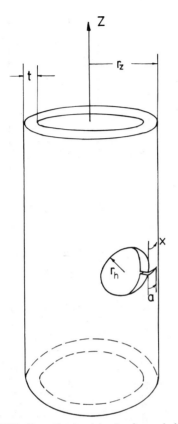

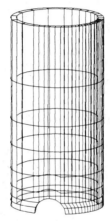

FIG. 12a—*Crack originating from a hole in the nail (sketch).*

FIG. 12b—*FEM model used for the calculation of the stress distribution and the stress-intensity factor.*

power series in (a/r). The constants in this power series have been determined by fitting Eq 1 to the FEM-displacement fields. The basic equation of the weight function method [5,6]

$$K = \frac{E}{K_r} \int_0^a \sigma(x) \frac{\partial u_r}{\partial a} dx \qquad (2)$$

reproduces K_r, if $\sigma(x) = \sigma_0 =$ constant is the reference load.
By this method the reference stress-intensity factor is

$$K_r = \left[E\sigma_0 \int_0^a \frac{\partial u_r}{\partial a} dx \right]^{1/2} \qquad (3)$$

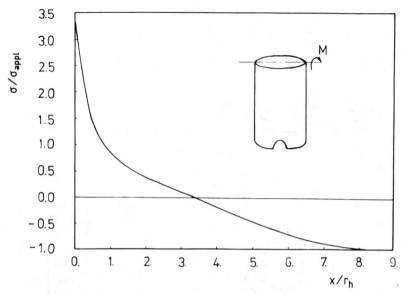

FIG. 13—*Circumferential distribution of axial bending stresses normalized on the externally applied maximum bending stress.*

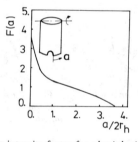

FIG. 14—*Normalized stress-intensity factor for physiologically directed bending load.*

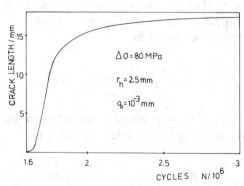

FIG. 15—*a(N) fatigue crack growth curve under the physiological bending load from Figs. 6 and 13.*

one side cracked   both sides cracked

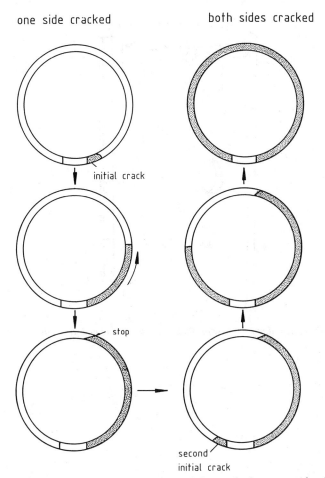

initial crack

stop

second
initial crack

FIG. 16—*Schematic sketch of crack growth history (only one considered).*

Using Eqs 1–3 for every axial stress $\sigma(x)$, the related stress-intensity factor can be computed by simple numerical integration. For the stress distribution from Fig. 13 the normalized stress-intensity factor $F(a) = K(a)/\sigma_{appl} \sqrt{\pi a}$ is calculated, where σ_{appl} is the maximum bending stress applied externally. Figure 14 shows high F-values for short crack lengths due to the notch stresses and, for increasing crack lengths, a continuously decreasing behavior of $F(a)$ down to negative values at which no further crack growth by fatigue is possible.

In order to calculate the number of physiological load steps up to crack arrest, a fatigue crack analysis has been carried out. The one leg stand (Fig. 6) was assumed to be representative for a walking step. Amplification of these loads by inertia effects has not been regarded because it is plausible that a

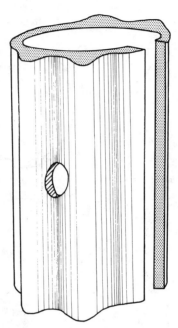

FIG. 17—*One possibility for a new profile of an interlocking nail with reduced notch stresses at the holes.*

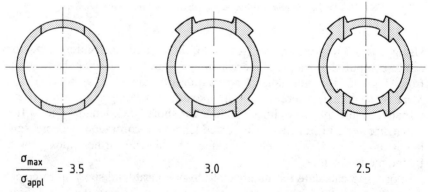

$$\frac{\sigma_{max}}{\sigma_{appl}} = 3.5 \qquad 3.0 \qquad 2.5$$

FIG. 18—*Successive reduction of notch stresses by use of the new profile with local wall reinforcement near the holes.*

person with a recently broken leg will neither jump nor walk very "dynam-ically." For the calculation the Paris law was used

$$\frac{da}{dN} = C(\Delta K)^n \qquad (4)$$

where

$C = 6.758 \times 10^{-12}$ and
$n = 3.726$

Note that the dimension of K is MPa$\sqrt{m}$ and a is given in m in Eq 4.

Starting from a very small initial crack of 10^{-6} m the $a(N)$-curve is shown in Fig. 15. After about 3×10^6 load steps no further significant crack growth can be seen. The crack growth has stopped. Nevertheless the cross section of the nail is now reduced by nearly half. It is possible that a second crack may originate from the opposite circumference of the hole, and this type of crack would not stop. Complete failure would occur. This is shown schematically in Fig. 16. Because of the fact that the cracks are initiated by notch stresses, the only way to avoid this crack initiation is to reduce these notch stresses by some constructive developments such as those described in the next section.

A New Nail Profile with Higher Fatigue Strength

In Ref. 7 a new nail profile is proposed. The main idea of this development is sketched in Fig. 17. The notch stresses result because the axial flow of forces is "stopped" at the hole, and this force flow has to pass the perforated cross section beside the hole, where higher stresses occur. These are the notch stresses. If the thickness of the nail wall is reinforced at both sides of the hole, over the full length of the nail, these notch stresses may be reduced. Figure 18 shows three different nail profiles loaded by simple axial tension. They are labeled with the ratio of maximum notch stress σ_{max} normalized on the externally applied stress σ_{appl}. It can be seen that the profile with reinforced wall thickness at the outer side of the cross section only, does not reduce the notch stresses as strikingly as the reinforcement of both the outside and inside wall thickness. The new profile will be man-ufactured in the near future.

Acknowledgments

The authors would like to thank Mrs. Burkhardt for her help in carrying out parts of the FEM-analysis as well as the surgeons Drs. Börner and Fux for many discussions on the clinical point of view.

References

[1] Beaupre, G., Schneider, E., and Perren, S. M., "An Investigation of the Mechanical Failure of a Partially Slotted, Femoral, Intramedullary Nail: A Finite Element Analysis," submitted 1983 to *Journal of Orthopedic Research*.

[2] Pauwels, F., *Gesammelte Abhandlungen zur funktionellen Anatomie des Bewegungsapparates* (in German), Springer, Berlin, 1965.

[3] Mattheck, C., Morawietz, P., and Munz, D., "Calculation of the Stress Intensity Factor of a Circumferential Crack in a Tube Originating from a Hole Under Axial Tension and Bending Loads," *Engineering Fracture Mechanics*, Vol. 22, No. 4, 1985, pp. 645–650.

[4] Petroski, J. and Achenback, J., "Computation of a Weight Function from a Stress Intensity Factor," *Engineering Fracture Mechanics*, Vol. 10, 1978, pp. 257–266.

[5] Bueckner, H., "A Novel Principle for the Computation of Stress Intensity Factors," *ZAMM*, Vol. 50, 1970, pp. 529–546.

[6] Rice, J., "Some Remarks on Elastic Crack Tip Stress Fields," *International Journal of Solids and Structures*, Vol. 8, 1972, pp. 751–758.

[7] Börner, M. and Mattheck, C., Patent-Anmeldung Nr. 100 259 7, 1984, Kernforschungszentrum Karlsruhe (in German).

Clare M. Rimnac,[1] *Timothy M. Wright,*[1] *Donald L. Bartel,*[1]
and Albert H. Burstein[1]

Failure Analysis of a Total Hip Femoral Component: A Fracture Mechanics Approach

REFERENCE: Rimnac, C. M., Wright, T. M., Bartel, D. L., and Burstein, A. H., **"Failure Analysis of a Total Hip Femoral Component: A Fracture Mechanics Approach,"** *Case Histories Involving Fatigue and Fracture Mechanics, ASTM STP 918,* C. M. Hudson and T. P. Rich, Eds., American Society for Testing and Materials, Philadelphia, 1986, pp. 377–388.

ABSTRACT: Failure analysis was conducted on 21 fractured femoral components from Trapezoidal-28™ total hip replacements performed at The Hospital for Special Surgery. These components were made from 316L stainless steel and had a trapezoidally shaped cross section in the neck and stem portions. Fractographic examination revealed multiple cracks on the medial side of the stems and a fracture surface with crack initiation sites from both the medial and lateral surfaces. Electron microscopy revealed striations on the fracture surfaces of both medial and lateral cracks, indicating that crack growth occurred by a fatigue fracture process in both cases. Analytical examination using curved beam theory with a bilinear stress-strain relationship predicted the medial cracks to initiate first, due to residual tensile stresses following an overload. In the presence of a medial crack, higher tensile stresses were predicted on the lateral side, making it the next most likely site for crack initiation. Fracture occurred when a lateral crack joined with a pre-existing medial crack. It was concluded this fracture mode was due to the combined effects of the trapezoidal stem design and the mechanical properties of 316L stainless steel.

KEY WORDS: total hip replacement, femoral component, failure analysis, fatigue, fracture mechanics, residual stress

Total joint replacement is an accepted treatment for diseased or damaged hip joints. Among the possible complications following replacement is fracture of the metal femoral component, though the incidence of fracture is reported to be less than 1.0% [1]. Component fracture has been shown to depend on material, design, surgical technique, and patient factors such as weight and activity level [2]. For example, in a clinical review of 120 cases

[1] Assistant scientist, associate scientist, associate scientist, and director, respectively, Department of Biomechanics, The Hospital for Special Surgery (affiliated with The New York Hospital and Cornell University Medical College), New York, NY 10021.

of fractured femoral components of a single design, a linear relationship between patient weight and time to fracture was found [3], with heavier patients experiencing fracture in less time.

In the early 1970s, the Trapezoidal-28™ (T-28) total hip was designed in an effort to improve range of motion, stability, and stem strength over previous hip designs. The Trapezoidal-28 was named for its trapezoidal stem and neck cross section and its 28-mm-diameter femoral head. A trapezoidal shape to the stem was considered preferrable to a diamond shape in maximizing the pressure of the acrylic grout in the femur when the prosthesis was inserted [4]. From 1973 to 1979, 805 patients received this implant at The Hospital for Special Surgery. Since that time, a minimum of 21 patients from this population have been treated for removal of a fractured femoral component (the number of additional patients treated elsewhere is unknown). Thus, a minimum fracture incidence of the T-28 femoral component in this series is 2.6%, about 4 times that reported for other femoral component designs.

We are currently engaged in a clinical review of the 805 patients with implanted T-28 hip components to identify the clinical factors (for example, height, weight, and diagnosis) which may indicate increased risk of fracture of the femoral component. Failure analysis of the retrieved femoral components, to determine the material and design factors which are contributing to the high incidence of fracture, is an important part of the overall investigation. The results of the failure analysis are reported here.

Materials and Methods

Twenty-one fractured femoral components were examined macroscopically. Table 1 provides pertinent patient and implant information. The fracture surfaces of the components were examined for crack initiation sites and for fracture markings indicating failure mode. In addition, the femoral components were examined for other cracks and for evidence of corrosion.

All T-28 femoral components are machined from surgical grade 316L stainless steel in the wrought and lightly cold-worked condition. Detailed failure analysis was conducted on 4 of the 21 femoral components using scanning and transmission electron microscopy (SEM and TEM). For TEM, two-stage replicas of the fracture surfaces were created, following standard techniques [5].

Results and Discussion

Experimental

Microscopic examination of the femoral components revealed minimal amounts of corrosion in the form of mild pitting, typically near the distal tip of the stem. Eighteen of the 21 components failed by fracture of the

TABLE 1—*Patient and implant data for the 21 fractured T-28 femoral components.*

Patient Weight, kg	Sex	Length of Service, months	Side	Implant Neck Size	Implant Stem Size	Multiple Cracks	Fracture Site[a]
48	F	58	L	medium	small	no	100
81	F	72	L	short	small	yes	75
53	F	76	L	medium	small	no	103
67	F	79	R	medium	small	yes	96
57	F	89	R	short	thin	no	90
79	F	69	L	medium	large	yes	107
76	M	74	L	long	large	yes	99
83	M	79	R	short	small	yes	88
74	M	96	L	medium	small	yes	100
100	M	64	R	medium	large	yes	91
98	M	76	R	long	large	yes	neck
100	M	85	L	medium	large	no	108
76	M	106	R	long	large	no	neck
58	F	109	R	medium	small	no	98
58	F	143	L	medium	small	yes	80
83	M	97	R	medium	large	yes	97
77	M	116	L	long	large	no	115
87	M	115	L	long	large	yes	106
83	F	105	L	medium	small	yes	102
70	M	113	L	medium	large	yes	114
92	M	97	L	long	large	yes	neck

[a] Distance in millimetres from distal tip.

stem, usually within the proximal third of the stem portion of the component (Fig. 1). The other three components failed by fracture of the neck [6]. Multiple cracks (as many as 7) were found on 14 of the 21 components. In every case but one, these cracks originated on the medial side of the stem (Fig. 2).

Two possible explanations for the multiple cracks were considered: stress corrosion cracking and fatigue. Stress corrosion cracking was unlikely, however, as it has been shown that 316L stainless steel is not susceptible to stress corrosion cracking in the physiologic environment [7]. In addition, since stress corrosion cracking occurs under tensile stress conditions, multiple cracks along the lateral aspect of the stem would have been also expected. Yet, only one of these cracks originated on the lateral side of the stem. Therefore, initiation and propagation of these cracks must have been associated with a fatigue fracture process.

Examination of the fracture surfaces by light microscopy indicated a crack initiation site in the posterior corner on the medial side of the stem (point A in Fig. 3). A major step in the surface was usually observed near the center of the cross section and appeared to be the boundary between the medial crack and a crack originating on the lateral side of the stem. The fracture surface appearance of the lateral crack often revealed clam shell

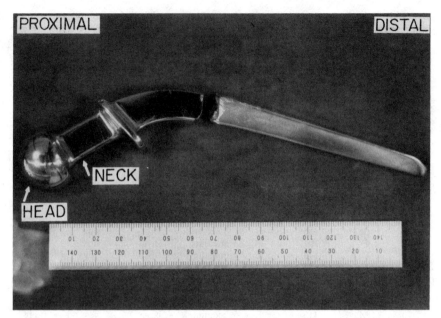

FIG. 1—*Fractured T-28 femoral component, showing proximal and distal aspects and femoral head and neck locations.*

markings emanating from the anterior corner, consistent with a fatigue failure process (point B in Fig. 3).

Examination of five of the fracture surfaces by SEM showed no evidence of gross metallurgical defects, such as porosity or inclusions, at the crack initiation sites on either the medial or lateral side. Thus, crack initiation is assumed to have occurred through a cumulative damage process at the surface of the component. Examination of one of the medial cracks using TEM revealed isolated regions of striations (Fig. 4), indicating a fatigue fracture mechanism. Similarly, TEM examination of the lateral cracks also revealed fatigue striations (Fig. 5). There were more striations on the lateral side than on the medial side. The striation spacing on the lateral side ranged from 7×10^{-5} to 3×10^{-4} mm and increased with increasing distance from the crack origin.

An estimate of the cyclic stress on the lateral side of a stem was made using nine measured striation spacings at various crack lengths, a, and from four different fracture surfaces (Table 2). In a study of fatigue crack propagation in 316L stainless steel weldment by Pickard et al [8], estimates of the microscopic crack growth rate (that is, striation spacing) were given for the parent metal at two cyclic stress intensity values ΔK as follows: when $\Delta K = 28$ MPa $\cdot$ m$^{1/2}$, the microscopic growth rate ≈ 0.52 μm/cycle and when $\Delta K = 50$ MPa $\cdot$ m$^{1/2}$, the microscopic growth rate ≈ 1.2 μm/cycle. Using these two correlations, we assumed a linear relationship between ΔK

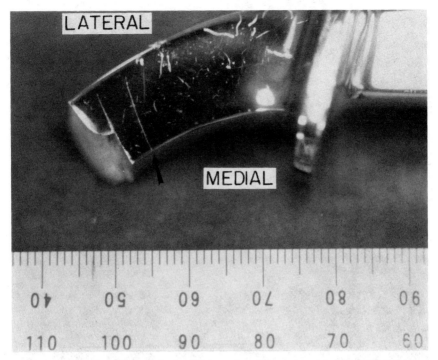

FIG. 2—*Light photomicrograph of multiple cracks originating on the medial side of the stem portion of a T-28 component* (arrow).

and the microscopic crack growth rate to estimate the prevailing ΔK on our fracture surfaces where striation spacings were measured. The cyclic stress ($\Delta\sigma$) on the stem was then determined from the stress-intensity solution for a semi-circular corner crack; $K = (1.12)^2(2/\Pi)\sigma(\Pi a)^{1/2}$ [9]. The estimated cyclic stresses for each crack length, a, are shown in Table 2 and range from 115 to 244 MPa.

Analytical

Fractographic analysis demonstrated that fracture initiated from medial cracks occurring by a fatigue mechanism. Fatigue crack growth requires tensile stress, yet femoral hip components are subjected to combined axial compressive and bending loads, resulting in compressive stresses on the medial side of the component where cracks initiate. To explain the initiation and propagation of these cracks, an analysis of residual stresses due to inelastic bending of the stem was conducted using curved-beam theory with a bilinear stress-strain relationship [10].

For this analysis, the stem cross section was divided into 484 quadrilateral subareas and a uniform, constant strain over each subarea was assumed. A bilinear approximation of the stress-strain relationship for 316L stainless

FIG. 3—*Light photomicrograph of fracture surface showing* (A) *crack growth from postero-medial corner and* (B) *clam shell markings emanating from anterolateral corner.*

steel was used (see following) which took into account the influence of strain hardening and the Bauschinger effect [11]. Loading was then specified, which consisted of bending in the lateral-medial plane and a superimposed compressive axial load applied at the centroid of the cross section.

The stresses due to a specified loading were determined as follows. First, an arbitrary strain distribution was chosen based on the deformation of curved beams under combined loading. The corresponding stress distribution was determined from the stress-strain relationship. The resultant bending moment and axial compressive load due to this stress distribution was then determined by integration. In general, the resultant bending moment and axial compressive load differed from the specified loading. Therefore, the strain distribution was modified automatically using an iterative optimi-

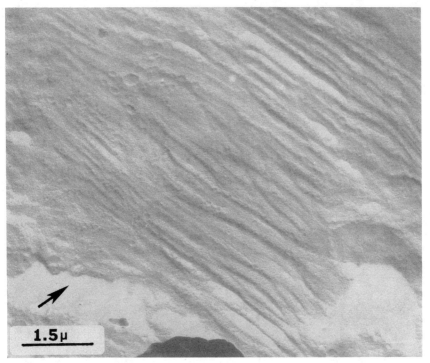

FIG. 4—*Transmission electron fractograph showing fatigue striations from fracture surface of a medial crack. Arrow indicates direction of crack growth.*

zation procedure until the resultant moment and load agreed with the specified moment and load. After agreement was obtained, the final strain distribution was saved to provide a starting distribution for the next specified loading. In this way, the stress distribution resulting from a sequence of applied loads was determined interactively. For example, the residual stress distribution due to an overload was determined by specifying the overload as the first load in the sequence and a zero load as the second load in the sequence.

The same analytical procedure was used to analyze stems with cracks on the medial side. A crack was modeled by modifying the stress-strain relationship so that material on either side of the crack could withstand only compressive stresses. When the strain in the region of the crack was compressive, the stress was determined from the normal stress-strain relationships. When the strains across the cracks were tensile, the stresses were set to zero.

The material properties (elastic modulus, yield strength, and post-yield modulus) were chosen so that load-strain curves predicted by the analytical model agreed with curves obtained experimentally. We had previously de-

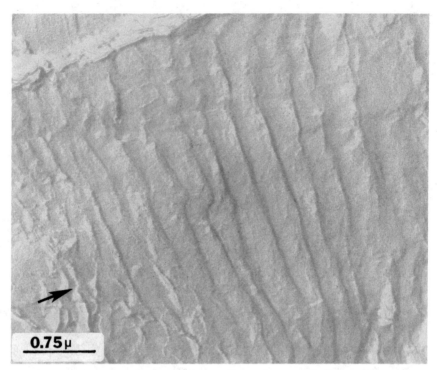

0.75μ

FIG. 5—*Transmission electron fractograph showing fatigue striations from fracture surface of a lateral crack. Arrow indicates direction of crack growth.*

termined load-strain data for T-28 components which had been loaded until strains on the medial side of the stem (measured using metal foil strain gages) exceeded the yield strain [12]. We used the analytical model to predict the strain at the gage location producing the largest strain reading. We then modified the yield strength and post-yield modulus of the bilinear stress-strain representation by trial and error until the load-strain relationship predicted by the model agreed with that determined experimentally. The properties obtained in this manner were: elastic modulus, 200 GPa; yield strength in tension and compression, 550 MPa; and post-yield modulus in tension and compression, 20 GPa. The strength values are reasonable for 316L stainless steel as used in the manufacture of T-28 components [13].

The load applied to the head of the femoral component of a total hip replacement varies depending upon the activity. For level walking, forces of 3.3 to 5.0 times body weight have been estimated [14]. Activities such as ascending or descending stairs and rising out of a chair tend to increase the forces across a joint. Sudden activities, such as stepping off an unexpected curb, can further increase these joint loads. In this analysis, we used a load of 3.5 times body weight to represent normal cyclic loading activity.

TABLE 2—*Fatigue fracture surface striation spacing measurements and estimated stress-intensity values and stresses for four components.*

Stem	Crack Length, a mm	Striation Spacing, mm	Estimated ΔK, MPa · m$^{1/2}$	Estimated $\Delta\sigma$, MPa
1	3	2.15×10^{-4}	18.1	233
	5	3×10^{-4}	20.9	208
	9	4.15×10^{-4}	25.0	186
2	7	7.73×10^{-5}	13.7	115
	9	1.8×10^{-4}	17.0	127
3	3	1.35×10^{-4}	15.6	201
	7.5	2.75×10^{-4}	20.0	163
4	1.5	6.7×10^{-5}	13.4	244
	11	2.6×10^{-4}	19.6	132

We used a load of 7.0 times body weight to represent an overload. This load is probably a lower bound on the loads that can occur under sudden loading conditions.

Stress distributions were determined for a sequence of loads at a cross section in the curved portion of the stem of a "large" T-28 femoral component where medial cracks were found. The sequence was to apply an overload, then a zero load, and then a load typical of those which would be cyclically applied during normal activities. The overload consisted of a bending moment of 110 Nm and a compressive force of 4.33 kN. This corresponds to a load on the femoral head of 7 times body weight for a 91 kg (200 lb) individual. Figure 6 shows the stress distribution, from the lateral

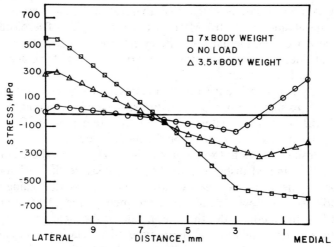

FIG. 6—*Calculated stress distribution through the cross section for three sequential applied moment and load histories.*

to the medial side, of the stem resulting from this loading condition. The stress on the medial side of the stem was -604 MPa and the stress on the lateral side was 556 MPa. Therefore, the stress on the medial side of the stem was much greater than the compressive yield stress (-550 MPa), while the stress on the lateral side was slightly greater than the tensile yield stress (550 MPa). When the load was removed (zero load applied), a residual tensile stress of 270 MPa occurred on the medial side and a residual tensile stress of 16.6 MPa occurred on the lateral side of the stem (Fig. 6).

The normal load following an overload consisted of a bending moment of 55 Nm, and a compressive force of 2.17 kN, corresponding to a load on the femoral head of 3.5 times body weight for a 91 kg (200 lb) individual [14]. This resulted in a compressive stress of -207 MPa on the medial side and a tensile stress of 289 MPa on the lateral side (Fig. 6). Thus, if the load on the femoral head were cycled between 0 and 3.5 times body weight following an overload, the cyclic stress range on the medial side would be -207 to 270 MPa, and on the lateral side it would be 16.6 to 289 MPa. The stress range on the medial side is more severe than that on the lateral side, which implies that crack initiation should occur first on the medial side. Note, too, that the applied stresses calculated from the measured striation spacings in the previous experimental section are in good agreement with the stress range calculated for the lateral side of the stem in this analysis.

Since the analysis predicts that crack initiation will occur first on the medial side, subsequent initiation of the crack on the lateral side was then examined by analyzing a stem with a 2.5-mm-deep medial crack. Again, a 7 times body weight overload was first applied. When the stem was unloaded to zero, the stress distribution across the entire stem cross section was near zero, except for a small residual tensile stress in the vicinity of the crack (Fig. 7). A 3.5 times body weight load then resulted in a stress of 286 MPa on the lateral side and -167 MPa on the medial side. Therefore, with a crack present on the medial side, the cyclic stresses were more severe on the lateral side, suggesting that the next most likely site for crack initiation would be the lateral side.

This analysis is consistent with the macroscopic observations that failed components exhibited a primary fracture surface with crack initiation sites on both the medial and lateral sides of the stem portion. The analysis suggests that cracks first occurred on the medial side of the stem due to the residual tensile stresses following an overload. However, these cracks only propagated a limited distance into the stem because the residual tensile stress on the medial side of the stem (due to compressive yielding from the overload) decreased with depth from the medial side. After a crack had occurred on the medial side, the lateral side of the stem was subjected to higher tensile stress and thus was more susceptible to fatigue crack initiation. Failure occurred when a crack which initiated on the lateral side eventually joined with a pre-existing medial crack.

Only a bending moment in the lateral-medial plane was assumed in this

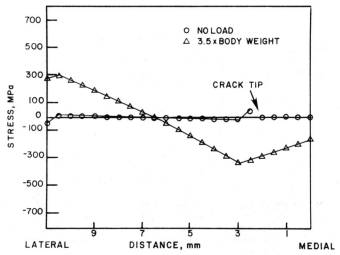

FIG. 7—*Calculated stress distribution through the cross section in the presence of a 2.5-mm-medial crack for two sequential applied moment and load histories.*

analysis; thus, crack initiation was assumed to occur at the midpoint of the medial side. However, fractographic examination of the stems indicated crack initiation occurred in the posterior corner of the medial side. This may be explained by an out-of-plane bending moment on the hip, due to a posteriorly directed compressive load [*14,15*]. The more severe loading condition on the posterior side would predict posterior corner crack initiation, as we observed fractographically.

Summary

Fracture of the Trapezoidal-28 femoral hip component is related to its design. The medial side of the stem is not as wide as the lateral side, causing residual tensile stresses on the medial side from a compressive overload. Other femoral hip components are designed with the medial side more narrow than the lateral side. However, only one of these designs is made from 316L stainless steel and the mode of failure of this design is similar to what we have found for the T-28 [*16*]. Therefore, it appears that the fracture mode of multiple fatigue cracks on the medial side of the stem, followed by crack growth on the lateral side, is due to the combined effects of the trapezoidal stem design and the mechanical properties of 316L stainless steel. This particular fracture mode is apparently more severe than other fracture modes of femoral hip components, as seen by the much higher incidence of fracture of the T-28.

Acknowledgments

The support of the Clark Foundation is gratefully acknowledged.

References

[1] Chao, E. Y. S. and Coventry, M. B., *Journal of Bone and Joint Surgery*, Vol. 63A, 1981, pp. 1078–1094.

[2] Galante, J. O., *Journal of Bone and Joint Surgery*, Vol. 62A, 1980, pp. 670–673.

[3] Wroblewski, B. M., *Acta Orthopaedica Scandinavica*, Vol. 53, 1982, pp. 279–284.

[4] Amstutz, H. C., *Clinical Orthopaedics and Related Research*, Vol. 95, 1973, pp. 158–167.

[5] *Metals Handbook*, 8th ed., Vol. 9, American Society for Metals, Metals Park, OH, 1974, pp. 24–26.

[6] Burstein, A. H. and Wright, T. M., *Journal of Bone and Joint Surgery*, Vol. 67A, No. 3, 1985, pp. 497–500.

[7] Williams, D. F., *Biocompatibility of Orthopaedic Implants*, D. F. Williams, Ed., Vol. 1, CRC Press, Boca Raton, 1982, pp. 197–229.

[8] Pickard, A. C., Ritchie, R. O., and Knott, J. F., *Metals Technology*, June, 1975, pp. 253–263.

[9] *The Stress Analysis of Cracks Handbook*, Tada, H., Paris, P. C., and Irwin, G. R., Del Research Corp., Hellertown, PA, 1973, pp. 8.1 and 26.1.

[10] Bartel, D. L. and Burstein, A. H., to be submitted, *Journal of Biomechanical Engineering*.

[11] Juvinall, R. C., *Engineering Considerations of Stress, Strain, and Strength*, McGraw-Hill, New York, 1967, p. 98.

[12] Reuben, J. D., Eismont, F. J., Burstein, A. H., and Wright, T. M., *Clinical Orthopaedics and Related Research*, No. 141, 1984, pp. 55–65.

[13] Disegi, J., Carpenter Technologies, Inc., Reading, PA, private communication, July, 1985.

[14] Crowninshield, R. D., Johnston, R. C., Andrews, J. G. and Brand R. A., *Journal of Biomechanical Engineering*, Vol. 11, 1978, p. 75.

[15] Brown, R. H., Davy, D. T., Heiple, K. G., Sr., Kotzar, G. M., Heiple, K. G., Jr., Berilla, J., Goldberg, V. M., and Burstein, A. H., *Transactions of the 31st Orthopaedic Research Society*, Las Vegas, NV, Jan. 1985, Vol. 10, p. 283.

[16] Mollan, R. A. B., Watters, P. H., Steele, R., and McClelland, C. J., *Clinical Orthopaedics and Related Research*, No. 190, 1984, pp. 142–147.

Adrian P. A. Demaid[1] and Alan Lawley[2]

The Markham Mine Disaster

REFERENCE: Demaid, A. P. A. and Lawley, A., **"The Markham Mine Disaster,"** *Case Histories Involving Fatigue and Fracture Mechanics, ASTM STP 918,* C. M. Hudson and T. P. Rich, Eds., American Society for Testing and Materials, Philadelphia, 1986, pp. 389–416.

ABSTRACT: The authors have applied linear elastic fracture mechanics to an official accident investigation of a fatigue failure. A single-line component in the mechanical braking system controlling the descent of a cage in a mine shaft fractured; the cage failed to slow down and eighteen miners were killed by the impact. It is shown that significant perceptions of the events prior to fracture are possible using a fracture mechanics approach. A Paris-Erdogan analysis is performed to highlight the constraints and limitations of a fracture mechanics fatigue analysis when applied to a real engineering problem. The Markham accident was a landmark case study in that it stimulated a new philosophy with respect to brake design.

KEY WORDS: ferrous metallurgy, strain gages, axial and bend stresses, bending theory, fatigue, fracture mechanics, estimating lifetime

This paper presents the results of an analysis of fatigue and subsequent brittle fracture of a critical component in the braking system of a coal mine winding mechanism. The case study developed from the application of a detailed analysis to an official report on the accident prepared in 1974. The purpose of the official report was sufficiently served by establishing that fatigue was the cause of the accident. This a metallurgist might do by microscopy alone. It was necessary, however, to measure the applied load on the brake rod in order to determine the origin of the critical loading pattern. This was done using strain gages which needed to be attached remote from the fracture site. It is perhaps surprising to realize that a fracture mechanics analysis of the load at the time of catastrophic failure would not be part of a normal investigation in 1974. In this study fracture mechanics suggests a different loading pattern to that resulting from an extrapolation of the strain gage readings while supporting the general con-

[1] Lecturer on Materials, The Open University, Faculty of Technology, Milton Keynes, United Kingdom.

[2] Professor, Department of Materials Engineering, Drexel University, Philadelphia, PA 19104.

clusion that the failure occurred by brittle fracture following a fatigue crack growth.

In this study, the bending stress analysis, the fracture toughness measurements and analysis, and the Paris-Erdogan analysis of the brake rod in cyclic loading are original. Strain gage readings and most of the fractography were contained in the official report of the investigation; their stress analysis was rudimentary.

Background

This case history documents events that occurred at the Markham coal mine (colliery) in Derbyshire, England in 1973. A descending mine (pit) cage failed to slow down as it approached the bottom of the mine shaft and 18 miners were killed by the impact. The shaft was sunk in 1886 and has a depth of 429 m.

There are two double-deck cages, each capable of carrying a maximum of sixteen persons per deck. In operation, the two cages are attached to opposite ends of a winding rope so that as one cage goes up, the other descends, Fig. 1. At the pit bottom, each cage rests on wooden baulks set into the walls of the shaft. The winding ropes are attached to the winding drum which is driven by a d-c motor (the winder motor). The armature of this motor is supplied by a generator set which is driven by an a-c induction motor. To begin a winding cycle, the engineman gradually increases the power applied to the winder motor while slowly releasing the mechanical brake. The cages are allowed to accelerate to a speed of about 6 m s^{-1}, and this speed is maintained until the cages approach the ends of the shaft. The engineman then gradually decreases the generator voltage to produce regenerative braking.

The arrangement of the mechanical brake and winding drum is shown schematically in Figs. 2 and 3. Basically the brake consists of a servospring mechanism that applies brake shoes to the underside of brake paths on the winding drum. These shoes are activated by the compressed spring nest which operates through a series of levers, Fig. 3. Force is transmitted from the spring nest to the main lever of the brake via the vertical brake rod which lies along the axis of the spring nest. Since the nest of springs is compressed, the brake rod is in tension with the brake in the "on" or "off" positions. The brake is released by using compressed air to counteract the force of the spring nest and so move the brake shoes away from the drum path.

In terms of safety devices, there is an automatic controller, driven from the shaft of the winding drum through a system of gears, which is designed to cut off the power supply to the winder motor and to apply the mechanical brake. There is also an emergency stop button which activates the "ungabbing" gear (see Fig. 2) that immediately disengages the engineman's brake

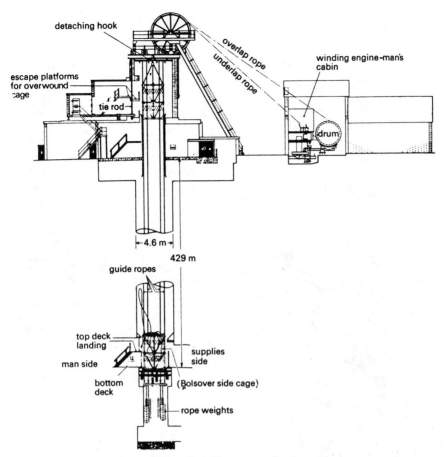

FIG. 1—*Shaft and winding system after the accident.*

control lever from the control valve on the servocylinder; the mechanical brake is then applied automatically.

The Accident

At approximately 6:20 a.m. on Monday 30th July 1973, 29 dayshift miners entered the cage at the surface of the mine. There were 15 men on the top deck and 14 men on the bottom deck. The other cage (at the bottom of the shaft) was empty at this time since all nightshift workers had been brought to the surface approximately 30 min earlier.

Winding proceeded normally with the two cages passing the midpoint in the shaft. The first sign of abnormality occurred with the start of regenerative braking. At this moment the descending cage was approximately 90 m from the shaft bottom and moving with a speed of about 6.0 m s^{-1}. The dayshift

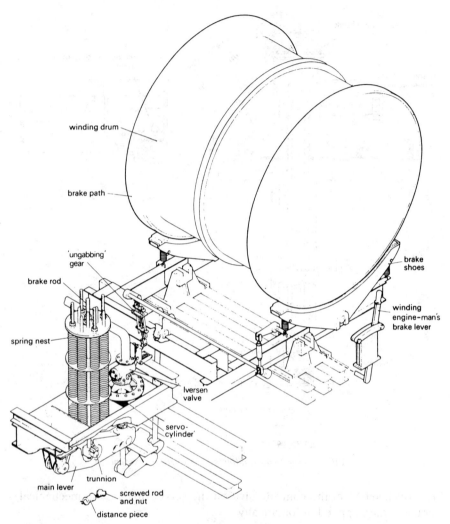

FIG. 2—*Mechanical brake and winding drum.*

winding engineman saw sparks under the brake cylinder and heard a loud bang. He immediately applied full regenerative braking and set the mechanical brake lever to the on position. Since this had no effect on the speed of the winding drum, he then pressed the emergency stop button. There was no response. The descending cage crashed into the pit bottom while the ascending cage struck the roof girders, dropped back, and finally hung suspended from the detaching hook, see Fig. 1.

The accident resulted in the loss of eighteen lives. Eleven other miners sustained serious bodily injury. Damage at the colliery included: severe

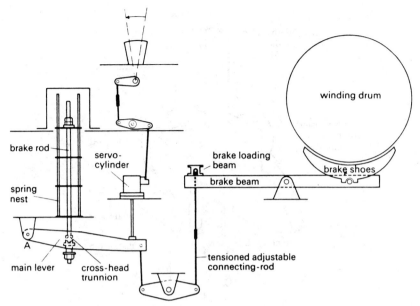

FIG. 3—*Schematic of the mechanical brake.*

distortion of the bottom deck of the cage (the speed of impact was estimated to be 12 m s^{-1}); a tearing of part of the side and brake path of the winding drum; and demolition of part of the brickwork, concrete, and roofing of the winding engine house as a result of the flailing underlap rope.

The Investigation

The accident was the subject of a formal investigation by the U.K. Department of Energy [1]. This encompassed all electrical and mechanical components of the winding system. It was confirmed that the ungabbing gear had operated, and a supply of compressed air was still available at the valve above the servo cylinder, see Figs. 2 and 3. The brake shoes, however, were about 3 mm clear of the brake paths. Thus, attention was directed to the mechanical brake system.

Metallurgical Examination

On inspection of this system it was immediately apparent that the brake rod within the spring nest had broken into two pieces, the fracture surface being located under the trunnion axle, Fig. 4. The brake rod broke in the threaded region 119 mm from the bottom where the rod passed through the "distance piece," below the crosshead trunnion axle. A polished and etched section of the rod, produced by cutting longitudinally along the axis

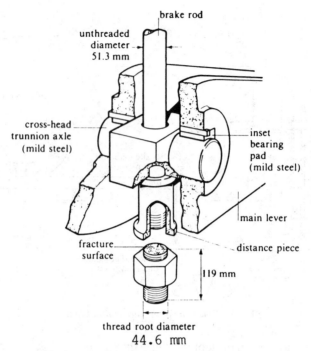

brake rod

unthreaded
diameter
51.3 mm

cross-head
trunnion axle
(mild steel)

inset
bearing
pad
(mild steel)

main lever

fracture
surface

distance piece

119 mm

thread root diameter
44.6 mm

FIG. 4—*Detail of the brake rod fracture and trunnion.*

of the smaller broken piece, revealed that there was a secondary crack at the root of the thread adjacent to that at which the rod broke, Fig. 5. Other secondary cracks of unknown depth and circumferential length were detected by the magnetic particle technique with a fluorescent inspection fluid. The secondary cracks, of various lengths, were located at the thread roots on either side of the fracture plane.

Drawings specified that the rod should be made from a medium plain carbon steel with the designation En 8, conforming to British Standard 970:1947. This was the prevailing standard at the time the brake rod material was purchased. A chemical analysis was carried out and the results appear in Table 1. It is seen that the material of the brake rod was within the specification demanded at the time of manufacture.

The microstructure of the brake rod was revealed by optical microscopic examination of a section parallel to the rod axis. It consists of an equiaxed grain structure of ferrite and pearlite. The rod was probably hot rolled in the single phase austenite region since inclusions visible in the microstructure are elongated parallel to the rod axis. There is no evidence of a distorted surface layer at the thread roots; thus, it is concluded that the threads were machined and not rolled.

Tensile properties were determined on two test specimens cut longitu-

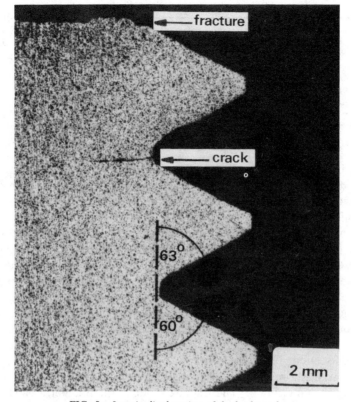

FIG. 5—*Longitudinal section of the brake rod.*

TABLE 1—*Chemical analysis of the brake rod and composition ranges permitted within the specification.*

Element	Brake Rod Steel, % by weight	BS 970:1947 En 8, % by weight
C	0.41	0.35 to 0.45
Mn	0.74	0.6 to 1.10
Si	0.24	0.05 to 0.35
S	0.032	0.06 max
P	0.007	0.06 max
Fe	97.4	...
Ni	0.39	...
Cr	0.057	...
Mo	0.01	...
Cu	0.21	...
Sn	0.02	...

TABLE 2—*Tensile properties of the brake rod.*

Property[a]	Specimen 1	Specimen 2	BS 970:1947 En 8
Yield stress, MN m^{-2}	351	346	262 min
Tensile strength, MN m^{-2}	599	600	540 min
Elongation, %	30	29	20 min
Reduction in area, %	47.5	46.5	...
Hardness (standard Brinell)	180	190	152 to 207

[a] Gage length 51 mm; gage cross section 1.61×10^{-4} m^2.

dinally from the rod at a distance 51 mm from the fracture plane. These results, together with hardness levels and the required specification for En 8 steel, are given in Table 2. It is seen that the measured values exceeded the minimum values.

Toughness was evaluated at three temperatures (0, 20, and 100°C) using the Charpy V-notch test on specimens cut from the rod; the resulting data are presented in Table 3. The 1947 specification does not include a minimum impact energy for the steel. However, it is clear from Table 3 that the steel has an extremely low toughness at ambient temperature.

The actual fracture is characterized by two distinct surface textures. There is a smooth, dull flat area covering approximately 65% of the fracture cross section, and a rougher, more reflective area occupying the remainder of the cross section, Fig. 6a. Both areas lie approximately perpendicular to the rod axis. Representative scanning electron micrographs from the two areas are shown in Fig. 6. The fracture mode in the rougher area (Fig. 6b) is characteristic of low energy brittle transgranular cleavage; note the faceted appearance of the fracture surface. The surface morphology of the larger flat area (Fig. 6c) appeared to be characteristic of fatigue since parallel striations are visible.

To fully understand the scenario that led to the Markham disaster, and to unambiguously interpret the two modes of fracture of the brake rod, it is necessary to determine the in-service loading of the brake rod. The stress levels reached during brake operation can be then calculated.

Loading Conditions of the Brake Road

Axial and Bending Loads—The broken brake rod was replaced with a new rod at the Markham site. Four strain gages were attached to the rod at a distance of 533 mm from the lower end with their active direction parallel to the long axis; the gages, numbered 1 to 4, were located 90° apart around the circumference of the rod, Fig. 7.

Gage readings were taken during both manual release and application of the brake, and during emergency application of the brake; the winding

TABLE 3—*Charpy V-notch data for the brake rod.*

Test Temperature, °C	Energy Absorbed, J	Brittle Fracture, %
0	7.05	100
20	10.00	100
100	46.00	35

drum was stationary in both cases. The results, converted to stress, are displayed in Fig. 8; stresses were found to be reproducible to within ±3 MN m^{-2}.

Dealing with the tensile stress first; this changes from an average value of 66 MN m^{-2} with the brake on to 73 MN m^{-2} with the brake off. This is precisely the result which would be expected from the kinematics of the braking system: slightly more tension with the brake off. What is surprising is the presence of large bending stresses. Indeed, the presence of the trunnion in the design suggests that the brake rod was not intended to bend when the main lever rotated during braking, Fig. 3. The major bending stresses are in the (2–4) plane and these show large changes when the brake is moved from on to off. There are also bending stresses in the (1 to 3) plane, but these do not change quite so much.

It was found that the mild steel bearing pad surfaces on the main lever (see Fig. 4) were badly worn. Similarly in the crosshead trunnion, the top surfaces of the steel axle in contact with the bearing pads were fretted and scored; this is illustrated in Fig. 9. The importance of the lack of lubrication of the trunnion was clearly demonstrated by lubricating the crosshead trunnion axle bearing surfaces with a graphite grease and then measuring the trunnion rotation relative to the main lever and the associated stress as a function of the number of on–off brake cycles. After only 13 brake cycles the alternating stress had increased to more than ±75 MN m^{-2}. It appears that high bearing pressures caused the lubricant to be progressively squeezed out of the gap between the trunnion axle and the bearing pads.

Returning to the evaluation of the bending stresses, the results displayed in Fig. 8 suggest that the stresses in the (2–4) plane are not necessarily the maximum bending stresses. Utilizing bending theory [2] and with the aid of Fig. 10, the bending stress at the arbitrary location A on the perimeter of the brake rod is given by

$$\sigma = \frac{M_x \, r \sin \theta}{I} + \frac{M_y \, r \cos \theta}{I} \tag{1}$$

where I is the second moment of area of the section on which the bending moment is M. Since σ and θ are the only variables, the angular position θ_m for maximum and minimum stress is given by $d\sigma/d\theta = 0$; hence

$$\tan \theta_m = M_x/M_y = \sigma_B/\sigma_C \qquad (2)$$

The maximum bending stress is then given by

$$\sigma_m = \sigma_B \sin \theta_m + \sigma_C \cos \theta_m \qquad (3)$$

Using the data in Fig. 8, the most highly stressed case occurs with the brake in the off position, when a bending stress of 103 MN m^{-2} and a tensile stress of 73 MN m^{-2} produce a combined stress of 176 MN m^{-2}. Maximum bending

(a)

FIG. 6—*Brake rod fracture surface:* (a) *complete fracture surface*, (b) *smaller rough region* (*SEM*), (c) *larger smooth region* (*SEM*).

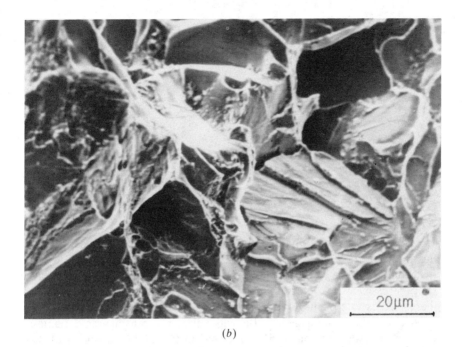

(b)

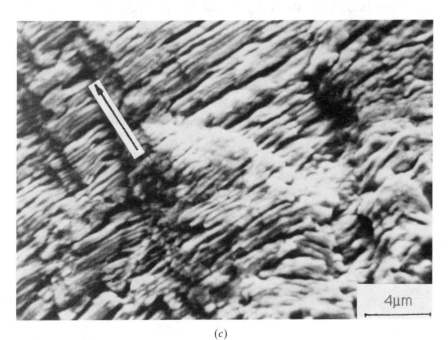

(c)

FIG. 6—Continued.

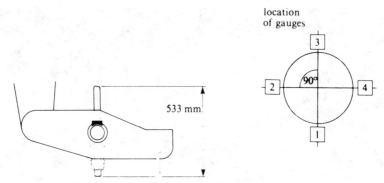

FIG. 7—*Location of the four strain gages on the replacement rod.*

stresses and the corresponding values of θ_m in the off and on positions are illustrated in Fig. 11.

Strains could not be measured at the fracture site because of the threads. Thus, stresses were determined from strain measurements taken at the nearest accessible position. To infer the stress at the fracture surface from the stress at the strain gage site, the rod will be taken as having a diameter in the unthreaded region of 51.3 mm and in the threaded region of 44.6 mm; these being given the suffices 0 and t, respectively, Fig. 12.

To relate the bending stress determined at the position of the strain gages to that at the fracture site it is necessary to model the loading of the brake rod. The simplest model assumes that, due to uneven loading on the nut, a moment is introduced at the top of the rod, Fig. 13. The moment will remain constant along the length of the rod until it is reacted by the nut bearing against the collar beneath the main lever. This model predicts a maximum bending stress (that is, with the brake off) at the fracture site of

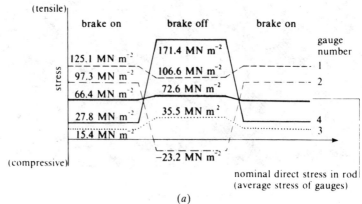

(a)

FIG. 8—(a) *Stresses on the replacement brake rod,* (b) *summary of stresses at the cross section through the gages,* (c) *axial and bending stresses.*

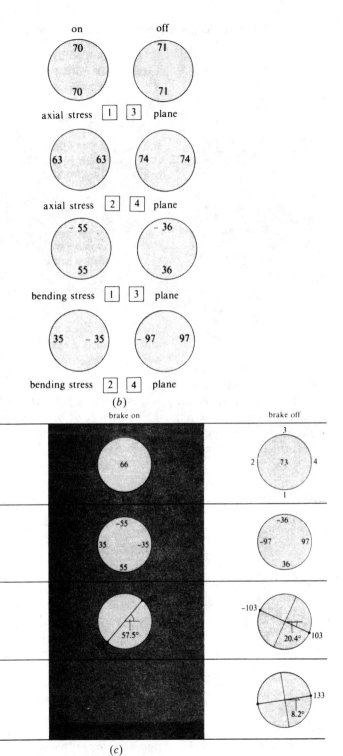

(b)

(c)

FIG. 8—*Continued.*

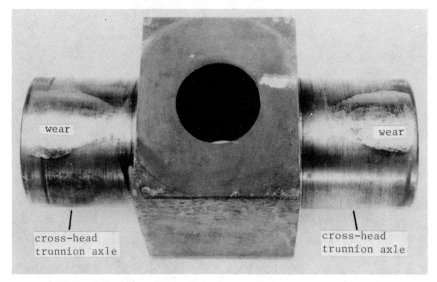

FIG. 9—*Cross-head trunnion showing wear on top surface of trunnion axle.*

157 MN m^{-2} and an axial stress of 97 MN m^{-2}, Fig. 14. The combined tensile stress is therefore 254 MN m^{-2}.

Based on yield stress, the factor of safety is 1.36, and on tensile strength, it is 2.36. Thus, the combined *static* effect of bending and tension is not sufficient to cause an overload failure of the uncracked brake rod. From Ref 3, the stress concentration factor (α_K) at the thread roots is about 4.

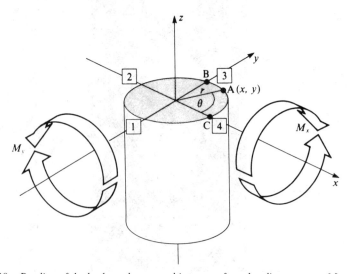

FIG. 10—*Bending of the brake rod expressed in terms of two bending moments* M_x *and* M_y.

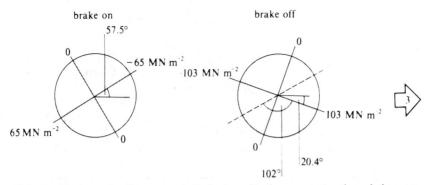

FIG. 11—*Maximum bending stresses in the brake rod in the cross section through the gages.*

Taking this into account, there will be some localized plastic deformation at the thread roots.

Fracture Toughness of the Brake Rod Steel—Fracture toughness[3] was determined on standard, notched, three-point bend specimens machined from the brake rod, Fig. 15. The method used was that given in B.S. 5447 [4]. A fatigue crack was first formed at the notch, and the load required to propagate this crack was measured.

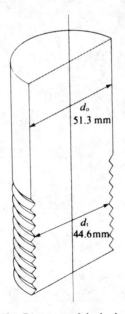

FIG. 12—*Diameters of the brake rod.*

[3] Courtesy of Professor J. T. Barnby, Aston University, Birmingham, U.K.

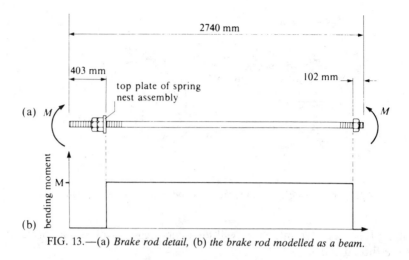

FIG. 13.—(a) *Brake rod detail, (b) the brake rod modelled as a beam.*

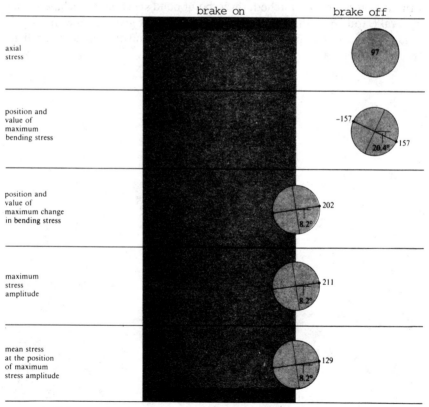

FIG. 14.—*Stresses in the brake rod at the fractured (threaded) section.*

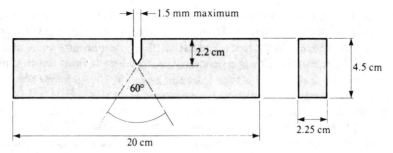

FIG. 15—*Fracture toughness three-point bend specimen.*

The apparent value of fracture toughness, K_Q, is given by

$$K_Q = P_Q Y_1 / B W^{1/2} \qquad (4)$$

Values of a/W and Y_1 for the brake rod tests appear in Table 4, together with the values of K_Q derived from them. Because of the brake rod dimensions [4]

$$B < 2.5 \, (K_Q/\sigma_y)^2 \qquad (5)$$

The values of K_Q are therefore not valid measurements of K_{Ic}. It is observed, however, that only a small area of the fracture surface exhibits shear lips. Thus, K_Q is probably close to the value of K_{Ic}. We will take the average value of K_Q, approximately 44.5 MN m$^{-3/2}$, as being K_{Ic} for the brake rod steel.

The general expression for stress intensity is of the form

$$K = Y\sigma(\pi\alpha)^{1/2} \qquad (6)$$

where Y is a dimensionless geometrical factor. At the moment of final fracture, the rod contained a deep fatigue crack, thereby becoming much

TABLE 4—*Fracture toughness data: brake rod steel.*

	Test 1	Test 2	Test 3
Thickness, B/mm	22.55	22.28	22.55
Width, W/mm	44.53	44.53	44.53
Average crack length, a/mm	22.63	22.78	26.65
Average crack length, a/W	0.508	0.511	0.598
Y_1 for above value of, a/W	10.89	11.00	14.94
Force P_Q for crack growth, kN	19.37	19.37	13.98
K_Q, MN m$^{-3/2}$	44.33	45.32	43.89
σ_Y, MN m^{-2}	348	348	348
$2.5(K_Q/\sigma_Y)^2$, mm	41	42	40

more flexible than it was when it was uncracked. This would have the effect of accommodating the small rotation of the main lever, and relaxing most of the bending stress in the rod. Therefore, it is assumed that the rod was under axial stress only. For the geometry of an edge crack of depth a in a circular bar of diameter D under a tensile stress σ, the dependence of Y on a/D taken from the British standards document [5] is shown in Fig. 16.

For the brake rod at the moment of final fracture, the crack length measured from the fracture surface was 27 mm and a/D was $27/46.6 = 0.58$. Extrapolating the curve in Fig. 16, the pertinent value of Y is 3.1. Hence

$$\sigma_{cr} = K_{Ic}/Y(\pi a)^{1/2}$$

and

$$\sigma_{cr} = 49 \text{ MN m}^{-2}$$

Thus, the rod with a 27 mm crack needs a stress of only 49 MN m^{-2} to initiate brittle fracture. This should be compared with the estimated stress at the fracture site, made from strain gage measurements on an uncracked rod, of 97 MN m^{-2}, Fig. 14. It is clear that all the load on the cracked rod was not reaching the fracture site. The only thing between the strain gage site and the fracture site is the crosshead trunnion axle and it is here that

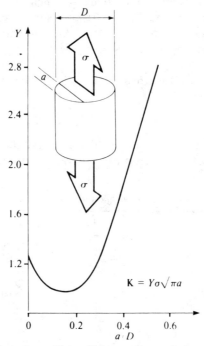

FIG. 16—*Stress intensity coefficient* Y *for an edge-notched round bar in tension* [5].

some of the load on the rod must have been reacted. Indeed, the investigators observed rubbing marks on the brake rod where it passed through the trunnion. Contact between rod and trunnion would be associated with a frictional force which would prevent some of the axial load from being transmitted to the lower end of the rod. This may always have been the case, if the rod was assembled to one side of the hole in the trunnion or if the elastic deflection was enough to cause contact, Fig. 17a. Alternatively, the crack may have caused the rod to become so flexible as to bring one side (Fig. 17b) or both sides (Fig. 17c) of the rod into contact with the sides of the trunnion block.

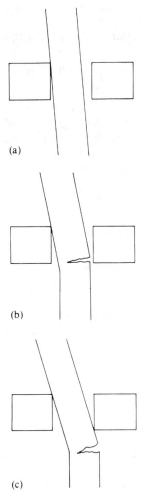

FIG. 17—*Possible modes of contact between the brake rod and the trunnion.*

Fatigue Analysis—The *change in axial* stress in the threaded portion of the rod is given by the change in axial stress in the unthreaded region (see Fig. 8) multiplied by the ratio of the square of the diameters

$$(73 - 66) (51.3)^2/(44.6)^2 = 9 \text{ MN m}^{-2}$$

We now have to estimate the *change in bending stress*. The maximum bending stress varies around the perimeter of the rod and the principal bending axis swings through some 102° between the brake on and brake off conditions at the site of the strain gages, Fig. 11. By reference to Fig. 10, the bending stress in both cases is given by

$$\sigma = \sigma_B \sin \theta + \sigma_c \cos \theta \tag{7}$$

Substituting in the values of σ_B and σ_c (see Fig. 8) for the brake off and brake on conditions

$$\sigma_{\text{off}} = -36 \sin \theta + 97 \cos \theta$$
$$\sigma_{\text{on}} = -55 \sin \theta - 35 \cos \theta$$

The *change in bending stress* $\Delta\sigma$ is given by the difference between the stresses with the brake off and the brake on, $(\sigma_{\text{off}} - \sigma_{\text{on}})$.

$$\Delta\sigma = (-36 \sin \theta + 97 \cos \theta) - (-55 \sin \theta - 35 \cos \theta)$$

which simplifies to

$$\Delta\sigma = 19 \sin \theta + 132 \cos \theta \tag{8}$$

The *maximum value of* $\Delta\sigma$ occurs when the value of θ is θ_m, which is given by

$$d(\Delta\sigma)/d\theta = 0 \tag{9}$$

which is

$$19 \cos \theta_m - 132 \sin \theta_m = 0$$

and

$$\theta_m = 8.2°$$

Using this value of θ_m in Eq 8 to give the *maximum change in bending stress*

$$\Delta\sigma_m = 133 \text{ Mn m}^{-2}$$

This result is significant in that it clears up an anomaly from the first analysis. From Fig. 11, the angular position of the maximum bending stress varied from $\theta = -122.5°$ with the brake on to $-20.4°$ with the brake off and yet the fracture origin was at roughly $0°$. The maximum change in bending stress at $8.2°$, is much more consistent with a fracture origin at $0°$.

To estimate the *maximum change in bending stress* in the threaded portion of the rod from that in the unthreaded segment, the bending stress is changed in proportion to the ratio of the diameters cubed

$$\Delta\sigma_m = 133 \ (51.3/44.6)^3 = 202 \ \text{MN m}^{-2}$$

To investigate the fatigue performance it is sensible first to check the material response to the stresses in the threaded section, assuming that all the load from the strain gage site is transmitted to the fracture site (a worst case analysis). To accomplish this we need to calculate the *mean stress* and the *stress amplitude* and plot these points on a Goodman diagram [6].

The *mean stress level* at the position of maximum stress amplitude has two components: the contribution of bending and the contribution of the axial stress. The *bending stresses* at the position of the maximum stress amplitude are given by

$$\sigma_{\text{off}} = -36 \sin \theta + 97 \cos \theta$$

$$\sigma_{\text{on}} = -55 \sin \theta - 35 \cos \theta$$

Evaluating these stresses at $\theta = 8.2°$

$$\sigma_{\text{off}} = 91 \ \text{MN m}^{-2}$$

$$\sigma_{\text{on}} = -42 \ \text{MN m}^{-2}$$

The mean of these two values is

$$\frac{91 + (-42)}{2} = 24.5 \ \text{MN m}^{-2}$$

The value of *mean bending stress* in the threaded portion of the rod is

$$(\sigma_{\text{mean}})_{bt} = 24.5 \ (51.3/44.6)^3 = 37 \ \text{MN m}^{-2}$$

As the *axial stress* is uniform over the section, its mean value is simply given by the average of the axial stresses with the brake on and brake off, corrected for the change in cross sectional area:

$$(\sigma_{mean})_{bt} = (66 + 73)(51.3/44.6)^2/2 = 92 \text{ MN m}^{-2}$$

The mean stress at the position of maximum stress amplitude is then simply given by the sum of the axial and bending components

$$\sigma_{mean} = (92 + 37) \text{ MN m}^{-2} = 129 \text{ MN m}^{-2}$$

The stress amplitude is again given by a contribution from *bending* and *axial* stresses. The change in stress due to bending was calculated to be 202 MN m^{-2}. In addition there is a change in stress due to axial loading, of 9 MN m^{-2}. The sum of these two contributions gives the total change in stress as

$$\Delta\sigma = (202 + 9) = 211 \text{ MN m}^{-2}$$

The stress amplitude is just half this value. We conclude that the cyclic stress of maximum amplitude is that shown graphically in Fig. 18.

The best data available for estimating the fatigue properties of the brake rod refer to an En 8 steel [7]. This material, in the normalized condition, was tested in the form of an electropolished specimen containing a notch with an elastic stress concentration factor, α_k, of 3.15. The tensile strength of the steel was 540 Mn m^{-2}, and its reverse bending fatigue limit was found to be ±120 MN m^{-2}. These data were used to plot a Goodman diagram, Fig. 19.

The Goodman diagram predicts that fatigue will occur in bending. This is an interesting result as it suggests that when the rod was new and the bearing seized then all the load above the trunnion axle was transmitted to the fracture site and, it was only after the crack had grown, probably to a significant length, that the rod was able to flex sufficiently to touch the trunnion block, thereby reducing the load on the fracture site.

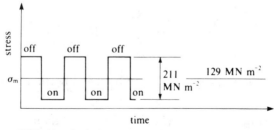

FIG. 18—*Cyclic loading condition of the brake rod.*

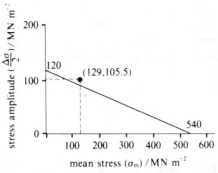

FIG. 19—*Goodman diagram for the brake rod.*

Estimating Lifetime

Fractography and stress analysis strongly suggest fatigue as the cause of failure of the brake rod and a number of important questions arise:

1. How many on–off cycles did the brake rod experience in its service lifetime of 21 years?
2. Was the crack already initiated when the brake rod was installed or was there an incubation period?
3. Once a crack had been formed, how would its length change with increasing cycles of braking?

About 1000 miners used the shaft in which the accident occurred, each making two trips per working day. This constitutes 2000 man trips a day. If the shifts of miners are lowered and raised independently of one another, each journey by the cages accomplished 32 man trips (because a full cage carries 32 men), and 2000 man trips are accounted for by 2000 per 32 journeys of the cage per working day. For each man a year will consist of about 230 working days (assuming a five-day week and that four weeks are spent on holidays and illness). The number of cage journeys per year will therefore be about $(2000 \times 230)/32$, and thus during the 21 year life of the brake rod approximately $(2000 \times 230 \times 21)/32$ journeys were made. For each journey the brake would be applied once so the estimate of the total number of brake applications is about 3×10^5. Evidence from the mine records suggested 7×10^5 cycles, and this is not in conflict with the estimate. The cages in this shaft were used also to transport materials other than coal, and limited quantities of stone which could account for the difference in total number of cycles.

Cracks were sought by nondestructive testing techniques only after the rod had failed and the age of those that were found is completely unknown. However, there is no evidence to suppose that these cracks were present in the as-manufactured rod, and it is likely that they were initiated by cyclic

loading. Since yielding is thought to have occurred due to the stress concentration at the thread roots, the incubation period for the cracks may
have been quite short. For example, yielding may have caused some of the
hard, brittle inclusions near the thread roots to become cracked, thereby
creating microscopic cracks in the first few loading cycles.

For low-strength ferrite-pearlite steels, the Paris-Erdogan equation is of
the form [8]

$$da/dN = 10^{-8}(\Delta K)^3 \tag{10}$$

where da/dN is given in millimetre per cycle and ΔK is measured in MN
m$^{-3/2}$. This equation produces the classic sigmoidal shaped da/dN-ΔK curve
suggested by Paris [9]. Provided that ΔK is not too small (say $\Delta K > 10$ MN
m$^{-3/2}$) and the value of K_{max} does not approach K_{Ic} (say $K_{max} < 0.7\ K_{Ic}$),
this equation can be applied irrespective of the mean K-value or the exact
identity of the steel [8]. In view of the age and unclean nature of the steel
the use of some of the more complicated crack growth/stress intensity factor
relationships was not justified.

The variation of stress-intensity range ΔK with crack length a for a rod
of circular cross section (diameter 44.6 mm) loaded in bending to a stress
range $\Delta\sigma$ is given in Fig. 20, taken from Ref 10. For a given crack length,
this graph specifies the particular value $\Delta K/\Delta\sigma$ for the rod. Multiplying this
value by $\Delta\sigma$ gives ΔK.

This approach is feasible provided the loading conditions are known for
all stages of crack growth. However, in this case the understanding which
we have developed so far suggests that the bending load on the rod decreases
as the crack grows. We start the analysis by assessing how the crack grows
to a *short* length of, say 3 mm and then take stock of the situation. The
maximum change in stress, $\Delta\sigma$, of 211 MN m^{-2} is composed of two con-

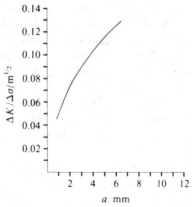

FIG. 20—*Variation of $\Delta k/\Delta\sigma$ with crack length for a rod loaded in bending* [10].

tributions, 202 MN m^{-2} from bending, and 9 MN m^{-2} from tensile loading. Since the crack is short, we can consider the value of $\Delta\sigma$ to be unchanged throughout the stages of crack growth. We also make the approximation that the total change in stress of 211 MN m^{-2} is all attributable to bending.

To integrate the Paris-Erdogan equation numerically, we first assume the presence of a crack; a crack of length 0.875 mm puts us in the range of $\Delta\sigma$ where the Paris-Erdogan relation is valid. From Fig. 20, for a crack 0.875 mm long

$$\Delta K/\Delta\sigma = 0.045 \text{ m}^{1/2}$$

Taking $\Delta\sigma = 211$ MN m^{-2}

$$\Delta K = 0.045 \times 211 = 9.5 \text{ MN m}^{-3/2}$$

This is approximately the lower limit of ΔK over which the growth law is valid [8]. Using this value for ΔK gives the rate of crack growth as

$$da/dN = 10^{-8}(\Delta K)^3 = 8.56 \times 10^{-6} \text{ mm per cycle}$$

If the crack grows at this rate over a small increment of growth, say from a crack length of $a_0 = 0.75$ mm to $a_f = 1.00$ mm (that is, linearly 0.125 mm, each side of the average crack length $a_{av} = 0.875$), then

$$\Delta N = (a_f - a_0)/da/dN = 0.25/8.56 \times 10^{-6} = 29.2 \times 10^3 \text{ cycles}$$

where ΔN is the number of load cycles involved in the increment of crack growth. This process can be repeated with successive increments of growth; the results are given in Table 5. The running total of load cycles $\Sigma\Delta N$ gives the total number of cycles needed to grow the crack from the beginning of the first increment to a length of 3 mm. In Fig. 21 the crack length is plotted against the total number of load cycles that have elapsed since the crack was 0.75 mm long.

This analysis shows that it took roughly 98 000 cycles to grow a crack from 0.75 to 3 mm long. There is probably not much point in carrying the analysis beyond this point because the modelling becomes uncertain due to the changing load. The result, however, does seem to be of the right order when compared with the officially estimated figure of 700 000 braking cycles to incubate and grow the crack to its critical length of 27 mm. Most of these cycles were spent in nucleating a crack and growing it to 0.75 mm long, but we have no analysis to estimate how many.

Linear elastic fracture mechanics presupposes that there is an existing crack, so it can account for the growth but not for the incubation of the crack. If the striations are clear and extend over the entire region of fatigue

TABLE 5—Data for increments of crack growth in the rods.

a_0/mm	a_f, mm	$a_{av} = (a_0 + a_f)/2$, mm	$\Delta K/\Delta\sigma$, m$^{1/2}$	$\Delta K = 211 (\Delta K/\Delta\sigma)$, MN m$^{-3/2}$	$da/dN = 10^{-8}(\Delta K)^3$, mm per cycle	$\Delta N = dN/da\,(a_f - a_0)$ × 10^{-3}, cycles	$\Sigma\Delta N \times 10^{-3}$, cycles
0.75	1.00	0.875	0.045	9.5	8.56×10^{-6}	29.2	29.2
1.00	1.25	1.125	0.052	10.97	13.2×10^{-6}	18.9	48.1
1.25	1.5	1.375	0.059	12.45	19.29×10^{-6}	13.0	61.1
1.5	1.75	1.625	0.065	13.71	25.8×10^{-6}	9.7	70.8
1.75	2.0	1.875	0.070	14.77	32.2×10^{-6}	7.8	78.6
2.0	2.5	2.25	0.078	16.46	44.58×10^{-6}	11.2	89.8
2.5	3.0	2.75	0.0865	18.25	60.8×10^{-6}	8.2	98.0

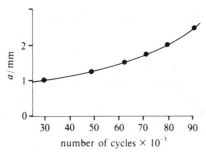

FIG. 21—*Crack length as a function of the number of load cycles occurring since the crack was 0.75 mm long.*

it should be possible to couple the crack growth analysis (at low *da/dN*) with fractography (at higher *da/dN*) where striations are probably more visible, to get some idea of the incubation period. Unfortunately, definite striations only occurred in patches on the fracture surface of the steel brake rod, and this precluded using the *da/dN* analysis to correlate with and extend the fractographic observations. Aluminum would have been the ideal material in this regard.

The preceding estimates of lifetime are incomplete. However, they do fully support the verdict of the investigation, namely, that the brake rod became progressively cracked by fatigue until finally the fatigue crack became unstable, causing brittle fracture. The agent of fatigue was an unforeseen bending which arose because the trunnion mechanism was ineffective.

The center rod in the spring nest is an example of a single-line component, in that the operation of the mechanical brakes on the winding engine depended completely on it. Not surprisingly, this type of system was taken out of service following the accident. The official report states that single-line components should be either eliminated or so designed as to prevent danger.

Conclusions

It is possible to identify fatigue as the cause of this failure and to highlight bending moments resulting from friction at the trunnion as the loading which caused this fatigue crack growth using straightforward techniques. The analysis presented here has taken this understanding further and identified a variation in principal bending axis, between the brake on and brake off positions. This produces a maximum change in bending stress—the critical fatigue loading—more in line with the fracture origin.

In order to measure loads in an uncracked brake rod it was necessary for the strain gages to be positioned away from the fracture site and the measurements to be extrapolated to the fracture site using simple analytical models. A fracture mechanics analysis using data measured at the fracture site showed that the loads measured at the strain gage site did not reach the threaded section where failure took place. Using this understanding it was possible to model the load paths through the structure during the growth of the fatigue crack. Such an approach was not possible using strain gage measurement alone.

The Paris-Erdogan analysis supports the overall understanding of the failure and illustrates the practical limitations in applying the technique.

At a more general level the case study illustrates the danger of single line components. The brake rod had lasted for 21 years despite being made from what, by modern standards, would be called a dirty steel and having a cut as opposed to a rolled thread.

Acknowledgment

The authors are indebted to Professor C. N. Reid for his interest and guidance throughout the development of this case history and the co-operation of Mr. H. D. Munson and his staff at SMRE.

Figures 1 to 5 are from a SMRE report and reproduced with the permission of the controller of HMSO.

References

[1] Calder, J. W., "Accident at Markham Colliery," Department of Energy, HMSO, Cmmd. 5557, 1974.
[2] Popov, E. F., *Mechanics of Materials,* Second Edition, Prentice-Hall, Englewood Cliffs, NJ, 1976.
[3] Neuber, H., "Theory of Notch Stress: Principles for Exact Calculation of Strength with Reference to Structure, Form and Materials," Office of Technical Services, Department of Commerce, Washington, DC, 1964, Translated from Kerbspannungslehre, Springer, 1958.
[4] British Standard, BS5447, "Plane Strain Fracture Toughness (K_{Ic}) of Metallic Materials," 1977.
[5] Daoud, O. E. K., Cartwright, D. J., and Carney, M., *Journal of Strain Analysis,* Vol. 13, 1978, pp. 83–89.
[6] Dieter, G. E., *Mechanical Metallurgy,* Second Edition, McGraw-Hill, New York, 1976.
[7] Woolman, J. and Mottram, R. A., "The Mechanical and Physical Properties of the British Standard En Steels," Vol. En 1–20, The British Iron and Steel Research Association, 1964.
[8] Barnby, J. T., *Fatigue,* Mills and Boon Ltd., London, 1972.
[9] Paris, P. C. in *Fatigue—An Interdisciplinary Approach,* J. J. Burke, N. L. Reed, and V. Weiss, Eds., Syracuse University Press, 1964, pp. 107–127.
[10] Cannon, D. F., and Allan, R. J., *Railway Engineering Journal,* Vol. 3, No. 4, 1974, pp. 6–23.

Summary

The Symposium on Case Histories Involving Fatigue and Fracture Mechanics showcased the most recent applications of fatigue and fracture mechanics analyses. In order to assemble this showcase, papers were solicited which described (1) how fatigue and fracture mechanics concepts were employed to design new structures and (2) how these concepts were used either to evaluate the integrity of structures containing defects or to determine why structures failed. The symposium was successful in attracting 24 papers which described the latest fatigue and fracture mechanics applications. This publication contains twenty-two of these papers. A brief summary of each follows.

Rahka described the cracking of a hydrogenerating pressure vessel. This cracking resulted from hydrogen attack of improperly-alloyed weld metal in the vessel. A detailed fracture mechanics analysis showed the existing cracks did not threaten the structural integrity of the vessel. Consequently, it was allowed to remain in service until a scheduled shutdown.

Smith explained that radial cracks frequently develop in the nozzle corners of nuclear pressure vessels. Generally, these cracks are initiated by thermal shock, and then propagated by subsequent pressure cycles. Small-scale pressure vessels have been fabricated and tested to study the initiation and growth of these flaws. Smith employed frozen-stress photoelastic methods to determine the validity of applying the results of tests on small-scale vessels to full-size vessels. The frozen-stress method showed the flaw shapes for the small- and full-scale vessels were basically similar. However, the stress-intensity factor distributions were as much as two times higher in the small-scale vessels than in the full-scale ones. Thus test results generated on small-scale vessels were conservative when applied to the full-scale vessels.

Kaplan, Willis, and Barnett analyzed the failure of a hatch cover on a pressurized cement barge. The hatch failed when it was partially unlocked to bleed air from the barge. The remaining locks on the hatch were unable to carry the pressure load, and the hatch failed. Detailed testing and analysis indicated the hatch design, hatch material, and barge design all contributed to the hatch failure.

Pearson and Dooman studied the explosion of a truck-mounted propane tank. An initial investigation located a crack in the tank. This crack was believed to be the cause of failure. However, fracture mechanics analysis showed the tank should not have failed at normal operating pressure even with the crack present. Further investigation showed (1) the relief valve on the tank was corroded shut; (2) the truck driver had overfilled the tank with liquid propane (a properly filled tank contains no more than 82.4% liquid); and (3) heat from the motor exhaust and the sun could raise the tank pressure to the bursting point. Factors (1) through (3), plus the crack in the tank, all combined to fail the tank.

Chow and Simpson described the sudden failure of a Zircaloy-2 pressure tube in a nuclear reactor. This tube was encased in a concentric Zircaloy-2 tube to insulate it from the reactor coolant. During construction, one of the spacers which maintain the concentricity of the tubes was displaced. Thus the two tubes contacted each other. The resultant heat sink led to the development of zirconium hydrides in the pressure tube. These hydrides led to cracking of the tube, and a reduction in the fracture toughness of the material. The crack subsequently propagated to a critical size and failed the tube. Fracture mechanics analyses and tests were conducted to confirm the sequence of events leading to the failure.

Reid and Baikie described the material selection process for the design of high pressure penstocks. Two steels were considered for these penstocks. A series of stress, fracture mechanics, and cost analyses were performed to establish which of these two steels would be most satisfactory. In addition, these analyses were used to establish defect acceptance standards and inspection intervals for the penstocks. This paper presents the specific procedures used in evaluating the two steels.

Christensen and Hill studied the impact of longitudinal weld-toe cracks on the structural integrity of pipeline steel. Their study indicated that the depths of these cracks were only one-third of the critical flaw depth. They confirmed their analysis with a full-scale pressure test on a cracked pipe section. A subsequent fatigue-crack-growth analysis showed 20 full-pressure cycles per day were required to propagate the existing toe cracks to failure in a 30 year period. Finally, tests confirmed that cathodic charging had no detrimental effects on the pipeline material.

Reemsnyder explained that fatigue cracks frequently developed at pin holes in the legs of off-shore oil drilling rigs. These cracks developed while the rigs were enroute from their fabrication sites to their installation sites. Studies showed the cracks resulted from a combination of high cyclic stresses and tensile residual stresses at the pin holes. To alleviate this cracking (1) the cyclic stresses were reduced by modifying the intransit configuration of the rigs thereby reducing vortex shedding, and (2) the residual stresses were reduced by thermal stress-relief following pin hole fabrication. No cracking has occurred at the pin holes since the vortex shedding and stress-relief steps were taken.

Nelson and Hampton investigated the failure of a large wind tunnel compressor blade. They found that a crack had initiated at a scratch in the shank of the blade and subsequently propagated to failure. Analysis indicated that crack propagation was aggravated by operation of the compressor at or near its resonant frequencies. Operation at these frequencies introduced high stresses in the blade. In order to prevent a recurrence of such a failure, (1) special procedures were developed for blade installation in order to reduce the risk of scratching the blades; (2) the compressor was required to pass through resonant frequencies as quickly as possible in order to reduce the number of high-stress cycles and (3) a regular inspection interval was established to detect cracks before they approached a critical length.

Cipolla, Grover, and Richman performed both one- and two-dimensional stress analyses of a burst compressor disk. These analyses showed a defect-free disk should not have burst in the short time the disk was in service. Further study showed the disk was subjected to nonroutine processing during fabrication. Normally, such disks are heated and quenched prior to final machining. Such processing frequently introduced quench cracks which were subsequently removed during final machining. However, investigation showed the burst disk was heat treated and quenched after final machining. Thus, quench cracks could have been present when the disk was initially put into service. The authors proposed that the growth of these quench cracks led to premature failure of the disk.

Selz and Peterson described the rationale used to develop inspection intervals for high-pressure heater vessels. This rationale involved (1) location of the high-stress areas; (2) determination of the maximum flaw size which could escape detection during inspection; (3) calculation of the number of cycles required to propagate the flaw in (2) to a critical length; and (4) selection of a safety factor which permitted location of the flaw prior to its reaching a critical length. The authors describe in detail the procedures used in completing steps 1 through 4.

Chang explained that welded structures frequently contain crack-like flaws. He described a computer code for predicting the growth of these flaws. This code, named EFFGRO III, was used to predict the growth of embedded flaws in welded aircraft structures. The effects of residual stresses were included in these predictions. The prediction results were compared with test results and good correlations were found.

Rich, Pinckert, and Christian reported that cracks were found in fastener holes in the compression-loaded spar caps of F-15 aircraft. Analysis indicated that the local stresses around the holes exceeded the material's compressive yield strength. Upon unloading, residual tensile stresses were introduced at the holes. The authors developed an analytical model for predicting the role of these tensile stresses in initiating and propagating cracks in the spar caps. This model was used to predict crack initiation and propagation to a predetermined size. The U.S. Air Force used these predictions to set overhaul intervals for their F-15 aircraft.

Howard reviewed the inflight failure of an aircraft horizontal stabilizer. The failure resulted from a fatigue crack in the top chord of the stabilizer. This crack propagated through the chord and subsquently failed it. Once the chord failed, the stabilizer separated from the aircraft and it crashed. Inspection of the fracture surfaces indicated the crack propagated by fatigue intermixed with a series static crack jumps. Fracture mechanics analysis techniques were used to explain the growth sequence of the crack.

Saff and Ferman reported that fuel-structure interaction dynamics significantly increased stresses in wet-wing aircraft. These high stresses produced drastic reductions in fatigue life as compared to the life of dry-wing aircraft. Combinations of vibration loads and internal pressure were found to cause even further reductions in fatigue life. They presented procedures for predicting fatigue lives under dynamic conditions. Laboratory tests were strongly recommended to validate the predictions.

Denyer explained how fracture mechanics techniques were used to track fatigue damage accumulation in the wing structure of the T-39 aircraft. These techniques were used to compute the remaining structural life, and to establish inspection intervals for each aircraft. Tracking was accomplished by analyzing flight records and then using the analytical results to predict the growth of flaws. The paper described the crack-growth procedures used in making these predictions.

Rich and Orbison described the failures of two dies. One die was used to swage fittings onto wire ropes, the other to mold gears in a powder metallurgy process. Detailed analysis and inspection showed the swaging die failed as a result of a combination of high stresses, low material toughness, and small fabrication cracks. The powder metallurgy die failed as a result of the growth of a fatigue crack to a critical size.

Clarke investigated the failure of an arbor on a lumber mill trimming saw. Visual inspection indicated the crack initiated at an abrupt change in the diameter of the arbor. Initially, the stress concentration associated with this diameter change was thought to be responsible for the failure. However, a detailed fatigue analysis showed the operating stresses were too low to initiate a fatigue crack. Further study of the lumber mill operations showed a piece of steel had recently contacted the saw and knocked off several teeth on the blade. The imbalance loads associated with the loss of these teeth were adequate to initiate a fatigue crack. Once the damaged blade was replaced, the service loads were adequate to propagate the crack to a critical size.

Tubby and Wylde investigated the effects of six parameters on the fatigue resistance of welded joints. They found that (1) the material's ultimate tensile strength and fracture toughness, and (2) weld residual stresses had relatively little effect on fatigue resistance. They further found that contouring the weld joint was somewhat beneficial, but was probably not cost effective. Finally, they reported that stress concentrations and material thickness had significant impacts on fatigue resistance. Both high stress concentrations and thicker materials had lower fatigue resistances.

Mattheck, Kneifel, and Morawietz described the fracture of intramedullary nails which are used to repair fractures in large bones. Fatigue cracks initiated at slots and screw holes in the nails. The growth of these cracks was predicted using fracture mechanics techniques. The procedures for making these predictions, and for predicting crack arrest, were presented. A new design for the intramedullary nails was proposed. This design was expected to have considerably lower stress concentrations, and consequently longer fatigue lives.

Rimnac, Wright, Bartel, and Burstein investigated the failure of twenty-one hip implants. They found that fatigue cracks initiated on both the medial and lateral surfaces of the implants and then propagated to critical lengths. Crack development was attributed to the basic design of the implant, the high applied stresses and the properties of the AISI 316L stainless steel used for the implant.

Demaid and Lawley described the failure of an elevator brake rod. Inspection showed this failure resulted from the initiation and growth of fatigue cracks at the root of screw threads on the brake rod. Detailed stress analyses and stress measurements defined the stress distribution within this rod. These analyses and measurements were then used to approximate the life of the brake rod using fracture mechanics procedures. The inherent vulnerability of single-load-path structures was most clearly manifest in this case history.

C. M. Hudson
NASA-Langley Research Center, Hampton,
VA 23665; cochairman and coeditor.

T. P. Rich
Bucknell University, Lewisburg, PA 17837;
cochairman and coeditor.

Author Index

Subject Index